For Reference

be taken from this room

ENERGY RECOVERY FROM LIGNIN, PEAT AND LOWER RANK COALS

COAL SCIENCE AND TECHNOLOGY

Series Editor:

Larry L. Anderson
Department of Fuels Engineering, University of Utah, Salt Lake City, UT 84112, U.S.A.

COAL SCIENCE AND TECHNOLOGY 13

ENERGY RECOVERY FROM LIGNIN, PEAT AND LOWER RANK COALS

edited by

DEBRA J. TRANTOLO
Adjunct Professor, Northeastern University, Boston, MA 02115, U.S.A.

and

DONALD L. WISE
Cabot Professor of Chemical Engineering, Northeastern University, Boston, MA 02115, U.S.A.

ELSEVIER, Amsterdam — Oxford — New York — Tokyo   1989

ELSEVIER SCIENCE PUBLISHERS B.V.
Sara Burgerhartstraat 25
P.O. Box 211, 1000 AE Amsterdam, The Netherlands

Distributors for the United States and Canada:

ELSEVIER SCIENCE PUBLISHING COMPANY INC.
655, Avenue of the Americas
New York, NY 10010, U.S.A.

ISBN 0-444-87335-X (Vol. 13)
ISBN 0-444-41970-5 (Series)

Printed in The Netherlands

FOREWORD

This reference text emerged from well over a decade of research and development work dealing with the conversion and/or recovery of fuels and organic chemicals from biomass, peat, lignite, and low rank coals. Note that throughout our work we have emphasized the investigation of organic materials that, by others, are considered as "wastes". Perhaps a bit of reminiscing is appropriate and fitting as a way of describing what we believe is an emerging and practical technology, namely, "gentle" or low severity treatment of these "wastes" to recover useful products.

Our early work involved dealing with the organic fraction of municipal solid waste, specifically, fuel gas production via anaerobic digestion. It was clear from our own work, and that of others getting into this field, that some form of pretreatment enhanced this bioconversion process. Because we learned that the lignin portion of biomass did not undergo anaerobic fermentation, we recognized that a breakdown of the lignin could occur if we applied technology common to the paper pulp industry, namely, aqueous alkaline hydrolysis. Coupled with this perception was the quest to apply this pretreatment technology to peat, lignite, and lower ranked coals. Thus evolved a research and development program in which an array of organic materials were considered as potential substrates for anaerobic digestion; provided the particular organic material under consideration could be broken down in molecular weight to such an extent that it could undergo methane fermentation.

As a result of this concept of a "gentle" or low severity pretreatment, a pretreatment sufficient only to prepare a lignaceous material for methane fermentation, we had the opportunity to initiate work on a peat biogasification process. Because this work was sponsored by the Minnesota Gas Company, we devoted our work only to fuel gas production from peat. However, we recognized several other aspects of this "gentle" or low severity pretreatment technology.

One insight based on our low severity pretreatment work was that we should be able to apply this technology to other "old" biomass lignin, in addition to peat, such as lignite and perhaps sub-bituminous coals. Also, we believed that following breakdown of the lignaceous feedstock it might be possible to recover some of the complex aromatics from the pretreated product. This recovered product we termed a "BTX-type" material, recognizing that it was not benzene, toluene, or xylene, but complex aromatics. However, discussion of a "BTX-type" liquid fuel was useful in a descriptive sense.

We also recognized, from our early work with peat, that we need not confine ourselves to methane fermentation. Closely-related work had revealed that "suppressed" methane fermentation did yield the aliphatic organic acids acetic through hexanoic. Moreover, Kolbe electrolysis experiments on these organic acids indicated that very desirable alcohols and alkanes could be produced from these organic acids. As a result, we began to explore a much broader overall process, namely, that of a "biorefinery".

This overall biorefinery concept includes as feedstock either biomass lignin (assuming the cellulose is used for paper pulp, rayon, or alcohol fuel), peat, lignite, or low rank coals. The first stage is the breakdown of this material by some gentle or low severity means, such as aqueous alkali hydrolysis. As an aside, and looking to the future, it may be that a lignase – an enzyme that cleaves lignin – will be used for the low severity pretreatment step. Regardless, the objective of this first step is to break down the lignaceous material to some smaller molecular weight fraction. This product may then be recovered or converted to a number of useful products such as a BTX-type fuel, be fermented direct to fuel gas, or undergo suppressed methane fermentation to organic acids which may then be recovered or subsequently converted to other fuels or organic chemicals.

During the course of this research and development program, we sought advice and recommendations of many colleagues with closely related goals and objectives. We found we were all considering alternatives to conventional processing technology. Much of our mutual interest related to the investigation of less severe approaches to the conversion and/or recovery of useful products from our lignaceous feedstocks. Under the funding of the U.S. Department of Energy, we had the opportunity to invite our colleagues to join an industrial/university review committee to comment on the progress of our work and, at the conclusion of this project, join us in a workshop on the general topic of the low severity treatment of biomass, peat, lignite, and low rank coal. This reference text is a formal result of this workshop.

The Editors

CONTENTS

Energy Recovery from Lignin, Peat and Lower Rank Coals, edited by D.J. Trantolo and D.L. Wise
Elsevier Science Publishers B.V., Amsterdam 1989 — Printed in The Netherlands

Chapter 1

FEASIBILITY OF A PEAT BIOGASIFICATION PROCESS*

M.G. BUIVID and D.L. WISE**
Dynatech R/D Company, Cambridge, MA 02139 (U.S.A.)
A.M. RADER
Minnesota Gas Company, Minneapolis, MN 55426 (U.S.A.)
P.L. McCARTY and W.F. OWEN
Stanford University, Palo Alto, CA 94305 (U.S.A.)

ABSTRACT

A feasibility study was conducted to determine if biogasification may be an economical process for the conversion of peat into pipeline quality methane (SNG). United States' peat resources, in terms of energy content, are greater than the energy recoverable from uranium, shale oil, or the combined reserves of petroleum and natural gas. Experiments and costing of SNG by the biogasification of peat were based on a two-stage process. In the first (assumed to follow wet harvesting of peat) an alkaline oxidation pretreatment of peat was employed to produce soluble, low molecular weight acids, wood sugars, and other soluble organic fragments. Unreacted peat solids were separated (to be processed as a potential boiler fuel) while the recovered liquid, containing the soluble organic material, was anaerobically fermented to methane and carbon dioxide in the second stage of the process. In this study, up to 26% of the energy value in peat was converted to methane in the two-stage process; more than 50% of the material from the first stage was fermented to methane in the second. More work is needed to optimize the first stage. Preliminary cost estimates are: a 79.2×10^3 GJ/day (75.0 billion BTU/day) plant, the total capital requirement is approximately $323 million (April 1978 costs). The annual operating cost is approximately $44 million. For a plant operating at a 90% stream factor, with peat delivered to the plant site as a 3% solids slurry at a cost of $0.0033/kg ($3/ton) dry peat, the average cost for SNG is estimated as $3.16/GJ ($3.34/million BTU). A significant advantage of biogasification appears to be that technical difficulties of dewatering, necessary for peat utilization in conventional gasification or direct combustion, are eliminated.

INTRODUCTION

It is well established that plant debris, primarily cellulosic and lignaceous in composition, will decompose when attacked by fungi and bacteria in the presence of oxygen. If, however, these debris fall into an oxygen deficient environment, such as a pond or saturated soil, the decomposition process is severely retarded and peat may be formed. The stoichiometry of peat formation may be described [1] as:

---

*Reprinted from *Resource Recovery and Conservation,* 5 (1980) 117-138
**Present address: Cambridge Scientific, Inc., 195 Common Street, Belmont, MA 02178 (U.S.A.

$$C_{72}H_{120}O_{60} \rightarrow C_{62}H_{72}O_{24} + 8CO_2 + 2CH_4 + 20H_2O \qquad (1)$$

This material will remain as peat unless fluctuations in the water table result in further decomposition or, if buried under additional sediments and subjected to heat and high pressure, ultimately converts into lignite or coal.

Peat is classified on the basis of its botanical origin, fiber content, moisture content, and ash content. The U.S. Bureau of Mines recognizes three types of peat: moss peat, reed-sedge peat, and peat humus. Any peat derived from moss is classified as moss peat. Peat derived from reed-sedge, shrubs, and trees is known as reed-sedge. Any peat whose origin is largely unrecognizable is classified as peat humus [2].

Recent estimates indicate that peat covers approximately 152×10^6 hectares (375 million acres) of the earth's surface [3]. Although the largest reserves — 62% — are located in the U.S.S.R., sufficient quantities exist in the United States to make this a potentially valuable national resource [5]. A comparison with other U.S. energy reserves [6] shows that peat reserves (1.5×10^{12} GJ or 1443 quads) are greater than the energy from uranium (1.2×10^{12} GJ or 1156 quads), shale oil reserves (1.2×10^{12} GJ or 1160 quads), and the combined reserves of both petroleum and natural gas (1.4×10^{12} GJ or 1408 quads); only coal reserves are greater ($5.3-10.6 \times 10^{12}$ GJ or 5,000–10,000 quads). Thus, available U.S. peat reserves represent an abundant and largely untapped energy source. Investigation of a practical process suitable for utilizing these reserves on a year-round basis and in an energy efficient, environmentally acceptable manner was the object of this study.

PROCESS CONCEPT

A two-stage biogasification process was studied for methane production from peat (Fig. 1). The first stage is an alkaline oxidation. The complex

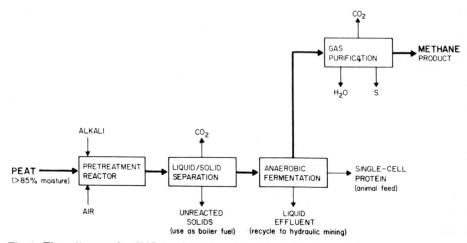

Fig. 1. Flow diagram for SNG production via the biogasification of peat.

benzene ring structure of the lignaceous fraction is broken down into water-soluble, low molecular weight aromatic acids and other organic compounds. The cellulosic fraction of peat is broken down into simple wood sugars. These reactions may be carried out with a high conversion of total carbon at temperatures of up to 150°C and pressures near 2.2 MPa. It is anticipated that a staged reactor would be used for this conversion. The second stage is anaerobic fermentation to produce methane. It had been previously established that the products of the alkaline oxidation may be converted to fuel gas [7—11]. The second or bioconversion stage takes place at a temperature of up to 60°C and at atmospheric pressure. As with all anaerobic fermentations, approximately 5% by weight of the organic feed, in this case the water soluble aromatic acids and sugars, are utilized by the microorganisms to grow additional cells. This biomass is a potentially valuable by-product; when dried it is a cream-colored powder of approximately 59% by weight protein.

The object of the first stage is to convert cellulosic and lignaceous fractions of peat, which may contain a small fraction of carbohydrates, to water-soluble compounds that could be fermented to methane. This conversion parallels other and previously reported experiments with wood and coal, especially lower rank coals.

Related background on the utilization of the cellulosic fraction of peat

There is much literature relating to acid hydrolysis of cellulose. Saeman [12] carried out a classic investigation of the hydrolysis of cellulose in dilute sulfuric acid and at high temperature. Later work by Harris et al. [13] showed that through acid hydrolysis organic acids could be created which were potentially fermentable to methane. Chemical changes in wood after treatment with alkali have been studied extensively, particularly by Tarkow and Feist [14]. The scission of the ester crosslinks has been demonstrated by increased calcium ion exchange ability of alkali-treated wood [15]. The overall effect of such treatments is to cause a "breaking down" of wood structure, allowing greater access by enzymes and, therefore, facilitate fermentation. More severe treatments produce significant destruction of the hemicellulose fraction of wood [16], possibly accompanied by formation of toxic by-products [17]. Overall, it appears that a fairly mild degree of alkaline pretreatment is effective, and that overtreatment may be either of no benefit or harmful.

Related background for aqueous alkaline oxidation of lignaceous fraction of peat

There is a background of experimental work on the aqueous alkaline oxidation of the type of lignaceous material found in peat to produce water-soluble organic compounds. Much of this relates to the aqueous oxidation of coal. Early work was carried out from the time of the First World War, in Germany largely, as well as in the United States, and up to the time of the Second World War [18—24]. Additional experiments on alkali oxidation of lignaceous materials were continued into the 1950's [26—28]. Apparently due to the overwhelming growth of the petrochemical business at that time, and to the impact of this growth on the potential for coal chemicals, the American work did not continue. However, during the 1960's, there was interest in Japan [30—32]. At the same time, and continuing through today with intensive interest, there has been Russian work in this area. A search in *Chemical Abstracts* from 1962 through 1978 found more than 35 Russian publications in this area. Areas of investigation include: (a) lower rank coals including lignite (but not peat) studied under aqueous alkaline oxidation conditions; (b) reaction conditions with the goal of optimizing production of benzene-carboxylic acids; and (c) reactor design concepts, specifically a two-stage plug flow reactor. In a recent article, Kukharenko [33] reports nearly complete (74 to 100%) conversion of coal into a soluble form. From 40 to 78% of the products are carboxylic acids, and up to 30% of this product acid mixture is oxalic acid. The two-stage process, using either NaOH or K_2CO_3, is conducted at temperatures up to 270°C. No reaction kinetics are presented on which to base reactor design and scale-up.

Several studies have also been made of lignin decomposition with gaseous oxygen under alkaline conditions [34,35]. The yields after the alkaline-oxidative reaction indicate the formation of acidic materials. Grangaard [36, 37] reports that various types of lignin can be almost completely degraded by oxygen in an alkaline solution to yield reactants digestible by anaerobic microorganisms.

Work on aqueous oxidation of lignaceous material without alkali has been conducted by Brink and coworkers [38—41]. Generally, the wet oxidation of particulate wood by the introduction of air into aqueous slurries from 160 to 220°C and at pressures from 1.6 to 3.4 MPa in staged reactors yielded an acid product, a portion of which was recycled to hydrolyze wood chips in the first stage. Similarly, McGinnis and coworkers [42—44] investigated the staged reaction of pine bark, but at somewhat more severe conditions (200 to 600°C). It would appear that the reaction products from the peat-like wood chips and bark, as reported by Brink and McGinnis, would be suited for anaerobic fermentation. These results may point to a peat processing technology which does not require the added cost of alkali.

Second-stage reaction: methanogenic fermentation of water soluble organic compounds

The second stage in the peat-to-methane conversion is the anaerobic fermentation of the benzenecarboxylic acids and wood sugars to fuel gas. Anaerobic fermentation is a biological process in which organic matter, in the absence of oxygen, is converted to methane and carbon dioxide. Peat has been anaerobically fermented directly without alkaline-oxidative pretreatment, but yielded relatively little methane [45]. Anaerobic fermentation studies of aromatic compounds, however, constitute a limited part of this background [46—55], compared to more conventional digestion work [55—58]. Tarvin and Buswell [59] reported the complete bioconversion to methane and carbon dioxide of benzoic, phenylacetic, hydrocinnamic, and cinnamic acids using a culture of microorganisms obtained from a sewage sludge anaerobic digester. Bioconversion of benzenecarboxylic acid was to 54.5% CH_4/45.5% CO_2 for the two experiments reported. The percentages of CH_4/ CO_2 in the gas mixture produced in the anaerobic fermentation of phenyl-acetic, hydrocinnamic, and cinnamic acids were 58/42, 60.5/39.5, and 58/42, respectively. Fina and Fiskin [60] later conducted detailed radioactive tracer experiments showing that benzoic acid was converted to methane and carbon dioxide in a fixed ratio with a conversion efficiency of greater than 95%.

EXPERIMENTAL

The experimental study combined two areas of investigation with the objective of testing the peat biogasification process. First, the peat pretreatment conditions were evaluated in terms of aqueous non-oxidative and oxidative reactions with and without alkali addition. Second, each type of pretreated peat and solubilized product was anaerobically fermented to produce methane. Two modes of fermentation were employed: in situ, an unstirred batch reactor, and a CSTR (continuous stirred tank reactor) with daily substrate addition and product removal.

Reed-sedge peat was obtained as directly wet-harvested by the U.S. Bureau of Mines from the Red Lake area of Minnesota. Delivered peat contained approximately 85% moisture. Chemical analysis of delivered peat is given in Table 1.

Total solids, organic solids, pH, total acids, nitrogen, and COD analyses were conducted according to *Standard Methods*, 1976 [61]. Gas analyses

TABLE 1

Analysis of peat used in this study

Element	Weight percent (oven-dry basis)
Carbon	50.84
Nitrogen	2.42
Oxygen	30.50
Hydrogen	4.75
Sulfur	0.43
Other	1.06
Ash	10.00

were performed with a Fisher—Hamilton Gas Partitioner, Model No. 29, using a helium carrier gas flow-rate of 40 mℓ/min.

Non-oxidative pretreatment

A slurry of peat was prepared in a Waring Blender® to produce a peat particle size range of approximately 10 to 100 Tyler® mesh and having the characteristics given in Table 2. "Heat treatments" were performed in bomb-type autoclaves of various capacities (Parr Instrument Co.) with a maximum size of 2.0 ℓ, as shown in Fig. 2. A Pyrex® liner was employed to minimize the introduction of heavy metals during heat treatment. The reactor was continuously stirred, and its temperature controlled to within ±5°C.

TABLE 2

Assay of initial pretreatment peat slurry

Total solids (TS)*	39.7 ± 0.1 g/ℓ
Volatile solids (VS)*	36.7 ± 0.1 g/ℓ
Chemical oxygen demand (COD)*	54.2 ± 1.1 g/ℓ
COD/VS ratio	1.48
Total Kjeldahl nitrogen	1.2 g N/100 g VS

*See Ref. 61.

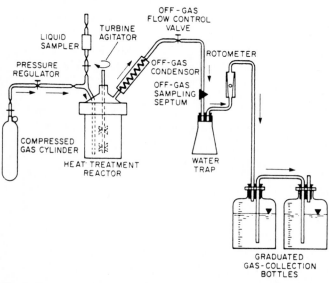

Fig. 2. Experimental apparatus for pretreatment of peat.

The heat treatment procedure is illustrated by the following example. A 150-mℓ sample of peat slurry was added to the reactor (600-mℓ capacity), assembled, purged with nitrogen gas for 15 min, the off-gas flow control valve closed, and heating commenced. The reactor took 20—35 min to reach the specified temperature. The reaction was held at temperature for one hour, then cooled to room temperature within approximately 20 min.

Non-oxidative pretreatments of peat slurries were also performed with the inclusion of various mixtures derived from the solubilization of ligno-cellulosic materials. Other additives investigated were: calcium hydroxide (at 115°C), potassium carbonate (at 170°C), and sodium hydroxide (at 25, 100, 150, 200, and 250°C). Concentrations were selected to leave 100 meq/ℓ after treatment at 250°C. For example, the initial concentration of sodium hydroxide used was 400 meq/ℓ.

The products from treatment with sodium hydroxide were anaerobically fermented exclusively in 125-mℓ batch in situ digesters. The other non-oxidative treatments, using calcium hydroxide and potassium carbonate, were performed in a similar manner but the products were not neutralized. They contained sufficient buffering capacity to be fermented in larger (50 ℓ) CSTR digesters.

Oxidative pretreatment

A slightly modified experimental procedure was followed compared to the non-oxidative heat treatment. The off-gas flow control valve was adjusted so that a continuous flow of oxygen could be passed through the slurry during treatment. After an alkali-treated peat slurry was added to the reactor, it was assembled and pressurized to 1.0 MPa with oxygen prior to heating. The oxygen was regulated to maintain the reactor at 1.0 MPa above saturated steam pressure to prevent excessive water loss.

The off-gas, flow rate controlled at 178 mℓ/min, was passed through a high-pressure condenser and flow control valve, and collected by displacement of water. The reactions were maintained for one hour at the specified temperature and operating pressure.

Peat was also oxidatively heat-treated at non-alkaline conditions. This was investigated to determine the effect of possible acid- or oxygen-catalyzed degradation of the lignocellulose fraction of peat and to evaluate the minimum requirements for chemical addition. These samples were acidic after heat treatment and were not neutralized before addition to the in situ digesters as adequate buffer was available to maintain a neutral pH for fermentation.

In situ anaerobic fermentation

Peat samples heat treated with sodium hydroxide were anaerobically fermented to methane using in situ fermentation digesters (125-mℓ bottles fitted with rubber serum caps). The digesters were initially purged with an oxygen-free gas mixture containing 70% nitrogen and 30% carbon dioxide. Volumes of pretreated peat (solids and associated liquids) were then added under anaerobic conditions. Two sample concentrations were investigated for each pretreatment product, 0.5 and 2.0 g/ℓ (based on total initial peat solids) so as to confirm that methane production was inhibited neither by the assay procedure nor by the presence of compounds toxic to anaerobic micro-organisms which might be formed during pretreatment.

Finally, each serum-bottle was inoculated with seed organisms which had been acclimated to the products from alkaline heat-treated lignocellulose. Total liquid volume was 50 mℓ; fermentation was conducted under meso-philic (35°C) conditions.

All sodium hydroxide heat-treated peats were anaerobically fermented in duplicate; experience has shown the precision of such fermentations to be within ±2%. Control fermentations contained only the nutrients and seed organisms. Methane production was monitored over a 31-day period.

Gas generated during fermentation was sampled with a syringe and analyzed. The methane produced from the peat products was computed by the differ-ence from the blanks. The conversion efficiency of peat to methane is refer-enced either to sample weight (mℓ CH_4/g sample TS) or to sample organic content (mℓ CH_4/g COD).

Peat treated with calcium hydroxide or potassium carbonate additions was also fermented (at 35°C) in the in situ mode using a 2000 mℓ Erlenmeyer flask digester. These digesters were connected to wet-test gas meters (Precision Scientific) and monitored daily for gas production and composition (CH_4/CO_2). The digesters were buffered with calcium carbonate, and the microorganism inoculum came from the effluent of the corresponding chemical pretreated peat CSTR digester.

CSTR anaerobic fermentation

The continuous stirred tank reactors (CSTR) were the largest operating digesters employed in this experimental program. CSTR digesters were 50-ℓ Nalgen® carboys, modified for anaerobic conditions, and continually stirred at ca. 100 rpm. Digesters were gas leak-tested before fermentation. The digesters were operated 24 hours/day, 7 days/week, for periods of 4 to 6 months. Product gas was continually monitored with wet-test gas meters for total gas output and sampled daily for CH_4/CO_2 composition using a gas partitioner.

Unlike in situ fermentation, CSTR operation required daily (7 days/week) product removal and substrate feeding. This was done via a feed port equipped with a liquid seal to reduce the amount of air introduced into the reactor during liquid transfers; 2 ℓ of effluent were removed and 2 ℓ of a peat slurry were added daily. For a 50-ℓ reactor, assuming well-mixed conditions, this resulted in an approximate 25-day hydraulic retention time.

Typically, the 2 ℓ addition consisted of 1 ℓ of a peat slurry containing

50 g dry peat solids, either as delivered (a control) or after various non-oxidative heat treatments, and 1 ℓ of primary sewage sludge. The sludge, containing 50 g dry sludge solids/ℓ, served as an inoculum of anaerobic microorganisms as well as a nutrient source. In order that the fermentation not be nutrient limited, 50/50 dry weight peat to sludge solids were used. A sewage sludge blank was also run for 25 days with a 2 ℓ feed (1 ℓ sewage, 1 ℓ H_2O). The primary sludge was obtained from the Nut Island Sewage Treatment Plant, Quincy, Massachusetts.

CSTR digesters were operated at both mesophilic (35°C) and thermophilic (60°C) conditions using the $Ca(OH)_2$ (115°C) and K_2CO_3 (170°C) peat pretreatments. Digester effluent pH was measured daily to assure the optimal fermentation conditions of pH ca. 6.8 for mesophilic and ca. 7.3 for thermophilic. Periodically, free ammonia nitrogen and dissolved orthophosphate in the digester effluent were measured to insure that the level of these nutrients was adequate for microorganism growth. Ammonia nitrogen was analyzed with an ion-selective ammonia electrode (Model 95-10, Orion Instruments, Cambridge, Mass.) and dissolved orthophosphate by a modification of the method of Fiske and Subbarow [62].

It is important to note that the anaerobic microorganisms used in these fermentations were not previously acclimated to peat or pretreatment products of peat. Ultimately, greater methane yields may be realized by employing a culture of anaerobic microorganisms that specifically utilize peat and/or derivatives of peat for conversion to methane.

EXPERIMENTAL RESULTS

The effects of different pretreatment conditions on conversion efficiency of peat to a methane product are shown in Figs. 3 and 4. Fig. 3 is based on the initial amount of peat submitted for pretreatment, whereas bioconversion in Fig. 4 is based upon the solubilized fractions of the peat product after pretreatment. After pretreatment, the products were either insoluble (to be separated for possible subsequent fuel for the biogasification) or soluble and feedstock for anaerobic fermentation. Some portion of the peat was oxidized to CO_2. In general, as increases in pretreatment temperatures with chemical addition were made, a greater amount of the initial peat was subsequently fermented to methane; also, a greater amount of solubilized peat was oxidized to CO_2.

In oxidative treatments, for a given reaction temperature, the addition of alkali decreased the amount of CO_2 produced. When organic substances are oxidized during wet oxidation, heat is liberated which can be used elsewhere in the process.

Alkali appeared to decrease the amount of COD lost due to volatilization and oxidation at both 150 and 200°C (23 and 18% less COD lost, respectively). Oxidative processing at 150°C increased the methane yield. At 200°C, the addition of alkali had two effects: (i) it increased the solubilization of the

10

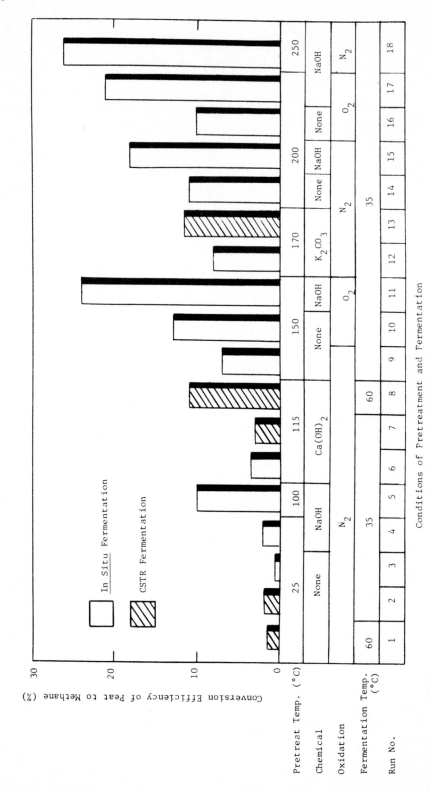

Fig. 3. Conversion efficiency based on initial amount of peat pretreated and subsequent fermentation.

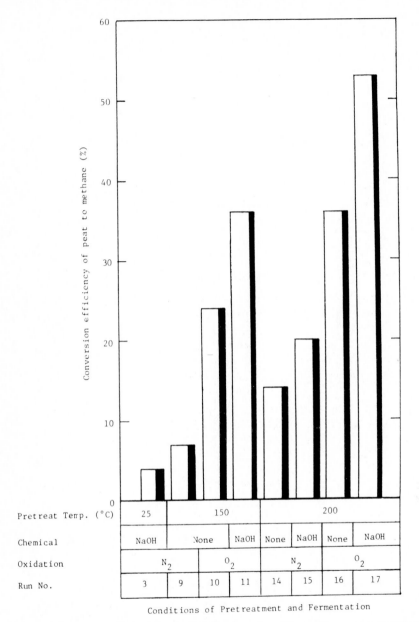

Fig. 4. Conversion efficiency of solubilized peat products after pretreatment and subsequent in situ fermentation at 35°C.

peat and subsequent methane yield; and (ii) it decreased the amount of COD lost by oxidation. However, the consumption of alkali was high, ranging from 1.7 meq/g total solids for peat soaked in alkali at room temperature for five hours (Exp. No. 4) to 70 meq/g TS for reaction at 250°C for one hour (Exp. No. 18).

Fig. 4 illustrates the conversion efficiency of the solubilized peat products to methane at various pretreatment and fermentation conditions. Note that wet alkali oxidation at 200°C produced the most fermentable soluble product (53%), but quite possibly many of the degradable organic materials produced were subsequently oxidized to CO_2 and water. The addition of alkali increased the amount of organic material solubilized and decreased oxidation to CO_2, thereby increasing the total yield of methane. Also, higher pretreatment temperatures possibly yielded soluble peat products inhibitory to anaerobic fermentation. However, these products were utilized after a longer period of microorganism acclimation.

In summary, untreated peat was virtually unused by anaerobic micro-organisms to produce methane. However, anaerobic microorganisms, not previously acclimated to the alkaline heat-treatment products of peat, utilized soluble organic material from the various pretreatment process conditions. In this study, up to 26% of the initial heat content of peat was biologically converted to methane when alkaline heat treatment was employed. Higher yields of methane from fermentation in a CSTR may be probable with recycling of microorganisms previously acclimated to the peat products. The amount of methane produced from soluble products of treated peat under alkaline conditions, both oxidative and non-oxidative, was generally about twice that of non-alkaline conditions.

Higher pretreatment temperatures increased methane yields but up to a limit. Unidentified products from more severe pretreatment conditions of alkaline heat-treatment of peat inhibited methane formation. Inhibition to the fermentation process was greatest in the solubilized peat products obtained at reaction temperatures higher than 200°C.

PROCESS DESCRIPTIONS

Few of the operational steps of the process have been dealt with at the necessary level of experimental sophistication and, in some cases, were outside the experimental scope of this present study. The present process design analysis incorporates experimental information and trends where available, along with assumptions and estimates from similar processes using other substrates. This procedure allows a wide range for each process variable, and a sensitivity study was conducted around a base-case design. The biogasification process postulated and analyzed incorporates conventional technology wherever possible. A conceptual process flow diagram for the base-case design was given in Fig. 1. The following is a discussion of a base-case process design to give a general understanding of the concepts involved.

Peat feedstock preparation

The postulated base-case is for a 79.2×10^3 GJ/day (75 billion BTU/day) methane plant. It requires 46.6 Gg (51,600 tons) of harvested peat feedstock/

day at 15% peat solids and 85% moisture (typical moisture in natural occurrence). For a 20-year plant life this requires harvesting an area of about 14,400 hectares (35,600 acres) of peat bogs with an average 2.1 meters (7 feet) depth. The abundance of water in peat bogs makes wet harvesting in slurry form by hydraulic mining/dredging techniques a logical choice. The peat slurry can be pumped but for limited distances at about 3% solids. This is a lower solids content for pumping than slurries of coal (up to 50% solids slurry) [63]. The type or cost of the harvesting/delivery equipment has not been determined and is still under investigation [64]. However, an estimated cost ($3/dry ton) of peat per unit weight delivered (at 3% slurry) to the plant is incorporated into the annual operating costs.

On arrival at the plant, the 3% solids slurry is mechanically dewatered or allowed to drain to 8% solids, which would eliminate about 65% of the water used for transport. The ultimate content of solids in the feed is critical to favorable process economics (discussed later). The excess water is recycled back to the hydraulic mining site. (Due to the weak cohesive strength of peat, hydraulic mining and slurry transport may effectively result in a degree of comminution that produces a material of fine consistency which may be pumpable at 8% solids.) The peat slurry is statically mixed with a suitable alkali and then pumped to an operating pressure of about 2.2 MPa.

Alkaline-oxidation reactor

An operating temperature of 150°C was selected for the base-case with a pressure of 2.2 MPa, a partial pressure of 1.7 MPa air. Maintaining this reaction pressure above saturated steam pressure prevents excessive water loss. Air is introduced into the reactor from reciprocating compressors; intercoolers are used to reduce the heat of compression sufficiently to obtain a 150°C delivery temperature.

Several oxidative reaction vessels are used in parallel, each with a 15-min reaction time. A countercurrent heat exchanger is used to preheat the peat feedstock before it enters the plug-flow reactor [65].

The composition of peat has been determined as: volatile matter, 65%; fixed carbon, 25%; and ash, 10%; having a gross heating value of 19.3 MJ/kg (8280 BTU/lb). As much as 100% of the fixed carbon and volatile material has been solubilized in other experimental studies [24]. For controlled wet oxidation this analysis assumes that 90% of the lignocellulose fraction of peat is solubilized to organic acids and simple sugars. It is important to consider that when organic materials are oxidized during wet oxidation heat is liberated, $\Delta H = -0.39$ MJ/mole; this can be used to satisfy the heat requirements of the pretreatment process. Specifically, it was calculated that when 15% of the available carbon is oxidized to CO_2, enough heat is provided to maintain the reaction temperature (150°C) and the thermophilic fermentation operation (60°C). The insoluble residue from the pretreatment contains about 10% of the heating value of the feedstock. The optimal conditions (temperature, alkali, time, air partial pressure) for wet oxidation of peat to produce high yields of soluble products requires further investigation.

Liquid—solid separation

The solids remaining after wet oxidation are separated from the liquor by high-volume clarification units. The total solids concentration is reduced from 8 to 1.5% by wet oxidation, and in the clarification step, the residue is removed as a sludge at 13% solids incorporating a loss of 10% of the organic liquor. Further dewatering is necessary in order to utilize the residue as a boiler fuel which may result in greater liquid recovery.

Fermentation

The thermophilic anaerobic fermentation unit is of conventional high rate design operating at a temperature of 60°C. Loading of 16 g/day-ℓ (1.0 lb feed/day-ft³) is assumed, based on data from peat and similar substrates (sewage sludge and municipal solid waste digestion units typically have loadings of 0.2 to 0.3 lb/day-ft³ [66, 57]. An average solids retention time of 3 days is based upon recent studies demonstrating rapid fermentations to methane [67, 68].

Microorganism considerations

Anaerobic digestion requires a balance between several different populations of microorganisms. Further experimental work is necessary to determine the effect of peat pretreatment products on each population. Important to microorganism growth is the availability of inorganic nutrients, in particular, nitrogen and phosphorus. The growth nutrient requirements are: 10% for nitrogen and 1% for phosphorus of the biological cells' organic solid weight. Reed-sedge peat, the substrate here, contains a sufficient amount of each.

Biomass recycle

Recycling the microorganisms maintains a high methane productivity in the fermenter. The effluent from the anaerobic fermentation system is continuously withdrawn and delivered to centrifuges. The supernatant liquids are treated to remove unfermented material or most likely recycled to the mining areas with no adverse environmental effects anticipated. Meanwhile, a portion of the resulting microbial cells is re-introduced into the fermenter to maintain as high a concentration of active organisms as is practical. Approximately 1.81 Gg (2000 tons) of new biomass (ca. 20% organic) is produced per day, utilizing about 6.5% of the total carbon entering the fermentation. The excess biomass (estimated composition $C_5H_9O_3N$) has an estimated value of 22¢/kg (10¢/lb). This may be used as an animal feed or fertilizer.

Pipeline gas preparation

The off-gas from the fermentation digesters is at 0.1 MPa, 60°C saturated with water vapor and contains 60% methane and 40% carbon dioxide. To meet pipeline gas quality specifications, a Benfield gas purification unit, which uses hot potassium carbonate solution as an absorption medium, was incorporated in this design [69].

COMPUTER-AIDED ENGINEERING AND ECONOMIC ANALYSIS

A computer model was prepared for using the process parameters to size the equipment and to calculate mass and energy balances, capital, annual operating, and average gas costs. Since the object of the present study was to evaluate the feasibility of a biogasification process, the computer model incorporated the significant parameters in a broad, comprehensive manner and was not intended to be all-inclusive. The base-case input data for process performance reflects values which were determined experimentally or which appear to be justifiable without being unduly conservative or over-optimistic. The base-case design specifies a plant operating continuously at a 90% stream factor (320 operating days/year).

Carbon and energy balance

A carbon material balance for the base-case is given in Table 3 in terms of carbon per day and as a percent of the input carbon. The overall process

TABLE 3

Carbon balance for peat biogasification. Basis: 1 day, 79.2×10^3 GJ/day (75×10^9 BTU/day) plant, base-case parameters

Input (%)	
Peat to process	100.0
Total	100.0
Output (%)	
Alkali-oxidation	
Degraded to CO_2	13.5
Liquid—solid separation	
Unreacted solids after oxidation	10.0
Soluble organic losses	7.6
Fermentation	
Utilized CH_4 product	32.4
Biomass	4.5
Lost CO_2	21.6
Unutilized soluble organics	10.3
Total	100.0

utilization of carbon in the feed peat to methane is 32.4%. An energy balance for the base-case is summarized in Table 4. The energy efficiency (including by-product energy credits) is 64%; the peat to methane energy efficiency is 53.4%. Note that conversion efficiencies are based on projected assumptions and these are presently several times greater than realized in laboratory experiments.

TABLE 4

Energy balance peat to SNG via biogasification process. Basis: as Table 3

Input (%)	
Peat to process	89.5
Energy requirements:	
Pumps	0.4
Air compressor	3.6
Digester mixing	0.2
Centrifuge	1.6
Gas purification: Mechanical	0.8
Steam	3.8
Total	100.0
Output (%)	
SNG	52.4
Biomass	2.6
Unreacted solids (for boiler fuel)	8.9
Heat value losses:	
Oxidation to CO_2	13.4
Other	22.6
Total	100.0

System performance calculation

Estimates of capital costs are as of April, 1978 [70]. Two process by-product credits were assumed. First, the recovered excess biomass produced was valued as an animal feed or a fertilizer. Second, solids remaining after wet oxidation were assumed to have a heating value as a fuel for power generation. The effluent from the anaerobic fermentation system, although of neutral pH and containing a small amount of unutilized organic material, was considered a process penalty. The calculation of the average unit gas costs was based on the public utility method developed in 1961 by the American Gas Association and modified in 1971 by the Panhandle Eastern Pipeline Company [71]. The base-case average gas cost was $3.16/GJ ($3.34/ million BTU) for the 79.2×10^3 GJ/day (75 billion BTU/day) plant, the total capital requirement was $323 million, and the net annual operating cost was $44 million.

Sensitivity evaluation

Because of uncertainties in the values used in the analytical description of the system, and because of unforeseen technical or economical problems or of possible technological breakthroughs ultimately resulting in cost reductions, it was reasoned that wide variations of the product cost from the base-case level might occur. The extent of the deviation of the product cost from the base-case level was used to establish the "sensitivity" of the system to the variation in input parameters.

A list of the parameters investigated, the limits within which they were varied, and the resulting cost range is given in Table 5. In order to evaluate the sensitivity of the methane cost to the variations in the different parameters, a cost sensitivity ratio, S, was defined for each input variable investigated, where S is the fractional increase in gas cost when the value of the variable is increased by a specific amount above its base-case level, i.e.,

$$S = \frac{\text{cost increase above base-case cost}/\text{base-case cost}}{\text{increase in variable above base-case level}/\text{base-case value of variable}}$$

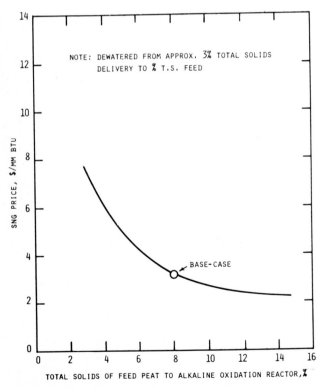

Fig. 5. Effect of total peat feed on methane cost.

TABLE 5

Summary of sensitivity study results including the cost sensitivity ratio, S. (Base-case SNG cost = $3.34/MMBTU)

Variable	Range of variable investigated			Cost range of variable investigated ($/MM BTU)		Cost sensitivity ratio, S
	Lower limit	Base-case	Upper limit	Lower limit	Upper limit	
A. Performance variables						
1. Total solids in feed (%)	3	8	15	7.75	2.38	-1.52
2. Heating value of dry peat (BTU/lb)	7500	8280	10,000	3.66	2.74	-0.86
3. Retention time in fermenter (days)[a]	3	3	10	3.34	4.00	+0.08
4. Unit efficiencies fraction $(A \times B \times C)$[c]	0.48	0.69	0.88	4.24	2.75	-0.64
B. Cost parameters						
1. Delivered peat cost ($/ton)	0.50	3.00	9.00	3.05	5.29	+0.30
2. Fermenter cost ($/ft³)[b]	1.50	3.00	4.50	3.17	3.50	+0.10
3. Total equipment cost ($MM)	81.5	171.8	208.7	2.13	3.84	+0.68
C. Credits and penalties						
1. Unreacted solids credit ($/MMBTU)[d]	0.0	1.75	2.50	3.64	3.17	-0.10
2. Fermenter effluent penalty ($/1000 gallon)	0.0	1.50	3.00	3.08	3.94	+0.17
3. Biomass credit (¢/lb)	0.0	10.0	15.0	4.42	2.76	-0.33

[a] Based on fermenter cost = $3/ft³.
[b] 3-day retention time.
[c] A = fraction ash-free peat solubilized, B = liquid/solid separation efficiency, C = fraction soluble organics to SNG.
[d] Removed from liquid-solid separation at 13% total solids and further dewatered to use as boiler fuel.

In other words, the ratio S for a system variable represents the fractional increase in methane cost when the value of that variable is hypothetically increased. When the value of S for some variable is positive and large, the methane cost increases substantially with an increase in the value of that variable investigated. When S is small, then the cost of the gas is only slightly affected by variations in the variable. When S is negative, the cost of gas decreases with an increase in the variable. Consider, for example, the effect of a cost change of delivered peat. The base-case value was $3.00/dry ton of peat delivered at approximately a 3% total solids slurry. The sensitivity ratio, S, for this variable was estimated to be +0.30 (Table 5). If the peat cost increases to $6.00/dry ton, then the cost of methane would be increased by 30% of the base-case value, i.e., by $0.94 to $4.28/MM BTU. Note that a knowledge of the ratio S is not sufficient to evaluate the impact which a variable may have on the process since it is also necessary to define the limits within which that variable might range. If, for example, a variable had a large S value (positive or negative) its impact on the process may be less important than that of another variable with a smaller S value but which might vary over a much wider range. Several of the variables, both technological and economical,

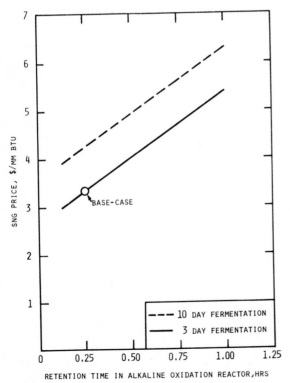

Fig. 6. Effect of retention time in alkaline-oxidation and anaerobic fermentation reactor units on methane cost.

which affect the methane cost to a significant degree are presented in graphical form in Figs. 5 and 6. Fig. 5 shows how the "most sensitive" factor, peat solids in the feed stream, affects gas cost. Fig. 6 shows how retention time in the fermenter, a key processing factor, affects gas cost.

ACKNOWLEDGEMENTS

This work was carried out at Dynatech R/D Company, with supportive experiments at Stanford University, under contract with the Minnesota Gas Company (Minnegasco), Minneapolis, Minnesota. The work was initiated November 1, 1977 and continued for a period of eight months. The authors are indebted to Mr. Paul W. Kramer, President, Minnegasco for the funding of this program and for his encouragement in developing peat resources. A comprehensive proprietary report of 148 pages was submitted by Dynatech to Minnegasco on July 25, 1978.

REFERENCES

1 Soper, E.K. and Osborn, C.C., 1922. U.S. Geological Survey, Bulletin 728.
2 American Society for Testing and Materials, 1969. Standard Classification of Peats, Mosses, Humus, and Related Products, D2607-69.
3 Tibbetts, T.E., 1968. Peat resources of the world — A review, Proc. 3rd Inter. Peat Cong., Quebec, Canada.
4 Hammond, R.F., 1975. The origin, formation, and distribution of peatland resources. In: D.W. Robinson and J.G.D. Lamb (Eds.), Peat in Horticulture, Academic Press, New York.
5 Farnham, R.S., 1967. Information compiled from U.S. Dept. of Agriculture.
6 Rader, A.M., 1977. Utilization of peat to produce Synthetic Natural Gas. Testimony before U.S. House of Representatives Sub-Committee on Environment, Energy, and Natural Resources, Sept. 29, 1977.
7 Clark, F.M. and Fina, L.R., 1952. Arch. Biochem. Biophys., 36: 26.
8 Evans, W.C., 1963. J. Gen. Microbiol., 32: 177.
9 Nottingham, P.M. and Hungate, R.E., 1969. J. Bact., 98: 1170.
10 McCarty, P.L. and Vath, G.A., 1962. Int. J. Air Water Pollut., 6: 65.
11 Rogoff, M.H. and Wender, I., 1962. Oxidation of aromatic compounds by bacteria. Bureau of Mines Bull. 602.
12 Saeman, J.F., 1945. Ind. Eng. Chem., 37: 43.
13 Harris, E.E., Belinger, E., Hajny, G.J. and Sherrord, E.C., 1945. Ind. Eng. Chem., 16: 869.
14 Tarkow, M. and Feist, W.C., 1969. A mechanism for improving the digestibility of lignocellulosic materials with dilute alkali and liquid ammonia. In: Cellulases and Their Applications, American Chemical Society Monograph No. 95.
15 Sjostrum, E., Janson, J., Maglund, P. and Enstron, B., 1965. J. Polymer Sci., 11C: 221.
16 TAPPI Monograph No. 27, 1964. The Bleaching of Pulp.
17 Kato, M. and Shibasaki, I., 1974. J. Fermentation Soc. Jap., 52: 177.
18 Howard, H.C., 1945. In: H.H. Lowry (Ed.), Chemistry of Coal Utilization, Vol. I, Wiley, New York.
19 Fisher, F., 1919. Gesammelte Abh. Kennt. Kohle, 4: 13.
20 Fisher, F. and Schrader, H., 1920. Gesammelte Abh. Kennt. Kohle, 5: 307.

21 Fisher, F., Schrader, H. and Treib, W., 1920. Gesammelte Abh. Kennt. Kohle, 4: 13.
22 Juettner, E., Smith, R.C. and Howard, A.C., 1937. J. Am. Chem. Soc., 59: 236.
23 Kasehagen, L., 1937. Ind. Eng. Chem., 29: 600.
24 Smith, R.C. and Howard, H.C., 1939. J. Am. Chem. Soc., 61: 2398.
25 Ruof, C.H., Savich, T.R. and Howard, H.C., 1951. J. Am. Chem. Soc., 73: 3873.
26 Franke, N.W., Kiebler, M.W., Ruof, C.H., Savich, T.R. and Howard, H.C., 1952. Ind. Eng. Chem., 44: 2783.
27 Montgomer, R.S. and Holly, E.D., 1956. Fuel, 35: 60.
28 Parker, F.G., Fugassi, J.P. and Howard, H.C., 1955. Ind. Eng. Chem., 47: 1586.
30 Kamiya, Y., 1961. Fuel, 40: 457.
31 Kamiya, Y., 1963. Fuel, 42: 347.
32 Kamiya, U., 1963, Fuel, 42: 353.
33 Kukharenko, T.A., 1972. Zh. Prikl. Khim., 45(1): 172.
34 Harris, G.C., U.S. Patent 2,673,148, 1954.
35 Dennis, J., 1976. The oxidation of lignin by gaseous oxygen in aqueous media. Ph.D. Thesis, University of Washington.
36 Grangaard, D.H., 1963. Decomposition of lignin, Part II. Oxidation with gaseous oxygen under alkaline conditions. Paper presented at the 144th meeting of the American Chemical Society, March, 1963.
37 Grangaard, D.H., Canadian Patent 611,507, 1961.
38 Merriman, M.M., Choulett, H. and Brink, D.L., 1966. Tappi, 49: 34—39.
39 Bicho, J.G., Zavarin, E. and Brink, D.L., 1966. Tappi, 49: 218—226.
40 Brink, D.L., Wu, Y.T., Naveau, H.P., Bicho, J.G. and Merriman, M.M., 1972. Tappi, 55: 719—721.
41 Schlager, L.L. and Brink, D.L., 1978. Tappi, 61(4): 65—68.
42 Fang, P., McGinnis, G.D. and Parish, E.J., 1975. Wood Fiber, 7: 136—145.
43 Fang, P. and McGinnis, G.D., 1975. J. Appl. Polym. Sci., Sym. 28: 363—376.
44 Fang, P. and McGinnis, G.D., 1976. In: F. Shafizadeh et al. (Eds.), Uses and Properties of Carbohydrates and Lignins, Academic Press, New York.
45 Melin, E., Norbin, S. and Oden, S., 1926. Ingeniorsventenskapsakademien, Handligar No. 53, Stockholm.
46 Buswell, A.M., 1934. Illinois State Water Survey Bull. No. 32.
47 Evans, W.C., 1969. In: D. Perlman (Ed.), Fermentation Advances, Academic Press, New York, pp. 649—687.
48 Cheng, K.J., Jones, G.A., Simpson, F.J. and Bryant, M.P., 1969. Can. J. Microbiol., 15: 1356.
49 Rogoff, M.H. and Wender, I., 1957. J. Bact., 73: 264.
50 Taylor, B.F., Cambell, W.L. and Chinoy, I., 1970. J. Bacteriol., 102: 430.
51 Taylor, B.F. and Heek, M.J., 1972. Arch. Microbiol., 83: 165.
52 Ponsford, A.P., 1966. Br. Coal Util. Res. Assoc., Mon. Bull., 30(1): 1.
53 Ponsford, A.P., 1966. Br. Coal Util. Res. Assoc., Mon. Bull., 30(2): 41.
54 Gossett, J.M., Healy, J.B., Stuckey, D.C., Young, L.Y. and McCarty, P.L., 1976. Heat treatment of refuse for increasing anaerobic biodegradability. Stanford University Report on NSF/RANN Contract AER-74-17940-A01.
55 Goodrich, R.D., Meiske, J.C. and Charib, F.H., 1972. World Rev. Anim. Production, 8: 4.
56 Toerien, D.F. and Hattingh, W.H.J., 1969. Water Res., 3: 385.
57 Wise, D.L., 1974. Fuel gas production from solid waste. Dynatech R/D Company Report on NSF Contract C-827, July 31, 1974. Cambridge, MA.
58 Wise, D.L., 1974. An engineering economic analysis of fuel gas production from solid waste. Paper presented at the Am. Soc. for Microbiology Conf., U. of Wisconsin, September, 1974.
59 Tarvin, D. and Buswell, A.M., 1934. J. Am. Chem. Soc., 56: 1751.
60 Fina, L.R. and Fiskin, A.M., 1960. Arch. Biochem. Biophys., 91: 163.

61 American Public Health Association, 1976. Standard Methods for the Examination of Water and Wastewater. 14th edn, Public Health Association, New York.

62 Fiske, C.H. and Subbarow, Y., 1925. J. Biol. Chem., 156(2): 375.

63 Wasp, E.J., 1979. Pipeline Gas J., 206(2): 50.

64 Personal communications with Messrs. S. Swan, B. Johnson and K.C. Stregig, U.S. Bureau of Mines, Twin Cities Mining Research Center, Minneapolis, Minnesota, 1978.

65 Myreen, B., 1979. The peat fuel process. Paper presented at Finn Energy '79 Symposium, Finnish Ministry of Trade and Industry, New York, January 23, 1979.

66 Cooney, C.L. and Wise, D.L., 1975. Bioeng. Biotech., 17: 119.

67 Bryant, M., 1976. Thermophilic anaerobic digestion of animal waste, unpublished manuscript, University of Illinois.

68 Finney, C.D., Evans, R.S. and Finney, K.A., 1978. Fast production of methane by anaerobic digestion. Natural Dynamics Report on DOE Contract EY-76-C-02-2900.

69 Ashare, E., 1977. Evaluation of systems for purification of fuel gas from anaerobic digestion. Dynatech R/D Company Report on U.S. ERDA Contract EY-76-C-02-2991, Cambridge, MA.

70 Fox, A.J. (Ed.), 1978. Engineering News Record, 200(18): 32.

71 Siegel, H.M., Kalina, T. and Marshall, H.A., 1972. Description of gas cost calculation methods being used by the synthetic gas-coal task force of the FPC national gas survey. Esso Research and Engineering Co. Report to the Federal Power Commission, June 12, 1972.

Energy Recovery from Lignin, Peat and Lower Rank Coals, edited by D.J. Trantolo and D.L. Wise
Elsevier Science Publishers B.V., Amsterdam 1989 — Printed in The Netherlands

Chapter 2

CONTINUED DEVELOPMENT OF A PEAT BIOGASIFICATION PROCESS*

D.L. WISE**, E. ASHARE, C.F. RUOFF

Dynatech R/D Company, 99 Erie Street, Cambridge, Massachusetts 02139 (U.S.A.)

and M.J. KOPSTEIN

U.S. Department of Energy, Washington, D.C. 20585 (U.S.A.)

ABSTRACT

Bench-scale experimentation and development of an economic process model were conducted for the conversion of peat to substitute natural gas (SNG). This work was a continuation of an earlier study [1]. The process consists of three steps: (a) hot (150°C) aqueous alkali solubilization of peat so that the complex ligneous structure is converted into single ring aromatic compounds with one or more 3 to 5 carbon side chains; (b) oxidation of these side chains, the exothermic reaction also providing the process heat; and (c) anaerobic fermentation of the solubilized/oxidized peat to fuel gas. While both solubilization and oxidation steps may be carried out in the same reaction vessel, these are two discrete processing steps.

Solubilization follows first order reaction kinetics with respect to peat and hydroxyl ion concentrations with $k_s = 2.30 \times 10^6 \exp(-3300/T)$ (l/h mol). The oxidation of solubilized peat was found to follow first order reaction kinetics with respect to solubilized peat and oxygen with $k_o = 7.5 \times 10^{14} \exp(-14560/T)$ (l/h mol), T in K. Methane fermentation of the pretreated peat followed first order kinetics; the rate constant was 0.015 day^{-1} under mesophilic (37°C) and continuous fermentation conditions. This constant is believed to be low primarily due to sodium ion inhibition from added soda ash. Methane fermentation of model compounds, assayed in the peat liquor, were much higher, for example, 0.1 day^{-1} for benzoic acid and 0.18 day^{-1} for syringaldehyde.

The fuel gas processing cost was calculated using the utility financing method approved by the American Gas Association and the kinetics and yields determined experimentally. A conservative "base cost" for pipeline quality methane was determined to be \$5.6/GJ (\$5.9/million BTU) for a 74×10^3 GJ/day (70 billion BTU/day) plant processing more than 7.3 Mg/day (8000 tons/day) (dry) of peat. A sensitivity analysis revealed that the alkali requirements and digester volume were most critical to the methane cost and that by further process development the average gas cost may be expected to be reduced to \$2.8—3.8/GJ (\$3—4/million BTU).

INTRODUCTION

The stoichiometry of the formation of several classifications [2] of peat may be described as [3]:

$$C_{72}H_{120}O_{60} \rightarrow C_{62}H_{72}O_{24} + 8\,CO_2 + 2\,CH_4 + 20\,H_2O$$

Sufficient quantities of peat exist in the United States to contribute sig-

---

*Reprinted from *Resources and Conservation*, 8 (1983) 213-231
**Present address: Cambridge Scientific, Inc., 195 Common Street, Belmont, MA 02178 (U.S.A.)

nificantly to domestic energy needs in the coming decades. The states with major resources, ranked as to energy potential, are Alaska (only peat outside the permafrost regions is considered), Michigan, Minnesota, Florida, Wisconsin, Louisiana, North Carolina, Maine and New York. Using three criteria that were employed by Farnham [4], the U.S. energy potential from these resources can be estimated. The criteria are: (1) an average peat deposit depth of 2.1 m, (2) an average bulk density of 240 kg/m^3 (for 35% moisture content peat), and (3) an average heating value of 14 MJ/kg (for 35% moisture content). On this basis, the total potential energy available from domestic supplies of peat is 1.5×10^{12} GJ thus ranking peat second in recoverable energy to the nation's coal reserve.

Numerous technical problems are associated with the large-scale production of energy from peat. The main obstacle in producing either SNG or electricity is the high moisture content of peat — it is almost 90% water. Dewatering is difficult and consumes large amounts of energy. The process described here, peat biogasification, bypasses this problem. In this process [1] the peat is assumed to be harvested wet, carried and treated through fermentation, without water removal, as described below.

PROCESS DESCRIPTION AND METHODS FOR EVALUATION

The peat biogasification process may be described in three steps. In the first step, peat and alkali (Na$_2$CO$_3$) are reacted at elevated temperatures to solubilize the peat by hydrolysis, producing many types of substituted phenols. In a second, discrete step, the solubilized peat is oxidized so that the side chains on the aromatic rings are converted into carboxylic and aldehydic functional groups. The oxidation is exothermic and provides sufficient heat for the first two steps.

The third step is anaerobic fermentation of the peat reaction products to methane and carbon dioxide. Anaerobic fermentation of aromatic carboxylic acids and aldehydes has been carried out experimentally [5—12]. Complete bioconversion to methane and carbon dioxide has been reported for benzoic acid, phenylacetic acid, hydrocinnamic acid, and cinnamic acid. The inoculum was anaerobic sewage sludge. Radioactive tracer experiments have shown that benzoic acid is converted to methane and carbon dioxide with a conversion efficiency greater than 9% [8].

PROCESS MODELLING

The present work was devoted to developing a process model that described the complex reactions occurring in peat biogasification. This model, using the three steps described above, was then used to determine process economics.

For the solubilization, it was assumed that the peat was solubilized by the following reaction:

$$C_P + C_B \rightarrow C_S,$$

where: C_P = concentration of insoluble peat; C_B = concentration of hydroxyl ion; C_S = concentration of soluble peat; then:

$$\frac{dC_S}{dt} = kC_P C_B,$$

where k is the absolute reaction rate constant. Upon integration, $C_S = kC_P C_B dt$. Analysis of data obtained followed this procedure:

- plot $C_P C_B$ vs. t;
- measure area vs. t, i.e. $\int C_P C_B dt$;
- plot C_S vs. area ($\int C_P C_B dt$) for each t;
- least squares fit gives slope which equals k.

This procedure was also followed for the oxidation and fermentation phases of the process model. For the numerous oxidation reactions, the following was assumed:

$$C_s + C_o \rightarrow C_F + C_R,$$

where: C_s = concentration of solubilized peat; C_o = concentration of oxygen; C_F = concentration of fermentable products; C_R = concentration of refractory products.

The analysis procedure was the same as described above, but there were four rate equations to be developed, i.e., dC_s/dt, dC_o/dt, dC_F/dt, and dC_R/dt. The development of the equations and the oxidation model was dependent upon measuring the concentrations of fermentable and refractory materials and their change with time.

The third model was used to describe the anaerobic fermentation phase. The following reaction was also assumed to follow first order chemical reaction kinetics:

$$C_F \rightarrow C_{CH_4} + C_{CO_2},$$

where: C_F = concentration of fermentable materials; C_{CH_4} = concentration of methane; C_{CO_2} = concentration of carbon dioxide.

Batch digesters were set up to yield data for total biodegradability and from this C_F was calculated. Continuous digesters were set up to give conversion rates.

METHODS AND MATERIALS

The data for the oxidation and solubilization phases of the process model were obtained using a two-liter Parr reactor. Reed sedge peat, obtained from the Red Lake area of Northern Minnesota (courtesy of the Minnesota Gas Company), was used in all cases. Peat slurry (3% total solids) was mixed in a blender and added to the reactor. The reactor was heated to temperature, and a known amount of alkali injected into the reactor with nitrogen gas. Liquid samples were withdrawn at regular intervals and measured for total solids, volatile solids, dissolved solids, dissolved volatile solids, and pH. Samples were retained for high pressure liquid chromatographic (HPLC) analysis.

Initially, the oxidation studies were carried out in a manner similar to the solubilization studies. Solubilized peat was added to the Parr reactor and heated to the desired temperature. Oxygen was injected and its concentration measured as a function of time. Liquid samples were analyzed as before.

Later, another set of oxidation experiments was run. Solubilized peat was added to the reactor, which was then heated to temperature. Air was bled into the reactor at a constant flow rate. A wet test motor recorded the flow rate output and gas compositions were measured periodically through the sampling septum. The amount of oxygen consumed was computed from the nitrogen/oxygen composition of the off-gas. Liquid samples were withdrawn periodically and total solids, volatile solids, dissolved solids, and volatile dissolved solids measured. Samples were analyzed by HPLC. The concentrations of refractory and fermentable products were monitored as a function of time.

In the third phase of the process, anaerobic digestion, two types of systems must be used to obtain the necessary data for the process model; batch digesters and continuous digesters. Batch digesters were used to determine the ultimate biodegradability of the material, while continuous digesters were used to determine the rate of fermentation at different retention times.

In the batch digesters, sewage sludge was mixed with the substrate, nutrients added, and the pH controlled by buffer addition. A control batch digester was charged with sewage sludge to determine gas production as a blank. Batch digesters were established by continuous culture transfers with the following aromatic compounds as substrates: benzoic acid, cinnamic acid, syringic acid, and syringaldehyde.

Batch digesters were established for peat at different pretreatment conditions. Sewage sludge and cultures acclimated to the aforementioned aromatic compounds were used as the inocula for digestion of the pretreated peat liquor. Nutrients were added in some cases and the pH was controlled by buffer addition. Continuous digesters were operated to determine a rate constant for the anaerobic digestion of pretreated peat. These digesters were started by adding 100 ml of cultures acclimated to benzoic acid, syringic acid, cinnamic acid, and syringaldehyde, 100 ml of sewage sludge, and 200 ml of pretreated peat. After significant gas production began, the digesters were drained and fed daily with pretreated peat. Two digesters were used, one on a twenty-day retention time and the other thirty-day; that is, the total volume of the digester (400 ml) was replaced in twenty days in one case and thirty days in the other case.

EXPERIMENTAL RESULTS

Solubilization

The kinetic analysis was carried out by fitting the solubilization data to

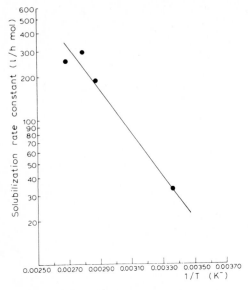

Fig. 1. Activation energy for solubilization of peat with alkali; here the value was found to be E_{act} = 27 kJ/mol (6.5 kcal/mol).

the proposed model as described earlier. It was assumed that the rate constants of solubilization at several temperatures could be described by the Arrhenius relation, $k_s = A \exp (E_{act}/RT)$, where A is a constant, E_{act} the activation energy, R the universal gas constant, and T absolute temperature in K. A plot of the results from the peat solubilization is given in Fig. 1. The slope of the line, obtained from least squares regression analysis, yielded an activation energy of 27 kJ/mol as shown in Fig. 1. The rate equation for peat solubilization was:

$$k_s = 2.30 \times 10^6 \exp (-3300/T) \qquad (1/\text{h mol})$$

This rate expression was used in the process model, described later.

Oxidation

In the initial oxidation experiments solubilized, filtered peat was added to the reactor, the reactor was heated to temperature, and oxygen injected. Oxygen consumption was measured as a function of time and plotted in Fig. 2. It is likely that the reaction occurs according to the proposed mechanism or hypothesis that as the reaction proceeds the concentration of oxygen remains proportional to the concentration of unoxidized peat. Thus a correlation may be developed solely in terms of oxygen concentration in which the reaction of solubilized peat (C_s) to products is

$$C_s + C_o \rightarrow C_F + C_R$$

28

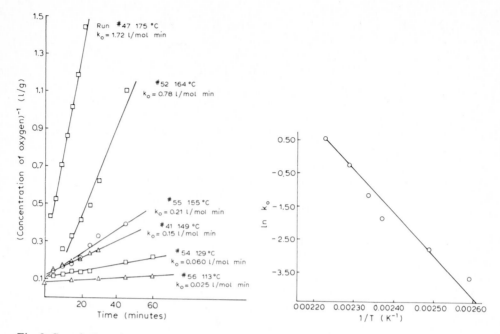

Fig. 2. Correlation of rate of oxidation of solubilized peat according to second order reaction kinetics (based on stoichiometric ratio of reactants).

Fig. 3. Activation energy for oxidation of solubilized peat; here the activation energy was
found to be E_{act} = 95 kJ/mol (22.6 kcal/mol).

The kinetic relationship, assuming oxygen (C_o) and solubilized peat (C_s)
are present in stoichiometric amounts, is

$$\frac{dC_o}{dt} = -k\,C_o^2$$

The results were analyzed and Fig. 2 shows the rate constants calculated for
each temperature assuming stoichiometric amounts of oxygen and solubilized
peat. As with the solubilization model, an Arrhenius relationship was assumed between rate constant and temperature. Fig. 3 shows a plot of ln k_o
versus 1/T; the slope gives E_{act} = 95 kJ/mol. The important item in the rate
constant obtained by this analysis is the activation energy which agrees well
with the oxidation tests discussed below.

The second type of oxidation experiment was conducted to monitor product formation with time using the HPLC. The HPLC analyses were conducted
by calculating the total area under the HPLC curve and calculating the percentage due to each fraction of peat, i.e., the hydrolyzed and oxidized fraction. The oxidized products were then measured as a function of time.

Due to differences in the gradient elution solvents used in the HPLC
analysis it was assumed that in the solubilized peat 40% of the material was
measured as the oxidized peak. As the oxidation proceeded, the fraction of

oxidized peat progressively increased. Thus the percentage of oxidized product could be determined as a function of time by measuring the area under the curve on the HPLC output. Implicit in this analysis is the assumption that both the solubilized and the oxidized product had similar extinction coefficients (at 254 nm). On this basis, a semilogarithmic plot of (100% conversion) versus time was made for each oxidation run (Fig. 4) following the rate expression for the oxidation mechanistic model of $dC_F/dt = k_o C_s C_o$. Here C_F refers to the oxidized portion of the HPLC chromatographs and C_s refers to the hydrolyzed portion.

Therefore, the material in the pretreated peat liquor is the $C_F + C_s$; thus by basing the analysis on percentages of the pretreatment mix, $C_F + C_s = 100\%$. Since C_o is constant throughout each continuous oxidation run, the integral of the rate expression becomes

$$\ln (100 - C_F) = -k_o C_o t + \text{constant.}$$

By plotting $\ln (100 - C_F)$ versus t for each temperature (Fig. 4) a slope of $k_o C_o$ was obtained; by dividing out the partial pressure of oxygen in the reactor for each test, a rate constant for oxidation was obtained for each temperature. Again, an Arrhenius relation was assumed between the rate con-

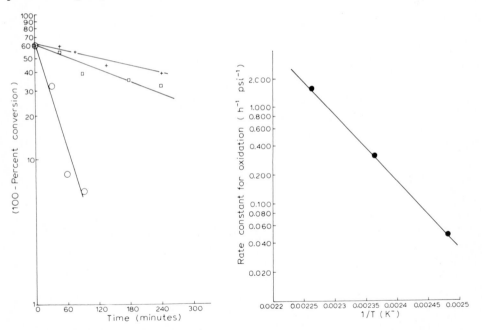

Fig. 4. Calculation of apparent rate constants for oxidation of solubilized peat. The results were found to be: + (temperature 130°C) : $k_o C_o = 0.122$ h^{-1} ; □ (temperature 150°C): $k_o C_o = 0.164$ h^{-1} ; ○ (temperature 170°C) : $k_o C_o = 0.167$ h^{-1}.

Fig. 5. Calculation of oxidation activation energy; here the activation energy was found to be $E_{\text{act}} = 121$ kJ/mol (29.0 kcal/mol) (1 psi = 6.895 kPa).

stant and the temperature. The rate constant was plotted semi-logarithmically versus inverse temperature in Fig. 5. From the calculations, as plotted in Fig. 5, an activation energy was obtained of 121 kJ/mol. This compares reasonably well with the initial set of oxidation trials discussed earlier, in which an activation energy of 95 kJ/mol was obtained. The rate constant expression for the oxidation was $k = 7.5 \times 10^{14} \exp(-14650/T)$ (l/h mol). The results of these two sets of rate experiments show that the mechanistic model proposed appears to describe the complex set of oxidation reactions occurring in the pretreatment.

Fermentation

Results from the acclimated batch digester using benzoic acid and syringic acid are shown in Figs. 6 and 7. First order kinetic constants were determined to be 0.13 and 0.18 day^{-1}, for benzoic and syringic acids respectively. Further, essentially complete conversion of the substrates was achieved. These values for the kinetics are higher than the values of 0.1 day^{-1} previously determined [13] for an array of cellulosic substrates such as straw, etc. On the other hand, a kinetic constant for the anaerobic digestion of whey was determined [14] to be 0.2 day^{-1}. It may be noted that acclimated cultures fermenting water

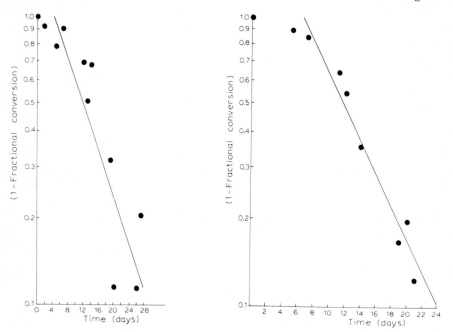

Fig. 6. Determination of benzoic acid rate constant; here the rate constant was found to be $k = 0.13$ days^{-1}.

Fig. 7. Determination of syringic acid rate constant; here the rate constant was found to be $k = 0.184$ days^{-1}.

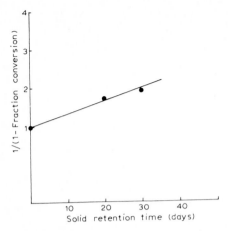

Fig. 8. Determination of fermentation rate constant; here the rate constant was found to be $k = 0.015$ days^{-1}.

soluble organic substrates may be expected to act more rapidly than those fermenting non-soluble cellulosic substrates.

Batch digestion of pretreated peat resulted in conversion of 60% of the total volatile solids in the feed. A first order kinetic constant for the continuous pretreated peat digester may be obtained by plotting 1/(1–fractional conversion) vs. SRT (solids retention time) as shown in Fig. 8, in which a value of 0.015 days^{-1} was determined.

This rate constant for fermentation of the pretreated peat or cooked peat liquor is extremely low and, as mentioned before, probably due to a high sodium loading ($\sim$3000 mg/l). According to [15,16], sodium toxicity in wastewater sludge digestion occurs in the 5000 to 8000 mg/l range. However, the concentration at which toxicity occurs is modified by antagonism, synergism, and acclimatization. It is thus probable that 3000 mg/l of Na$^+$ is inhibiting the digestion of the pretreated peat. Another possibility is the lack of nitrates or electron acceptors [17]. It appears that some sort of electron acceptor is essential for the anaerobic degradation of benzoic acid. In the batch fermentations of solubilized and oxidized peat, the addition of KNO$_3$ to the digesters increased gas production by a substantial amount ($\sim$5–10%). This is another possible inhibition factor that needs to be studied more thoroughly. Without this inhibition, it is anticipated that the rate constant can be increased.

ECONOMIC ANALYSIS

An economic analysis was carried out with several considerations, as follows.

Slurry concentration

In the slurry concentration step, the peat slurry — assumed to be obtained by wet harvesting — is dewatered from 3 to 8% solids. The efficiency of this process step has been assumed to be 100% (i.e., no peat solids pass through the filter). The efficiency must be determined, but it is expected to be about 90%.

Pretreatment reactor

The pretreatment step consists of two stages, solubilization followed by oxidation. A two-step plug flow reactor concept is assumed. Reaction for typical model compounds may be:

CH_2OH
$-CH$
$HO-CH$

$+ 2\frac{1}{2} O_2 \longrightarrow$ CHO $+ 2 CO_2 + 2 H_2O$

OCH_3 OCH_3
O OH

PEAT   VANILLIN

Here the heats of formation of peat and vanillin are approximately 23.9 and 25.2 MJ/kg respectively. The heat of reaction for this reaction is, therefore, about -4.3 MJ/kg solubilized peat reacted. The amount of heat produced daily is more than twice the heat required for the sensible heat difference between the influent and effluent stream and is, therefore, sufficient to provide the heat for the process.

Another energy requirement in the pretreatment step is the energy necessary to compress the oxygen to the desired pressure. For every unit weight of peat solids, the required alkali is 0.3 weight units of Na_2CO_3 and the required oxygen is 0.3 weight units. If oxygen were supplied by compressed air, then 1.05 weight units of N_2 would be supplied to the oxidation reactor which would have to be removed, preferably after the fermentable product stream has been cooled to eliminate significant water and energy loss. For the base line design it is assumed that 90% of the solubilized product is oxidized to fermentable products. The assumptions are based on the results presented earlier.

The "base case" preliminary process design was followed by an economic process model evaluation and sensitivity analysis. A process flowsheet for the base case is shown in Fig. 9, and assumptions are given in Table 1. A summary of this cost analysis follows.

For a pretreatment temperature of 150°C (the rate constant is 944 l/h mol), a conversion of 90%, and an initial peat concentration of 8% solids ($\sim$0.4 mol/l), the retention time for solubilization is 0.024 hours. The size of the

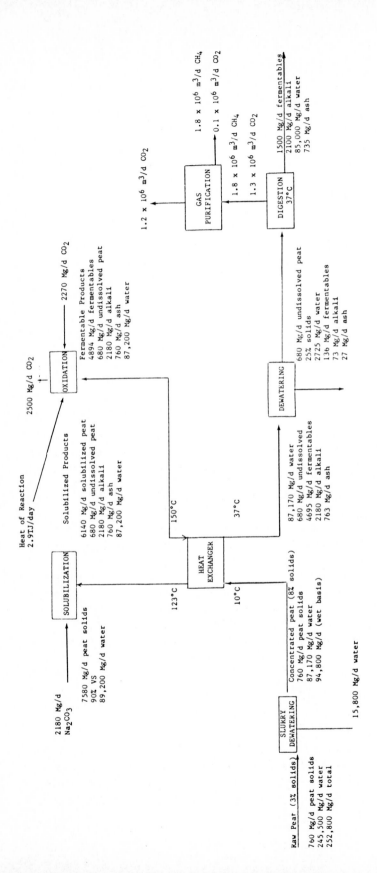

Fig. 9. Overall flowsheet for base case design.

TABLE 1

Base case design process assumptions

Process step	Assumptions
Slurry preparation	3% Solids feed 8% Solids output
Pretreatment—solubilization	Plug flow reactor 90% Conversion 0.4 kg O_2/kg Solubilized product converted 150°C
Heat exchanger	10°C Inlet slurry temperature 37°C Exit temperature for feed to digester
Liquids/solids separation	Undissolved solids separated to 25% solids
Digestion	CSTR 37°C 20 Day retention time 67% Conversion of fermentable products Gas composition, 60% CH_4/40% CO_2
Gas purification	3% CO_2 Concentration in purified gas 6.8 MPa Output pressure Power requirement per day of 7.90 W/m³

solubilization pretreatment reactor is strongly dependent on slurry concentration, temperature (which affects the rate of reaction), and desired conversion. The volume of the pretreatment oxidation reactor was based on 90% conversion, $k_0 = 4.4^{-1}$ l/h mol, and a calculated retention time of 0.5 h. For a 0.1 MPa oxygen partial pressure and oxygen supplied as compressed air, the pretreatment reactor must be designed for 1 MPa operating pressure. This reactor size is strongly dependent on solids concentration, reaction temperature, and conversion.

Another important aspect of the pretreatment step is the energy requirement. The major energy requirements are found by computing the sensible heats of the inlet and exit streams and heat of reaction for the oxidation reactor. This heat may be provided by the heat of reaction of the peat oxidation step. The heat of reaction for oxidation can be estimated from the differences in higher heating values of solubilized peat and fermentable products. For the base case design, it is assumed that the oxygen partial pressure is 0.1 MPa. The total system pressure is dependent on reactor temperature and oxygen partial pressure. For a reactor temperature of 150°C, the steam pressure will be about 0.47 MPa, so if pure oxygen were supplied, the total system pressure would be about 0.57 MPa. If air were used to provide oxygen the total system pressure would be about 1 MPa. The power required to compress air from 0.1 to 1.0 MPa to provide the needed oxygen can be estimated (assuming isothermal compression) by: watts = $3.25 P_1 Q \ln (P_2/P_1)$

where Q is the inlet gas flow. If pure oxygen were used, the energy requirement would be greater [19].

Digester

Digestion system design is a function of yield, kinetics, temperature, fermentable solids content, and product gas composition. For an assumed rate constant of 0.1 day^{-1} and twenty-day retention time, the fractional conversion of fermentable solids is $k/(1 + k) = 0.67$. The size of the digester is dependent on assumed values for slurry concentration (8%), retention time (20 days), and the throughput. The major capital cost component of the base case design is the digestion step. Other more efficient digestion concepts are necessary to reduce this capital cost. These concepts include the anaerobic contact process (which uses cell recycle), an example of which is the anaerobic packed bed filter. The anaerobic contact process is applicable to soluble, fermentable substrates with retention times of less than one day. This would reduce the digester volume to about 1/50 of the volume of a continuous stirred tank reactor (CSTR).

Gas purification

A gas purification step is necessary to provide pipeline quality gas. The digester gas must be treated to remove CO_2, H_2O, and other impurities. The digester gas will consist of approximately 60% CH_4 and 40% CO_2 (on a dry basis) [20].

The energy requirements for the process are for electric power for dewatering, mixing, and gas purification and compression. This power requirement, assuming 32.5% conversion efficiency, is about 24% of the gross energy output. The net energy output is 33% of the energy content of the peat feed.

Cost information

The "base case" capital costs are presented in Table 2. The costs were estimated from literature values and were updated to December 1980 costs. The costs presented in Table 2 indicate a total plant investment of $266 × 10^6. The total equipment costs are $192 × 10^6, of which 50% can be attributed to the digester costs and 35% to gas purification. Any major changes in process design should be directed at the digestion step, since this is where the greatest reduction in capital costs can be achieved. The gas purification capital costs were developed for an existing process, and it is not expected that any major reduction in this component cost can be achieved. In the sensitivity analysis, given later, the effect of reducing digester retention time, and hence digester volume and costs, will be examined.

The operating costs for the base design are presented in Table 3. For this

TABLE 2

Base case capital costs

	Cost ($)
Slurry preparation	100,000
Pretreatment — solubilization	100,000
— oxidation	1,300,000
Heat exchanger	2,300,000
Liquid/solids separation	8,100,000
Digesters	97,000,000
Gas purification	6,800,000
Pumps, piping, electrical, instrumen- tation	15,800,000
Total equipment costs	131,500,000
Supporting facilities	6,600,000
Total capital investment	138,100,000
Contractor's overhead & profit	13,800,000
Engineering & design	6,900,000
Subtotal plant investment	158,800,000
Contingency	23,800,000
Total plant investment	182,600,000
Interest during construction	32,900,000
Start-up	23,000,000
Working capital	3,600,000
Total capital requirements	242,100,000

TABLE 3

Base case annual operating costs

	Cost ($)
Raw materials	
Peat ($3.3/Mg or $3/ton)	9,100,000
Na_2CO_3 ($66/Mg or $60/ton)	52,600,000
Utilities	
Electric (5¢/kWh)	28,100,000
Labor	
Operating	3,100,000
Maintenance	4,000,000
Supervision	1,100,000
Administration and overhead	4,900,000
Supplies	
Operating	900,000
Maintenance	4,000,000
Local taxes and insurance	7,200,000
Gross operating cost	115,000,000

analysis, it was assumed that the peat cost is $3/Mg(dry), electric power is
5¢/kWh, and the operating labor requirement is 250 people (at $6/h).

The annual operating cost is estimated to be $115 million. The major con-
tributions to the annual operating costs are the cost of alkali and electricity
(46% and 25%, respectively). (It is assumed that the oxygen is supplied by
compressing air; this cost is included in the electric cost and amounts to about
10×10^6. If oxygen were purchased at $0.077/kg, the annual costs would
be about 50×10^6, or five times greater.) This suggests that the potential
areas for process improvement which could significantly decrease the unit
gas cost are alkali requirements and power requirements. The effects of
changing these variables on the unit gas cost are discussed in the sensitivity
analysis.

The unit gas cost for the base case design was calculated to be $5.61/GJ.
About 79% of this cost is due to operating costs, so that the alkali cost contributes more than 36% to the unit gas cost.

Sensitivity analysis

A sensitivity analysis was carried out to show the effects of changing process and economic variables on the unit gas cost. The effects of some of the
key variables are shown in Fig. 10. A discussion of the effects of these process variables on cost of pipeline quality methane appears below.

Slurry composition and peat cost
In the slurry preparation step it was assumed that the slurry would be dewatered to 8% solids before it was fed to pretreatment reactor. A change in
the slurry concentration will affect the required equipment size and cost. The
energy balance in the pretreatment step will also be affected since the volume
of water flow would change and, therefore, the sensible heat in the streams
would change. The results of changing slurry composition on unit gas cost
are shown in Fig. 10, for a concentration range of 4 to 12% (base case = 8%).
Over this range, the unit gas cost decreases by 35% when the concentration
increases from 4 to 12%. The sensitivity at lower concentrations is greater
since there is a greater percentage change in liquid volume, i.e., reducing the
concentration from 8 to 4% results in doubling the liquid volume, whereas
increasing the concentrations from 8 to 12% decreases the liquid volume by
one-third. In the base case economic analysis, the cost of peat was assumed
to be $3.3/Mg. This contributed about $0.33/GJ to the unit gas cost. The
sensitivity of the unit gas cost to peat cost is about $0.11/GJ per $1.1/Mg
peat cost.

Pretreatment reaction
The pretreatment reactor conditions included a plug flow reactor at 150°C
with 90% conversion for solubilization and 90% conversion for oxidation.
This resulted in retention times of 0.024 h and 0.5 h for solubilization and

38

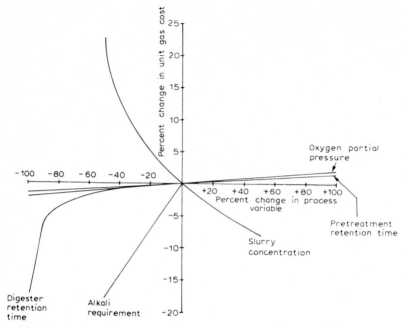

Fig. 10. Sensitivity of unit gas cost to process variables.

oxidation, respectively, when the kinetic relationships developed from the experimental data were used.

Changes in pretreatment conditions, such as temperature, amount and type of alkali, or oxygen partial pressure, will change the pretreatment retention times. (The effect on unit gas cost of varying retention time is not specified; the results are presented to determine whether a change in retention time significantly affects unit gas cost.) The results indicate that there is little sensitivity of unit gas cost to changes in this retention time. An increase in retention of an order of magnitude (e.g., from 0.5 to 5.0 h) results in only a three percent increase in unit gas cost.

Another factor which could significantly affect unit gas cost is the cost of pretreatment chemical. For the base case design, it is assumed that 0.3 kg Na_2CO_3/kg peat is required with no recycle of Na_2CO_3. If a recycling process were utilized, the amount of Na_2CO_3 addition would be reduced. Fig. 10 shows the effect on unit gas cost of changing the amount of Na_2CO_3/kg peat added. For the base case, 0.3 kg Na_2CO_3/kg peat is added, whereas with 50% recycle, only 0.15 kg Na_2CO_3/kg peat will be added. The sensitivity of this change is significant, with 50% recycle resulting in a 20% decrease in unit gas cost.

Since the pretreatment chemical requirement is a highly sensitive variable, it is one area which should be considered for potential process change. In particular, pretreatment should be investigated to determine conditions which would give the same yield of solubilized product but require significantly less chemical addition. This would probably result in an increased pretreatment retention time, but since the sensitivity to retention time is insignificant, the net result would be a decreased unit gas cost.

For the base case design, the oxygen partial pressure was assumed to be 0.1 MPa and the oxygen was supplied as compressed air. If the oxygen partial pressure were changed, there would be a change in compression requirement and hence electricity cost. The effect of varying oxygen partial pressure on unit gas cost is shown in Fig. 10, assuming the only effect is on electricity cost. There is very little sensitivity of unit gas cost to oxygen partial pressure, and if a change in oxygen partial pressure were to result in a significant change in pretreatment retention time (which is also a low sensitivity variable), unit gas cost would not change significantly.

Digestion

Digester cost contributes about 70% to the capital costs of the system when a 20 day retention time is used. It is apparent that reducing digestion retention time would significantly reduce cost. It has been shown [21,22] that anaerobic contact system, such as anaerobic packed bed filter, is capable of converting a soluble substrate to methane with a retention time of the order of 4—10 hours and giving a conversion efficiency of more than 75%. This type of system has not been used with solubilized, oxidized peat product, so the applicability can only be hypothesized. The effect on unit gas cost using a high rate system is shown in Fig. 10. The analysis uses the assumption that the conversion efficiency is the same as for a CSTR, but retention time is reduced. Costs for an anaerobic packed bed filter type methane fermenter include both the cost for the reactor plus the cost for the packing (assumed to be 50% packing volume and 50% voids). This cost was about 10 times the cost for a conventional CSTR type digester. The results indicated that unit gas cost is strongly dependent on digester retention time. High rate systems could potentially give higher yields which would result in even lower unit gas costs. Since a high rate system has the potential to reduce the unit gas cost significantly, and since application of such a system to soluble peat-fermentables has not been experimentally verified, it is recommended that such a process be investigated.

Other techniques can also be used to reduce digestion retention time, such as operating at thermophilic conditions or using a contact process (cell recycle). These techniques would also result in lower unit gas costs; this effect can be estimated from Fig. 10.

It is apparent from this sensitivity analysis that the variables which have the greatest influence on unit gas cost are the digester retention time and alkali pretreatment chemical cost.

CONCLUSIONS AND RECOMMENDATIONS

The results continue to indicate that biogasification of peat to pipeline quality methane is technically feasible but further research and development is required. Specifically, three major areas of continued experimental work are recommended:

(a) carry out the continuous solubilization/oxidation of peat to fermentable products — batch cooking is necessary to develop an understanding of the practical processing characteristics;

(b) conduct experiments on alternative alkali oxidation systems including alkali recycle as well as use of other alkalies;

(c) develop a high rate digestion process, using a packed bed or anaerobic filter system, for fermentation of the cooked peat liquor to fuel gas.

It is to be noted that high rate anaerobic systems are being operated at full-scale [21,22] with hydraulic retention time being reduced from upwards of 20 days down to a period of as low as four hours. Coupled with these necessary experimental tasks, which may be carried out in parallel, should be a strong supportive process engineering.

ACKNOWLEDGEMENT

This work was carried out under sponsorship of the Minnesota Gas Company (Minnegasco), Minneapolis, Minnesota, the Northern Natural Gas, Omaha, Nebraska, and the U.S. Department of Energy. The authors are grateful for the assistance and encouragement of Mr. Arnold M. Rader, Minnegasco, and Mr. Russell L. Pargett, Northern Natural Gas.

REFERENCES

1 Buivid, M.G., Wise, D.L., Rader, A.M., McCarty, P.M. and Owen, W.F., 1980. Feasibility of a peat biogasification process. Resource Recovery and Conserv., 5: 117.
2 American Society for Testing and Materials, D2607-69, 1969. Standard Classification of Peats, Mosses, Humus, and Related Products.
3 Sopor, E.K. and Osborn, C.C., 1922. U.S. Geological Survey Bull. 728.
4 Farnham, R.S., 1979. Peat resources, classification, properties, and geological distribution. Paper presented at the Conf. on Management Assessment of Peat as an Energy Resource, Arlington, VA, July 22—24, 1979.
5 Clark, F.M. and Fina, L.R., 1951. Arch. Biochem. Biophys., 36: 26.
6 Evans, W.C., 1963. J. Gen. Microbiol., 32: 177.
7 Evans, W.C., 1969. Fermentation Advances. Academic Press, New York, pp. 649—687.
8 Fina, L.R. and Fiskin, A.M., 1960. Arch. Biochem. Biophys., 91: 163.
9 Nottingham, P.M. and Hungate, R.E., 1969. J. Bacteriol., 98: 1170.
10 McCarty, P.L. and Vath, C.A., 1962. Air Water Pollut., 6: 65.
11 Tarvin, D. and Buswell, A.M., 1934. The methane fermentation of organic acids and carbohydrates. J. Amer. Chem. Soc., 56: 1751—55.
12 Healy, J.B. and Young, L.Y., 1979. Applied and Environ. Microbiol., 38: 86.
13 Ashare, E., Wentworth, R.L. and Wise, D.L., 1979. Resource Recovery and Conserv., 3: 359—386.
14 Wise, D.L., 1981. Resources and Conservation, 6: 295—319.
15 Kugelman, I.J. and Chin, K.K., 1970. Toxicity synergism and antagonism in anaerobic waste treatment processes. Presented before Division of Air, Water, and Waste Chemistry, A.C.S., Houston, Texas.

16 Kugelman, I.J. and McCarty, P.L., 1965. Cation toxicity and simulation in anaerobic waste treatment. J. Water Pollut. Control Fed., 37 (1): 97—115.
17 Williams, R.J. and Evans, W.C., 1975. The metabolism of benzoate by moraxella species through anaerobic nitrate respiration. Biochem. J., 148: 1—10.
18 Perry, R.H. and Chilton, C.H., 1973. Chemical Engineer's Handbook, McGraw-Hill, New York.
19 Borne, J.G., Ashare, E. and Habert, R., 1979. Evaluation of oxygen enrichment processes used to improve combustion efficiency, Dynatech Report No. 1815, Dynatech R/D Company, Cambridge, MA.
20 West, C.E. and Ashare, E., 1980. Feasibility study for anaerobic digestion of agricultural crop residues — 1980 update. Report on SERI Subcontract No. NR-9-8175-1, Dynatech R/D Company, Cambridge, MA.
21 Lettinga, G., Van Helsen, A.F.M., Hubma, S.W., DeZee, U.W.W. and Klapwijk, A., 1980. Biotechnol. Bioeng. 22: 699—734.
22 Obayashi, A.W., Stensel, H.D. and Kominek, E., 1971. Chem. Eng. Prog. 77: 68—73.

Energy Recovery from Lignin, Peat and Lower Rank Coals, edited by D.J. Trantolo and D.L. Wise
Elsevier Science Publishers B.V., Amsterdam 1989 — Printed in The Netherlands

Chapter 3

THERMAL PROCESSING OF PEAT

John W. Rohrer, BSME, MBA
Vice President, Engineering, Signal Cleanfuels, Inc.
Hampton, New Hampshire 03842, USA

Professor Bertel M. Myreen, PhD
Consultant, J.P. Energy OY, Helsinki, Finland

ABSTRACT

It is not generally recognized that peat represents the United States' second largest fossil energy resource. While coal reserves exceed peat reserves, peat reserves are several times the estimated reserves of oil and gas. To date, however, primary problems preventing widespread peat utilization center around finding economic and environmentally acceptable ways to harvest peat and reduce its water content to useful levels. Peat is utilized extensively in Russia, Finland, and Ireland for the production of electric power and district heating. It has not been suitable for widespread industrial use in the United States, however, because its low bulk density makes storage, shipment, and combustion of sun-dried peat uneconomical in most locations. Wet harvesting of peat promises to be more economical and more environmentally acceptable than sun-dried, milled or sod peat. Because of the high water content and poor filterability of most fuel grade peats, however, wet harvesting has not been previously used in commercial operations.

Ra Shipping and Jaakko Poyry of Finland have developed a successful process for creating a high density, moisture resistant solid fuel out of wet harvested peat. The fuel has approximately one-fifth the ash content of coal and is essentially sulfur-free. Signal Cleanfuels, Inc. is a North American licensee of this process and is exploring the feasibility of building the nation's first peat fuel plant. This plant is expected to provide significant cost advantages over coal, oil, and gas in many U.S. markets. An example of a proposed project in Maine is presented.

1. <u>U.S. PEAT RESOURCES – A MAJOR ENERGY OPPORTUNITY</u>

It has been estimated that approximately 1% of the earth's land area or approximately 400 million acres are covered by surface peat deposits. The U.S. has the second largest peat resources, behind the Soviet Union (Figure 1), but ahead of Canada and Finland. The U.S. Department of Energy has estimated domestic deposits at 1,443 quads, about eighteen times total annual U.S. energy consumption. While peat reserves are substantially less than our recoverable coal reserves, peat exceeds our recoverable reserves of natural gas, petroleum and shale oil (Figure 2). Perhaps more importantly, U.S. peat reserves are located in regions of the country which are not

COUNTRY	ACRES (MILLIONS)
Soviet Union	228.0
United States*	52.6
Finland	35.6
Canada**	34.0
East and West Germany	13.1
Sweden	12.7
Poland	8.6
Ireland	7.3
Great Britain	5.8
Indonesia	3.3
Norway	2.6
All Others	5.2
TOTAL	408.8

*   Estimate Includes Non-Permafrost Peatlands of Alaska.
**   Estimate Does Not Include Artic Canada Peatlands.

Source: "Peat Prospectus," U.S. Department of Energy, July, 1979, p. 14.

Fig. 1. World Peat Resources

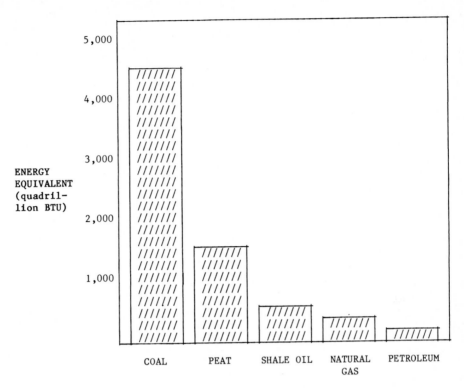

ENERGY
EQUIVALENT
(quadril-
lion BTU)

Coal, Petroleum, Natural Gas, and Shale Oil Are Reported As Proved And
Currently Recoverable Reserves While Peat Is An Estimate Of The Resource.

Source: "Peat Prospectus," U.S. Department of Energy, July, 1979.

Fig. 2. Total U.S. Fossil Energy Reserves

proximate to our coal reserves (Figure 3). Despite the magnitude of this
U.S. energy resource, to date it has been almost totally non-utilized as a
fuel or chemical feedstock.

      This is particularly surprising when one observes the physical
properties of dry peat. Sulfur contents are very low, generally below .3%.
Ash contents can vary significantly from one deposit to another, but many
extensive deposits have ash contents in the 3 - 6% range, substantially
lower than coal.   Heating values generally depend on ash content and the
degree of peat decomposition. The lower heating value of dry peat on a
moisture and ash-free basis generally ranges from 9,000 to 10,000 Btu's per
pound.   This is slightly above wood (8,500 to 9,000 Btu's per pound), but

STATE	ACRES (millions)	QUANTITY (billions of tons)	POTENTIAL ENERGY (10^{15} BTU)
Alaska	27.0	61.7	741
Minnesota	7.2	16.5	198
Michigan	4.5	10.3	123
Florida	3.0	6.9	82
Wisconsin	2.8	6.4	77
Louisiana	1.8	4.1	49
North Carolina	1.2	2.7	33
Maine	.77	1.8	21
New York	.65	1.5	18
All Others	3.66	8.4	101
TOTAL	52.58	120.3	1443

Source: "Peat Prospectus," U.S. Department of Energy, July, 1979, Page 16.

Fig. 3. United States Peat Resources and Potential Energy

somewhat less than bituminous coal (12,000 to 14,000 Btu's per pound). In combustion and chemical conversion, volatile content and chemical reactivity are generally more important properties than heating value. Here, peat has major advantages over various coals. Peat's volatility and chemical reactivity make it faster burning than coal and easier to convert into other synthetic fuels via gasification and liquefaction techniques (ref. 1).

Because peat exists entirely as surface deposits, it also has advantages over coal and wood in harvesting or mining costs. In 1977, U.S. bulk peat sold at an average value of $12.25 per ton (approximately 5% moisture) or about $1.40 per million Btu's (Figure 4) (ref. 2). Even escalating this figure by 30% for ensuing inflation places bulk peat at $1.80 per million Btu's. U.S. utility/industrial coal prices range from $1.40 per million Btu's at mine mouth to over $2.50 per million Btu's in New England and on the West Coast, making peat prices competitive in most regions in which it is found. While primary wood waste from mills is occasionally available for under $1.00 per million Btu's, it is widely distributed in small quantities and is quickly becoming unavailable in most regions. Green whole tree chips, which like harvested peat are about 50% moisture, currently cost $2.25 per million Btu's ($18.00 per green ton).

USE	IN BULK		IN PACKAGES		TOTAL	
	QUANTITY (short tons)	VALUE (thou- sands)	QUANTITY (short tons)	VALUE (thou- sands)	QUANTITY (short tons)	VALUE (thou- sands)
Soil Improvement	195,871	$2,425	324,263	$6,559	520,134	$ 8,984
Seed Inoculant	826	$ 6	3,859	$ 107	4,685	$ 113
Packing Flowers, Shrubs, Etc.	27,186	$ 229	4,648	$ 114	31,834	$ 343
Ingredient for Potting Soil	68,109	$1,025	62,541	$1,358	130,650	$ 2,383
Mushroom Beds	5,909	$ 99	2,649	$ 223	8,558	$ 322
Earthworm Culture Medium	5,142	$ 57	21	$ 1	5,163	$ 58
Mixed Fertilizers	14,717	$ 94	--------	------	14,717	$ 94
Other	6,985	$ 33	3,086	$ 190	10,071	$ 223
TOTAL	324,745	$3,968	401,067	$8,552	725,812	$12,250

Fig. 4. U.S. Peat Sales by Producers in 1979, by Use

Finland and Russia produce substantial quantities of bulk peat via the milled peat method for fuel use. In Finland, Vapo, the state-owned peat production company, sells milled peat for approximately $2.00 per million Btu's, comparable to the U.S. figure above. Thus we can see that in regions where it can be produced, milled peat is competitive with both wood and coal and has the additional benefit of avoiding costly sulfur removal systems required by coal users (reference Figure 5 for comparative prices).

2. PEAT - THE PROBLEMS

Based on peat's widespread availability, attractive physical properties, and competitive economics, why then is peat not extensively utilized in the U.S.? Conventional harvesting techniques yield a product with a very low storage or transport density (whether sod or milled peat). Unfortunately, most U.S. fuel markets are remote from major peat deposits. The low transport density of peat produced makes movement of this peat to major fuel markets impractical. Even for those near peat deposits, however,

48

INDUSTRIAL FUELS:	HEATING VALUE NET MMBTU/TON	SHIPPING DENSITY LBS/CU. FT.		$/MM BTU	SULFUR (%)	ASH (%)
10 Fuel Oil (1)	36	60	$5.00	($30/DI.)	.2 - 2	.1
Coal (Medium Sulfur) (2)	24	45	$2.50	($60/TON)	1.0 - 5	10.0 - 1
Electricity	--	--	$11.71	(4¢/KWH)	--	--
Peat-Derived Fuel ("PDF")	20	45	$2 - 3		.2	3
Raw Milled Peat (40% WC)	10	8	$2.00	($20/BULK TON)	.2	3
Round Wood (50% WC)	8	44	$2.25	($50/CORD)	.1	2
Green Wood Chips (50% WC)	8	18	$2.25	($18/GREENWOOD)	.1	2
Green Wood Waste (Bark, Sawdust - 50% WC)	8	18	$1.00	($8/GREEN TON Where Available)	.1	2
Wood Chips (Predried to 10% WC)	15	12	$3.20	($18/GREEN TON & 15% Dryer Loss & $10/Dry Ton Extra Cost)	.1	2
Pelletized Wood Fuel From Wood Waste	16	40	$2.50	($8/GREEN TON & 15% Dryer Loss & $10/Dry Ton Extra Cost)	.1	2
Pelletized Wood Fuel From Wood Chips	16	40	$3.63	($18/GREEN TON & 15% Dryer Loss & $20/Dry Ton Extra Cost)	.1	2
RESIDENTIAL FUELS:						
12 Fuel Oil	36	60	$8.00	($1/GALLON)	0	0
Natural Gas (Where Available) (1)	--	--	$4.00	($4/TCF)	0	0
Electricity	--	--	$14.65	(5¢/KWH)	--	--
Coldwood (Seasoned @ 20% WC)	13	27	$4.50	($100/CORD)	.1	2
Peat-Derived Fuel ("PDF")	20	45	$3 - 4		.2	3
Anthracite Coal (Sized Bulk)	25	45	$4.00	($1/TON)	1.0	10
Pelletized Wood Fuel (Bulk)	16	40	$5.00	($80/TON)	.1	2

Footnotes: (1) Being decontrolled, prices will increase substantially.
(2) Federal and State laws require extensive sulfur emissions control.

Fig. 5. Comparison of New England Industrial and Residential Fuels

other problems exist. The low density and dusty nature of the product necessitates specialized fuel handling and burning equipment, which is more than twice as expensive as oil and gas fired units and substantially more expensive than coal units, if sulfur controls are not required. The variable moisture content of air-dried peat necessitates the use of auxiliary fuels, such as oil or gas, to stabilize combustion and boiler steam flow. Additionally, the dusty nature of the product creates an ever present danger of dust explosion in storage silos and fuel-handling systems. Most U.S. peat deposits are in northern climates where the harvesting season is short and heavily dependent on weather conditions. Peat production can vary, therefore, by more than 50% from season to season and requires the use of seasonal labor forces, which are often difficult to obtain in countries with high living standards like the U.S. Despite these limitations of conventional air-dried peat, some industrial and utility fuel users proximate to peat deposits may find it more attractive than scrubber equipped coal plants and, therefore, some utilization in this form will undoubtedly occur over the next decade. Use of this form, however, will probably be insignificant in comparison to the quantity of resource available. Some have proposed thermal drying and densification of conventional field dried peat. While this can improve transportation costs and provide uniform moisture content, these steps add significant cost to the field harvested peat (an additional $1.50 to $2.00 per million Btu's), and the product is still subject to regaining moisture and must, therefore, be stored under cover.

Conventional peat production techniques also present serious environmental problems which might make it unacceptable in the environmentally conscious U.S. Field drying techniques remove only a small portion of the total peat deposit in any one year, and must, therefore, clear the entire deposit surface to provide adequate drying area for production. After clearing the surface of vegetation, the bog is ditched and drained. Because drainage water differs from normal surface runoff in terms of pH, suspended, and dissolved solids, such ditching can have impacts on local rivers, streams, and lakes. Ditching is performed to drop the water table on the deposits somewhat, thus reducing the amount of moisture which must be evaporated by the air and sun. The peat production fields, sometimes unaffectionately referred to as "brown deserts", are particularly vulnerable to producing fugitive dust emissions (especially in windy areas) and are subject to bog fires which are extremely difficult to extinguish. Conventional production techniques, which remove only a small layer of the deposit each year, cannot initiate reclamation efforts until the entire deposit has been removed. Thus promises of reclamation often appear empty when postponed for twenty or thirty years until harvesting operations have been completed.

Alternative bog uses, such as horticultural peat production and perhaps more significantly preservation areas, are also a major environmental concern. U.S. horticultural users consume less than .5 million dry tons per year. U.S. reserves are estimated at approximately 85 billion dry tons, thus horticultural peat production cannot have a significant effect on the quantity of reserves available. It would be environmentally irresponsible to remove virtually all peat deposits from a given region. Some peat deposits contain unique wildlife habitats and ecological systems. Others are particularly sensitive areas where peat production could have major environmental impacts. Fortunately, however, a major portion of all deposits could be put into commercial production, environmental and energy interest should responsibly determine which bogs should be set aside and how bogs should be reclaimed when production is completed.

Many of the environmental and seasonal harvesting problems outlined above could be circumvented if peat were harvested in its wet state, rather than requiring labor, time, and land-intensive air drying. Wet harvesting could achieve much higher production rates and much lower cost. Deposits could be mined vertically rather than horizontally, such that only a small portion of the bog need be disturbed in any one year, and could subsequently be reclaimed the following year after production was complete (i.e. rather than removing six inches of an entire fifteen foot deep bog for thirty years, one would remove fifteen feet out of one-thirtieth of the bog area in one year). Pre-ditching and draining of a substantial bog area would not be required. Dust and fire problems would also be eliminated.

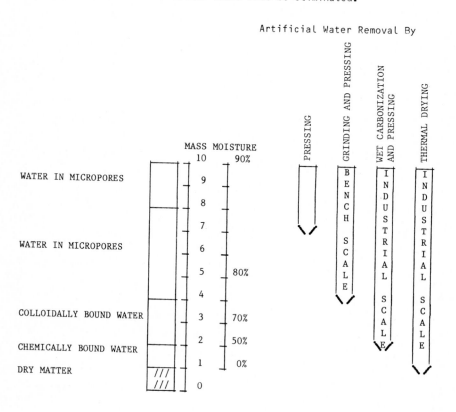

Source: Myreen, Bertel, "Industrial Refining of Peat By Wet Carbonization."

Fig. 6. Water Distribution in Raw Peat

Wet harvested peat, however, contains about 90% moisture. A major portion of this moisture is held in micropores or is colloidally bound within the peat structure and is, therefore, not removable by filtering or pressing. Tests conducted by J.P. Energy OY and numerous others have shown that it is not possible to mechanically remove sufficient water from fuel grade or hemic peat (which comprises the bulk of most peat deposits) to a sufficiently low level that the balance of the moisture content can be economically removed by thermal drying (Figure 6).

Numerous alternatives to mechanical water removal have been and are being pursued. The U.S. Department of Energy is currently sponsoring research aimed at extracting this colloidal and micropore moisture via the use of solvents which dissolve in water (but hopefully not into peat), displacing some of the moisture. The solvent can be flashed off the peat with lower energy consumption than is necessary with water. The solvents are subsequently condensed or separated from the water and reused. Favorable solvent balances and process economics have not yet been developed. An alternative dewatering approach involves biological treatment. It has been observed that some biological digestion (either anaerobic or aerobic) of peat slurry will improve filterability (anaerobic digestion can also produce some methane directly). To obtain significant biological digestion, however, a substantial portion of the peat must first be solubilized. Acid hydrolysis and alkali cooking liquors have both been utilized. Solids remaining after biological activity tend to be more easily dewatered, although significant and costly thermal drying is still required. In research to date, however, only minor portions of the solubilized peat have been digested by microbes (into CO_2 or methane) leaving a major water treatment task remaining. Efforts toward finding more effective microbes are currently being sponsored by the U.S. Department of Energy and Minnegasco.

3.   THE J.P. ENERGY WET CARBONIZATION PROCESS - A SOLUTION

It was recognized around the turn of the century that the dewatering properties of peat slurries could be substantially improved by wet carbonization. Wet carbonization might be defined as the heating of an aqueous peat slurry under sufficient pressure and temperature to produce some carbon loss (as gaseous and/or dissolved products). It was discovered that wet carbonization at temperatures above 180°C appears to break down the colloidal suspension of peat and to permit the mechanical expression of water from the micropores, thereby permitting mechanical dewatering to greater than 50% solids content. The first commercial wet carbonization plant was operated in Dumfries, Scotland between 1910 and 1920 following their previous pilot plant effort. The plant utilized batch rather than continuous operations and was plagued by corrosion and scaling problems and heat exchanger fouling problems. In the late 1950s, a peat carbonization plant was started at Boksitogorsk, Russia. This plant, however, processed previously air-dried peat via a batch process for use as several chemical feedstocks. The Swedes, who later evaluated the process, felt it was not economic under conditions prevailing in Western Europe at that time. Beginning in 1951, the Swedish Peat Company and the Swedish Government funded the development of wet carbonization in that country leading to the construction of a one-tenth commercial scale pilot plant is Sostala, Sweden. The plant commenced operation in 1954 and completed experimental runs totaling approximately 1,000 hours. Problems in the plant focused on fouling and scaling of heat exchangers and erosion damage to pumps, pipes, and other equipment due to the abrasive nature of the slurry. The advent of low cost petroleum in the late 1950's resulted in abandonment of further work.

Renewed work on wet carbonization was resumed in Finland in 1974 with the advent of drastically increased oil prices. This work was privately sponsored by Ra Shipping OY and was aimed at overcoming difficulties encountered in the Swedish work which proceeded. Ra Shipping's development efforts were focused on further improving the heat economy of the wet carbonization plant and in overcoming certain equipment difficulties previously experienced. Unit operations were tested on commercially available equipment, and a pilot plant was subsequently constructed and commenced operating in May of 1977. Work resulted in a new wet carbonization plant configuration (U.S. Patent No. 4153420) and in various equipment patents (i.e. U.S. Patents Nos. 4184540 and 4200600).

The pilot plant equipment was sized for approximately 20,000 tons per year of production but was not configured to operate for sustained periods at this rate (due to storage capacity, water capacity, etc.). Perhaps the most significant benefit of the pilot efforts was the hands-on experience gained in running the carbonization plant with modern equipment. Ra Shipping formed a joint venture with Jaako Poyry Engineering, a major international engineering and construction company headquartered in Helsinki. This joint venture is called J.P. Energy OY. Recently, several other Finnish companies including Neste, the state-owned oil company; Kemira OY, a major Finnish chemical company; and Imadran Vioma OY, formed a consortium to advance the technology in Scandinavian countries. This consortium holds the Finnish license to the process and is currently sponsoring enlargement of the pilot plant and additional pilot runs, as well as complimentary programs,

including work aimed at peat wet harvesting techniques and the use of
peat-derived fuel in oil and gas boilers and furnaces. Figure 7 is an
illustration of peat-derived fuel in pelletized form. For gasifier feed and
other on-site applications, pelletization and/or final thermal drying are
often not required. Figure 8 compares peat-derived fuel with the raw peat
from which it was produced. Note that 72% of the dry matter was retained
while its Btu content was increased. Nitrogen and oxygen content were
reduced, thereby increasing the carbon content slightly. Primary material
losses include carbon dioxide, and water soluble and volatile organic
matter. Most of the changes in the peat, however, are physical rather than
chemical, resulting in coagulation of colloids and the opening of micropores, thus making mechanical dewatering of the dry matter possible.

Because peat can enter the carbonization plant at up to ten parts water
per one part dry matter, heat economy is the primary design consideration of
the plant. Careful consideration also had to be given to corrosion and
fouling of heat exchangers, however, as this proved to be a major problem in
prior efforts.

Fig. 7. Peat-Derived Fuel Pellets

CHEMICAL CHANGES IN RAW PEAT BY WET CARBONIZATION	RAW HEAT	PRODUCTS OBTAINED BY WET CARBONIZATION				
	Dry Matter	Dry Matter	Water Soluble Organic Matter	Water	Carbon Dixoide	Volatile Organic Matter
Mass Reaction	100	72	11	8	7	2
Elemental Analysis						
C%	53.1	62.1	45.2	--	27.3	76.6
H%	5.4	5.2	5.5	11.1	--	6.2
S%	0.2	0.3	--	--	--	--
N±0%	36.3	25.5	49.3	88.9	72.7	17.0
Ash%	5.0	6.9	--	--	--	--
Gross Calorific Value						
BTU/LB.	9,060	10,950	7,280	--	---	14,700

Source: Myreen, Bertel, "Industrial Refining of Peat by Wet Carbonization."

Fig. 8. Wet Carbonization of Peat

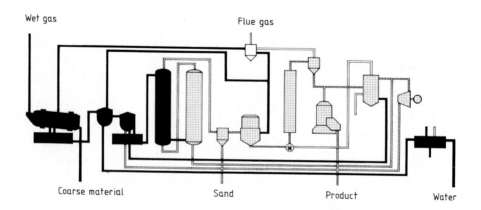

Fig. 9. The Peat Fuel Process

55

Figure 9 shows the simplified process flow diagram of the J.P. Energy
Peat Fuel Process. The process consists essentially of raw peat slurry
preparation, preheating of peat up to reaction conditions by recycling the
heat contained in the reacted products, carbonization of peat in the pre-
heaters and reactors, separation of the carbonized peat from filtrate by
pressure filters, and subsequent thermal drying and pelletization of the
carbonized peat where desired. In addition to a proprietary two stage
preheater/heat recovery system employing both surface and flash heat
exchange, the process also utilized oversized material from slurry
preparation (mostly wood) in its steam generating boiler. Press filtrate is
utilized as a scrubbing solution for the steam generating boiler and thermal
dryer, further improving the heat economy of the plant and preventing the
loss of fine material. Figure 10 shows the overall heat economy of the
process. While approximately 72% of the product is recovered as useful fuel,

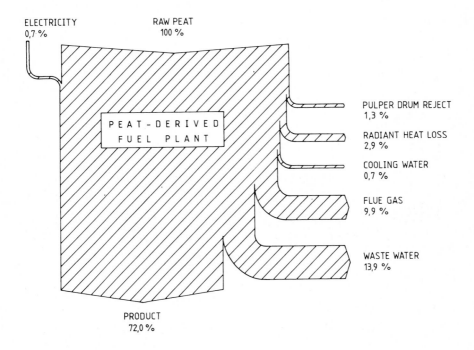

Fig. 10. The Energy Balance of the Wet Carbonization Plant

it should be remembered that the process also utilizes the junk wood found in most peat bogs which often can exceed 10% of peat deposits by weight.

The use of higher severity carbonizing conditions has recently been suggested by other experimenters in the U.S. Why has J.P. Energy opted for the low severity processing conditions utilized in their process? J.P. Energy has found, and other researchers have more recently confirmed, that solubilized and gaseous losses (CO_2 and volatile hydrocarbons) increase substantially with increased processing severity (time or temperature) (ref. 3).

These more severe processing conditions do not result in a commensurate improvement in dewatering capability which has already been reached at the milder process conditions utilized in the J.P. Energy process. Devolatilization of the peat does improve the Btu content of the final product somewhat (by increasing carbon content), but this devolatilization substantially deteriorates both combustion properties of the fuel (burning rate) and chemical reactivity, so useful for hydrogasification or liquefaction. Milk processing conditions substantially reduce water treatment requirements on the filtrate. Severe processing conditions also substantially increase plant capital costs, lower plant heat economy, and necessitate the use of undeveloped equipment. All of the equipment utilized in the J.P. Energy process is currently available in comparable alternative service and scale.

4. PEAT-DERIVED FUEL MARKETS

PDF retrofits of both oil burning utility and industrial facilities and coal capable units now burning oil are prime markets. A major portion of our utility and industrial boiler stock was originally designed to burn coal. In 1979, FERC identified approximately 30,000 megawatts of coal capable generating capacity, which is currently burning either oil or gas. Most of this capacity was located in the eastern industrialized states with approximately 6,000 megawatts in New England. Likewise, many industrial boilers are also capable of burning solid fuels. A portion of this boiler stock will be converted back to coal under either voluntary conditions or under FERC reconversion orders.

Often, reconversion is impossible, however, because of economic or environmental considerations. Many air quality regions do not permit the burning of sulfur containing fuels without sulfur controls. Many coal capable plants either lack the space or the remaining useful economic life to justify expenditures in sulfur controls. Other sites have major problems with ash disposal, due to reduced landfill capacity and stricter disposal standards. The low sulfur and ash characteristics of peat-derived fuel make it well suited for many applications where coal reconversion is not possible.

Approximately 30% of the utility boiler stock and more than two-thirds of our nation's industrial boilers are designed to burn only oil or gas. Some of these facilities lack the room to handle solid fuels on site, but most of them are precluded for solid fuel usage due to design restrictions, such as inadequate furnace site or boiler tube spacing. Strenuous efforts are being made toward reducing oil consumption in such units by the use of coal/oil mixtures (COM). While some COM conversions appear possible, such mixtures are not feasible for most oil-fired boilers because the high ash content creates furnace slagging and boiler tube erosion problems. Peat-derived fuel has substantially lower ash content than coal, as well as lower sulfur and, therefore, is better suited for slurry with oil. The rapid burning characteristics of fully dried peat also holds forth promise that peat-derived fuel can be burned in pulverized form in boilers originally designed for fuel oil without derating. Several large Finnish industrial boilers, originally designed only to burn oil, are currently firing thermally pre-dried peat alone at full boiler rating.

Interest has also been expressed in the use of peat-derived fuel as an auxiliary fuel for industries firing green wood waste, whole tree chips, or air-dried peat. All of these fuels have variable and high moisture content (40 - 60%) and, therefore, require oil or gas auxiliary fuel for combustion stabilization. In addition to the utility and industrial retrofit market, the sulfur control requirement placed on new boiler plants by the Clean Air Act amendments and resulting New Source Performance Standards makes the use of sulfur-free fuel especially attractive. The cost of sulfur control equipment on moderate sized industrial boiler plants can equal or exceed the cost of the balance of the boiler plant. The uniform fuel properties of PDF and its low ash content also make it more desirable than coal for new boiler installations. Fuel storage, conveying, and pulverization equipment is also less costly than that required for coal burning.

Another promising fuel market for PDF is residential, commercial, and institutional space heating. The high cost of fuel oil and natural gas have created a major move to solid fuels (coal and wood) in many regions of the country. In several states (i.e. Maine, Vermont, Minnesota, and Wisconsin) up to 25% of all households currently use solid fuels as their primary source of space heating. This was virtually unheard of five years ago but the price differentials, as shown in Figure 5, now provide sufficient motive to switch to solid fuels. Lower sulfur and ash content of PDF and its uni-form fuel properties promise to reduce the sulfur, particulate, and unburned hydrocarbon emissions which are particularly troublesome for these small scale burning installations where pollution controls are not practical. The uniform size, lack of dust, and fines found in PDF make it more compatible with automatic storage and feeding equipment than coal, wood chips, or other solid fuels.

For both large and small scale installations where burning of solid fuels or slurries is not possible due to boiler design, the use of close-coupled gasifiers offers considerable promise. Within the last three years, several dozen such units have been put into commercial operation in the U.S. They appear to be less costly than new solid fuel replacement burning plants.

The advantage of peat over coal for hydrogasification and liquefaction are well documented but not generally recognized. Tests, for example, conducted by both the Institute of Gas Technology and Rockwell, on their respective high Btu gasifiers, show that lower hydrogen partial pressures and shorter reaction times were required with peat than with coal, and that the proportion of methane made directly in the gasifier was substantially increased, reducing the expense of subsequent methanization and additional hydrogen production (ref. 4).

Higher liquid yields also appear probable when peat is used as a feed-stock in direct liquefaction processes, although little work has been done to date in this regard. At the recent International Peat Society Symposium, held in Duluth, Minnesota, both Swedish and Canadian representatives indicated programs to develop pilot plants for direct peat liquefaction. New pilot plants are really not necessary since the U.S. has numerous coal liquefaction pilot plants capable of running peat-derived fuel feedstock. Another U.S. pilot facility which looks particularly promising is the Wood-to-Oil liquefaction pilot plant in Albany, Oregon, currently operated by Signal's Rust Engineering Company for the U.S. Department of Energy. This plant is compatible with aqueous feeds and therefore would not require final thermal drying after carbonization and filtering. We are hopeful of utilizing existing U.S. liquefaction facilities to gain the necessary information for proceeding to commercial plants. The production of synthetic gas from peat-derived fuel is also quite promising. The synthesis gas could subsequently be converted to liquids via Fischer-Tropsch (Sasol) or by methanol synthesis. Ammonia synthesis for fertilizer production is also possible both here and in third world countries so desperately in need of fertilizers.

J.P. Energy has also successfully produced coke out of peat-derived fuel. The coke produced has extremely high strength and, of course, a very desirable low ash content.

6. COMMERCIALIZATION OF THE PEAT-DERIVED FUEL PROCESS - PROPOSED PEAT PROJECT

With the pilot plant work completed and markets for peat-derived fuel and feedstock identified, the appropriate next step is construction and operation of the first commercial scale PDF plant. It is now possible that this first plant might be built in the U.S. rather than in Finland. Signal Cleanfuels, Inc., as the North American Licensee of the J.P. Energy PDF process, is investigating a possible 290,000 ton per year plant to be located in the State of Maine. At the same time, the consortium of Finnish companies is pursuing construction of a similar plant in that country. We have elected to investigate a fuel plant, rather than a feedstock plant (i.e. PDF to methanol), because it permits a smaller first plant, and one which can be constructed in a shorter period of time. Primary uncertainties regarding this project are not technical, but rather involve siting and environmental issues because no precedents exist in Maine (or elsewhere in the U.S.) for major scale peat utilization.

Maine was picked as an ideal site for the first American PDF plant because it has little pipeline gas distribution, extremely highly priced coal (due to long distance transport) and is, therefore, very dependent on imported oil for utility, industrial, and consumer fuel requirements. Peat is the only major fossil energy resource located within the New England states. The only other major fuel resource in this region is wood. The forest product industries, however, are major employers in the region and significant utilization of merchantable wood for fuel may drive up their raw material costs to the point where New England mills are not competitive with those in other regions of the country, especially in the South where forest re-growth occurs at much faster rates. Forest products companies in this state, therefore, were amongst the first to endorse this project. A major portion of the peat reserves are on their land. Such a project not only offers relief from high oil prices, but can permit reclaimed bogs to be replanted as productive forest land.

The project has received strong support from the New England Congressional Caucus, the New England Energy council, the Maine State Office of Energy Resources, and regional planning groups in major peat resource counties. The Maine Audubon Society also supports the feasibility study and has stated it is appropriate to utilize a portion of the State's peat resources for energy production, if this can be done in an environmentally responsible manner.

Output from this plant will be distributed to several classes of users, even though the largest fuel oil and coal distributor in the region (Sprague Energy Inc.) has expressed a willingness to take the entire plant output. Output will go to industrial boilers as stoker fuel, pulverized fuel, and possibly in slurry form with oil. It will also be distributed to the residential, commercial, and institutional markets where it is projected to be less costly than oil, coal, or wood.

Several Signal divisions are involved in the company's PDF efforts. Our Rust International Company, the largest engineering and construction company serving the pulp and paper industry, will use their related expertise in pulping and slurry pumping, waste water treatment, dewatering, and solid fuel industrial boilers. Signal's Swenson Division will supply the heat recovery systems and reactors. Signal's UOP Process Division will operate the plant. They recently constructed the slurry heat exchange system for a wet carbonization pilot unit built at the Institute of Gas Technology for the U.S. Department of Energy. The PDF efforts are being coordinated by Signal Cleanfuels Inc.

While the proposed project at 290,000 tons per year is relatively small, it is sufficient to provide enough residential heating fuel for approximately 42,000 households, or about 20% of all Maine residences. During its lifetime, the plant will utilize less than 1% of the State's peat resources. Signal envisions future plants in other contiguous U.S. peat states as well as in Canada and Alaska where peat-derived fuel and metallurgical coke could be produced for shipment to the U.S. West Coast and to the Orient.

7. ENVIRONMENTAL CONSIDERATIONS

Maine's recent economic decline is in large part caused by heavy dependence on increasingly expensive imported fuels and energy (oil, imported electricity, and more recently, coal). Maine's greatest economic development opportunity lies in solving its own energy problem with indigenous energy resources. Their $1.6 billion imported energy bill could create 80,000 jobs (15% of the Maine labor force) if this money could be retained in Maine's economy. This economic solution would not require the influx of massive industry into Maine to offset the dollar drain caused by imported energy.

Many people have recognized that a deteriorating economic situation is not conducive to progress in the area of environmental protection and

improvement.   Economic decline is often associated with declining public
sentiment for maintaining or advancing environmental standards and the
inability to afford maintenance or achievement of such standards.

In formulating a policy on peat, we would hope that environmental
groups carefully evaluate the alternatives which are likely to be utilized.
Coal and wood are likely to be the primary alternatives to peat in Maine as
price increases in oil and gas drive Maine energy users to less expensive
fuels. Coal is substantially dirtier than peat and sulfur controls are
affordable only by the largest potential energy users. Large-scale wood
utilization affects approximately 100 times more acreage (i.e. 20 dry tons
per acre versus 2,000 dry tons per acre). Wood, if properly harvested, can
make a valuable energy contribution but is not a total solution.   Major
increases in wood utilization could affect the price of saw logs and pulp
wood, thus affecting the competitive position of Maine's forest products
industries in relation to other regions of the country where forest growth
is faster and energy less expensive. Peat and wood are the State's only
indigenous fuel resources and both should be utilized in an environmentally
responsible way to help resolve the State's energy and economic problems.
(Solar, hydro, wind, and conservation are desirable and renewable sources of
energy, but are not fuels per se).

Several years ago, at the peak of the energy crisis, there was much
speculation that Maine's peat bogs were threatened by a development stampede
which has not materialized. Signal believes that traditional peat harvesting
methods relying on air-drying as practised in the U.S.S.R., Finland, and
Ireland cannot compete with wood and coal on an economic basis in Maine.
Air-dried peat requires specialized combustion equipment not available in
Maine; it is costly to transport to customers due to its low bulk density,
and has environmental disadvantages versus wet excavated peat. Our studies
have concluded that the Finnish PDF process is perhaps the only one which
can produce a peat fuel readily substitutable for coal in northern climates
at a price competitive with coal. We do not intend to undertake additional
projects until this first one has proven itself as an environmental and
economic success.

Peatlands appear to cover approximately 770,000 acres in Maine, or
about 3% of the State's land area. The State's Geologists Office and prior
surveys have confirmed approximately 94,000 acres of peatlands with
"commercial potential". They estimate that up to 125,000 acres maximum, or
approximately 20% of Maine's peatlands, may have commercial potential. This
is significant in that it ensures that at least 80% of the State's peatlands
will remain in their undisturbed state.

Figures 11 to 18 review the impact of the proposed peat process in Maine. The proposed project would utilize only 3,000 to 5,000 acres of peatlands over a twenty year project life, or only 150 to 250 acres of peatland affected by mining each year. While this is a very small portion of the State's total peat resources, the proposed 290,000 tons per year of plant output would provide sufficient space heating fuel for approximately 20% of all Maine homes over its project life. In other words, less than 3% of the State's peatlands could produce all of the State's residential space heating requirement for twenty years. This is significant in that it minimizes the potential for siting conflict between bogs truly worthy of preservation and those which are promising for energy development.

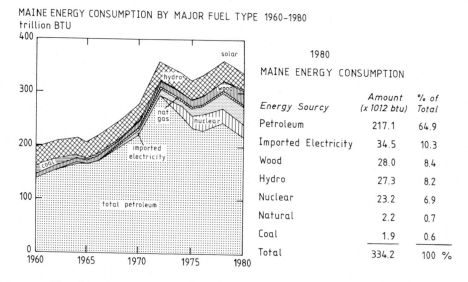

MAINE ENERGY CONSUMPTION BY MAJOR FUEL TYPE 1960-1980
trillion BTU

1980
MAINE ENERGY CONSUMPTION

Energy Sourcy	Amount (x 1012 btu)	% of Total
Petroleum	217.1	64.9
Imported Electricity	34.5	10.3
Wood	28.0	8.4
Hydro	27.3	8.2
Nuclear	23.2	6.9
Natural	2.2	0.7
Coal	1.9	0.6
Total	334.2	100 %

Fig. 11. Maine Energy Consumption by Major Fuel Type, 1960-1980

Despite the insignificant impact of a single project on the State's overall peatlands, there are bound to be some local objections as there would be with any major industrial development project. There are also those who are philosophically opposed to any natural habitat alteration and/or industrial development, regardless of its economic or environmental impact. It is important that a set of objective criteria be established and

MAINE ENERGY CONSUMPTION BY SEKTOR, 1960-1980

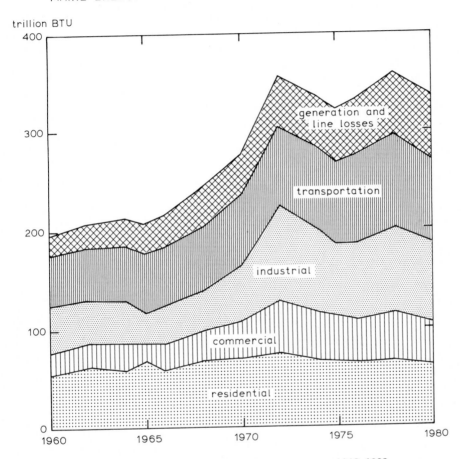

Fig. 12. Maine Energy Consumption by Sector, 1960-1980

adopted which can be applied by potential developers, State agencies, and environmental groups in a non-arbitrary way to determine whether a partic- ular peat bog is truly rare or unique or contains habitat truly worthy of preservation. An arbitrary set of preservation standards can only create confusion, misunderstanding and ultimately prove counterproductive to the preservation of peat bogs which truly deserve it.

It is important to understand the distinction between traditional harv- esting methods utilizing air-drying and the proposed project which utilizes wet excavation of peat with subsequent water removal under controlled conditions in a dewatering plant. With air-drying, an entire peat reserve must be cleared of surface vegetation, ditched and drained to allow for the removal of two to three inches of peat each year. (A typical deposit depth might be ten feet). With wet excavation only a small portion of the total peat reserve is affected in any year. All of the peat, with the exception of that left for reclamation, is removed. Water is removed from peat in the process plant with that water being fully treated prior to re-discharge. Other peatlands remain undisturbed until required for production. Each plot is reclaimed in the year after use as wildlife habitat, forest land, or agricultural land.

SOURCE	$ DRAIN FROM ECONOMY (million)
PETROLEUM 38 Million Barrels @ $35/Barrel	1,330
IMPORTED ELECTRICITY 3.3 Billion kwh @ $0.05/kwh	165
NUCLEAR 2.2 Billion kwh @ $.015/kwh (Fuel Only)	33
NATURAL GAS 2.2 Million cf @ $4/Thousand cf	9
COAL 80,000 Tons @ $75/Ton	6
TOTAL MAINE DOLLAR DRAIN	1,543

Maine Job Loss @ $20,000/Primary Job = 77,150 Jobs

Fig. 13. Impact of "Imported" Energy on Maine Economy (At Approximate 1982 Wholesale Prices)

COUNTRY	PEATLANDS ha X 10^6	PEAT PRODUCTION – TONNES X 10^3		
		FUEL PEAT	MOSS PEAT	TOTAL
Canada	170.0	––	488	488
U.S.S.R.	150.0	80,000	120,000	200,000
U.S.A.	40.0	––	800	800
Indonesia	26.0	––	––	––
Finland	10.0	3,100	500	3,600
Sweden	7.0	––	270	270
China	3.5	800	1,300	2,100
Norway	3.0	1	83	84
Malaysia	2.4	––	––	––
United Kingdom	1.6	50	500	550
Poland	1.4	––	280	280
Ireland	1.2	5,570	380	5,950
West Germany	1.1	250	2,000	2,250
TOTAL	417.2	89,771	126,601	216,372

Fig. 14. World Peat Resources (Kivinen and Pakarinen, 1980)

Initial developers recognize that the burden of proof will be on them to establish that a particular deposit does not contain rare or endangered wildlife or botanical species, and that it is not of a unique geomorphological type. Obviously all peat bogs, like other land forms, are unique in some way. If and when a peat energy industry is actually established in Maine, the State might then be well justified in spending public money to identify those bogs worthy of preservation. Responsible developers will also participate in identification and preservation of truly unique peatlands. Given the multitude of potential project sites in Maine it would

66

AREA	ACRES
Total Maine Wetlands	2,000,000 (approx.)
Total Peatlands/Organic Soils (Approx. 3% of State)	770,000
Commercial Potential Confirmed by Physical Surveys[1]	94,000
Major Peat Project Lifetime Requirement[2]	3,000 – 5,000
Acreage Excavated and Reclaimed/Year	150 – 250

[1] Per Maine Geological Survey Criteria
[2] 1/3 Million Ton/Year Briquette Fuel Plant or 50 Million Gallons/Year Methanol

Fig. 15. Maine Peatlands in Perspective

be unreasonable to delay an initial peat energy project until the inventory of all sites had been completed.

The proposed project will no doubt involve peatlands habitat alteration. Reclamation, however, can render the site more productive than its original use. Peatlands can be reclaimed as a more productive and varied wildlife habitat. There are numerous examples in Maine and New Brunswick where peatlands and other wetlands were over-flooded by fish and wildlife agencies or other groups to improve water fowl habitat or fisheries.

While peatlands are not renewable in the conventional sense (they regenerate less than one-half dry ton per acre per year), reclaimed peatlands can be utilized to grow renewable energy crops at up to 100 times this regeneration rate. Hybrid, alders, willows, and cattails, are examples of high yield energy crops which could be grown on reclaimed peatlands.

Signal selected Maine as the potential site for an initial North American peat fuel plant because of Maine's high dependence on imported fuels and the high cost of coal delivered to Maine. We believe the project is environmentally sound and can contribute to the resolution of Maine's economic and energy problems.

☐ Annual Effected Area Reduced
  Twenty-to-Fortyfold

☐ No Fugitive Dust Emissions

☐ No Untreated Drainage and Runoff

☐ All Water Discharges Treated

☐ Immediate Bog Reclamation and Use

☐ Clean Burning Processed Fuel

Fig. 16. Environmental Advantages of Wet Versus Dry Peat Harvesting

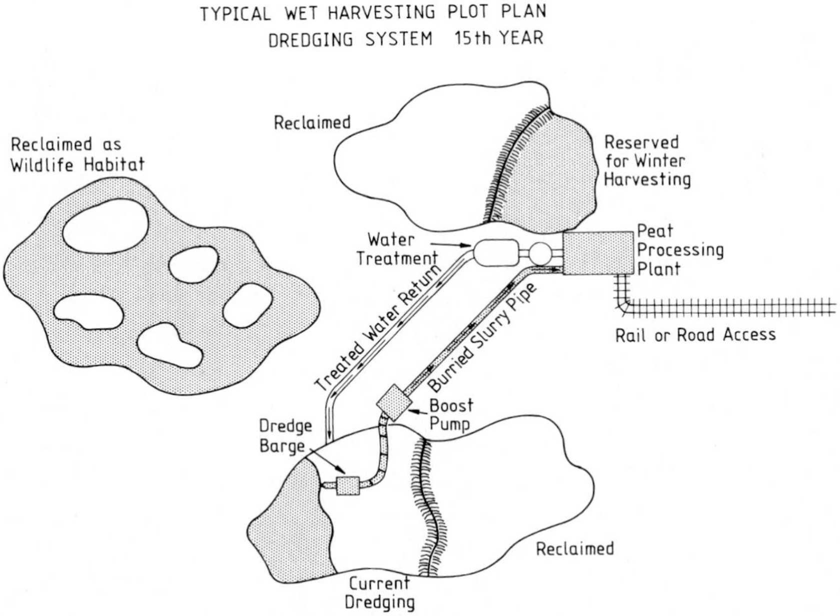

TYPICAL WET HARVESTING PLOT PLAN
DREDGING SYSTEM 15th YEAR

Fig. 17. Typical Wet Harvesting Plot Plan Dredging System, 15th Year

68

```
┌─────────────────────────────────────────┐
│ Oil and Gas                               │
│                                           │
│   High Fuel Cost                          │
│   Maine Dollar Drain                      │
├─────────────────────────────────────────┤
│ Coal                                      │
│                                           │
│   High Sulfur (0.7 to 5%)                 │
│   High Ash (6 to 15%)                     │
│   SO₂ Scrubber Sludge                     │
│   Highly Capital Intensive                │
│     (W/SO₂ Controls)                      │
│   Maine Dollar Drain                      │
├─────────────────────────────────────────┤
│ Nuclear                                   │
│                                           │
│   Limited to Electricity Only             │
│   Unresolved Waste Disposal and           │
│     Decommissioning Issues                │
│   Highly Capital Intensive                │
├─────────────────────────────────────────┤
│ Peat                                      │
│                                           │
│   Habitat Modification/Improvement        │
│   Low Sulfur - 0.1 to 0.2%                │
│   Low Ash - 2% to 4%                      │
│   Low Creosote and Aromatics              │
│   Nonrenewable but Indigenous             │
└─────────────────────────────────────────┘
```

Fig. 18. Environmental and Economic Impacts of Alternative Maine Fuels

8. CONCLUSION

Peat represents the nation's second largest fossil energy resource, but has been almost totally non-utilized to date. Its low sulfur and ash content and high chemical reactivity give it major advantages for both synthetic fuels and chemical feedstock uses. Surface deposits hold promise for lower production costs than other fuels. Environmental, transportation, and utilization difficulties associated with conventionally harvested air-dried peat, however, will seriously restrict use of this resource in the U.S. The Peat-Derived Fuel Process developed by J.P. Energy OY has completed a successful pilot program and is based on the use of existing commercial equipment. This process, in conjunction with the wet harvesting of peat, may prove to be the key which unlocks the use of this promising U.S. resource for both fuels and feedstocks. The market for this product has been substantiated and efforts have been launched toward to the first commercial PDF plant in the U.S. to be sited in Maine. It is quite possible that this project could be the first commercially operating modern peat wet carbonization plant worldwide. The economic, environmental, and improved physical properties of wet harvested PDF could make this first major scale U.S. peat energy project an exemplary one.

REFERENCES

1. D.V. Punwani and F.C. Schora, Peat, An Energy Alternative, Paper presented at Energy in the Third World Conference, July 28, 1980.
2. Joel Davis and Gregory K. White, Production and Utilization of Maine's Peat Resources, August, 1979.
3. M.J. Kopstein and D.V. Punwani, Peat Beneficiation and its Effect on Dewatering and Gasification Characteristics, Paper presented at American Chemical Society 179th National Meeting, March 23, 1980.
4. M.J. Kopstein, S.A. Nandi, D.V. Punwani, and S.A. Weil, Peat Hydrogasification, Paper presented at 176th National Meeting of American Chemical Society, 1978.

Energy Recovery from Lignin, Peat and Lower Rank Coals, edited by D.J. Trantolo and D.L. Wise
Elsevier Science Publishers B.V., Amsterdam 1989 — Printed in The Netherlands

Chapter 4

DEVELOPMENT OF A BIOCHEMICAL PROCESS
FOR PRODUCTION OF ALCOHOL FUEL FROM PEAT

Donald L. Wise*
Cambridge Scientific, Inc., 195 Common Street, Belmont, MA 02178, USA

EXECUTIVE SUMMARY

This chapter relates progress in the development of a process for production of mixed alcohol fuel from peat. The process has four steps — pretreatment of peat to promote biodegradability, anaerobic fermentation to produce organic acids, electrolytic oxidation of organic acids to olefins, and hydration of the olefins to alcohols. Since production of alcohols by hydration of olefins is an acknowledged technology, the development program focuses on demonstrating technical feasibility of the other three steps.

Known domestic reserves of peat total over 1,400 quads of energy. The primary obstacle to its widespread utilization is its high moisture content and the difficulty of drying it. Wet processing alternatives include wet oxidation with heat recovery, wet carbonization, and fermentation. In the process under discussion, partial wet oxidation is used to enhance bio-degradability in anaerobic fermentation. Organic acids (C_2 - C_6) are produced in the anaerobic fermentation. After concentration by liquid-liquid extraction, these acids are electrolytically oxidized to olefins, which may be recovered by phase separation. Subsequent hydration of the olefins pro-duces the alcohol product. Figure 1 shows the processing concept explained in greater detail in Section 1.

The experimental program involved laboratory scale work in three process steps: peat pretreatment, fermentation, and electrolytic oxidation. Each of these areas was approached independently to establish the technical feasibility of the operation and to determine process operating conditions and yields. Conditions were found in which the pretreatment of peat increased its biodegradability. Solubilization of peat in the presence of alkali (0.004 g moles/g peat), oxidation temperature of 250°C, and supply of

*Dr. Wise is also the Cabot Professor of Chemical Engineering, Northeastern University, Boston, MA 02115, USA

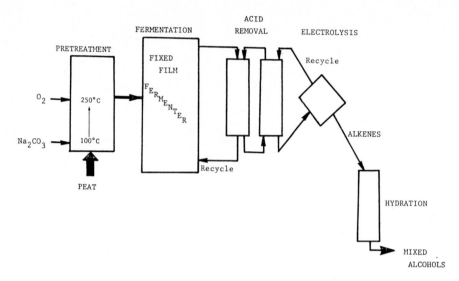

Fig. 1. Processing Concept

approximately 50% of the oxygen required for complete combustion were found to be optimal conditions. Half of the peat volatile solids were oxidized in this step and half of the remaining material was biodegradable (total biodegradation of 25% of input material). Lower yields were observed when the peat solids concentration was increased. These experiments are reported in detail in Section 3.

Fermentation experiments were carried out using benzoic acid (model compound) and pretreated peat as substrates. For both substrates, cultures of microorganisms were adapted from sewage digester contents. These adapted cultures anaerobically produced acetic acid. While some higher acids appeared, their sustained production has not yet been demonstrated. This is a key point remaining to be demonstrated in future work performed in this area. The first-order rate constant for conversion of pretreated peat to methane in a CSTR fermenter was measured as 0.25 day^{-1}. In a packed bed type reactor, complete conversion was achieved at 1.5 day retention time. Work on fermentation is reported in Section 4.

Electrolytic oxidation of organic acids was investigated to develop a mechanistic understanding of these conversions and provide insight into the control of product distribution by manipulation of operating parameters. Increased reactor pressure and presence of caproic acid in the reacting solution resulted in increased yield of alkane dimer (Kolbe product).

Increased temperature and use of a carbon anode increased olefin and alcohol yields. A working hypothesis was developed which explains the data presented in Section 5.

The engineering design, presented in Section 6, includes all material flows and process energy requirements and specifies the size and material of construction of all major equipment. The weight yield of olefins from the process as described in the base case is 4.6% of the peat organic solids fed into the pretreatment. The most significant material losses are in the pretreatment and fermentation steps. The unit fuel cost, calculated using a discounted cash flow method with 15% return on investment, is $52.20 per million Btu for the base case design. The effect of process performance (yields) and economic parameters on product cost were examined and are summarized in Figure 2. The product cost is most sensitive to ultimate yield. Doubling the product yield will reduce the unit product cost by about 40%.

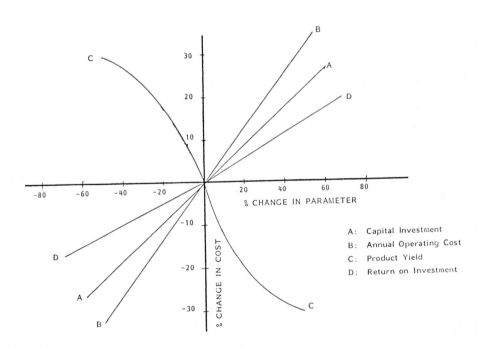

Fig. 2. Summary of Sensitivity to Changing Parameters

74

The experimental program has successfully demonstrated the technical feasibility of the pretreatment, liquid-liquid extraction, and electrolytic oxidation steps of the proposed process. The fermentation step shows probable feasibility. Production of acetate from peat was demonstrated; now cultures must be developed for production of the higher molecular weight products necessary in the process. Experimental data were sufficient to estimate the material flows and operating conditions for the full-scale process design. The economic analysis of the process at the present stage of development indicates that mixed alcohol fuel cannot be produced at a cost competitive with the present-day cost of liquid motor fuels refined from petroleum. The most direct approach to reducing product cost is increasing the yield, with particular emphasis on the pretreatment and fermentation steps.

Altering the process by reducing the number of steps is another approach to improving the yield. Two process alternatives for achieving such improvement can be suggested. Recovery of the pretreatment product for direct use as a fuel is one potential solution in which a liquid fuel (though substantially different from the alcohols produced in this process) would be produced. Fermentation of the pretreated peat to methane, instead of to organic acids, with eventual production of electricity or gas clean-up to pipeline quality is another processing alternative in which peat energy can be utilized as a non-liquid fuel.

1. INTRODUCTION AND BACKGROUND

The diminishing world-wide petroleum reserves and rising crude oil prices have created the need to find other raw materials for the production of fuels and feedstock chemicals. Private industry and the state and federal governments are undertaking a substantial research and development program to utilize other readily available resources to bridge the gap between supply and demand. The sizeable deposits of peat in the north central and coastal sections of the United States may provide raw materials for the production of significant quantities of fuels and chemicals.

It is a well established fact that plant debris, which are primarily cellulosic in composition, will decompose into water, methane, and carbon dioxide when attacked by fungi and bacteria in the presence of oxygen. If however these debris fall into an oxygen deficient environment such as a pond or saturated soil, the decomposition process is severely retarded and peat may be formed. Soper and Osborn (1922) described the stoichiometry of peat formation:

$$C_{72}H_{120}O_{60} \rightarrow C_{62}H_{72}O_{24} + 8CO_2 + 2CH_4 + 20H_2O$$

Peat is classified on the basis of its botanical origin, fiber content, moisture content, and ash content. The U.S. Bureau of Mines recognizes three types of peat: moss peat, reed-sedge peat, and peat humus. Any peat derived from moss is classified as moss peat. Peat derived from reed-sedge, shrubs, and trees is known as reed-sedge. Any peat whose origin is largely unrecognizable is classified as peat humus.

Recent estimates indicate that peat covers approximately 152×10^6 hectares (375 million acres) of the earth's surface (Tibbetts 1968). Sufficient quantities of peat reserves exist in the United States to make this a potentially valuable natural resource. The U.S. Bureau of Mines has compiled a map of the location and quantity of peat reserves in the U.S. (Figure 1.1). A comparison with other energy reserves shows that peat reserves (1,442 quads) are greater than the energy from uranium (1,156 quads), shale oil reserves (1,160 quads), and the combined reserves of petroleum and natural gas (1,408 quads) (Rader 1977). Only coal reserves (5,000 - 10,000 quads) represent a greater energy reserve in the U.S. The quantity of peat makes it a vast and largely untapped potential energy source, an alternative to imported petroleum.

The combustion of dry peat will release the greatest amount of energy (10,000 Btu/lb). Since peat is harvested from bogs, its moisture content is high (85 - 95%). Combustion requires pre-drying, a costly and energy-intensive process. Wet processing provides distinct advantages over any use which requires pre-drying. Recovery of a portion of the heat content of peat may be accomplished by wet oxidation, a process in which the peat is oxidized in its wet state (Othmer 1978). Wet oxidation has been developed and utilized in the treatment of waste streams which are high in organic matter (Teletzke 1964; Randall and Knopp 1980). These reactions are typically run at elevated temperatures (250 - 300°C) at the highest feasible solids loading.

There are a variety of options for the wet processing of peat, as depicted in Figure 1.2. Alkaline oxidation may be carried out exhaustively producing steam or, as proposed in this process, under conditions of limited oxygen producing a dissolved substrate for anaerobic acetogenic bacteria. The lignin portion of peat is represented in Figure 1.3 (Before). It consists of substituted aromatics linked by alkyl chains through ether bonds. The lignin polymers vary in size from 3,000 to 10,000 molecular weight. Peat also contains a carbohydrate fraction varying between 5 and 40 percent of its dry weight (Waksman 1930).

The pretreatment (alkaline oxidation) of peat is intended to render it more susceptible to microbial attack by disrupting its native structure. The

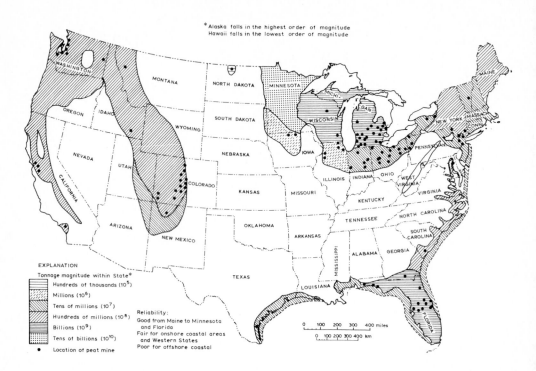

Fig. 1.1. Peat Mines in Operation in 1971 and Estimated Resources in Short Tons of Air-Dried Peat Suitable for Commercial Use

products of the pretreatment are substituted in single-ring aromatics, carboxylic acids, and methanol as shown in Figure 1.3 (After). These water soluble products are all potential substrates for anaerobic bacteria.

The pretreated material may then be fed to an anaerobic fermenter in which microbially mediated conversion to methane can proceed. Under controlled conditions, conversion to aliphatic organic acids may be encouraged. The organic acids can then be removed from the fermenter by liquid-liquid extraction and concentrated as the final product, or electrolytically oxidized to form other products – linear alkanes, organic acid esters, or alkenes. The product spectrum of the electrolytic oxidation is determined by the conditions under which the reaction occurs. Further processing of the electrolysis product, such as hydration of olefins (alkenes) to alcohols, represents another option.

The current program involves the production of a mixed alcohol fuel from peat. The process is shown schematically in Figure 1.4. Following the

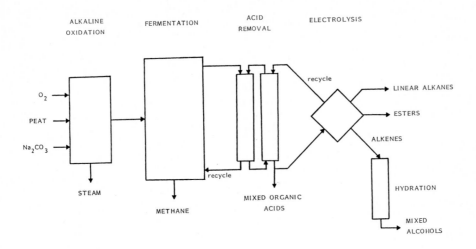

Fig. 1.2. Potential Options for Peat Wet Processing

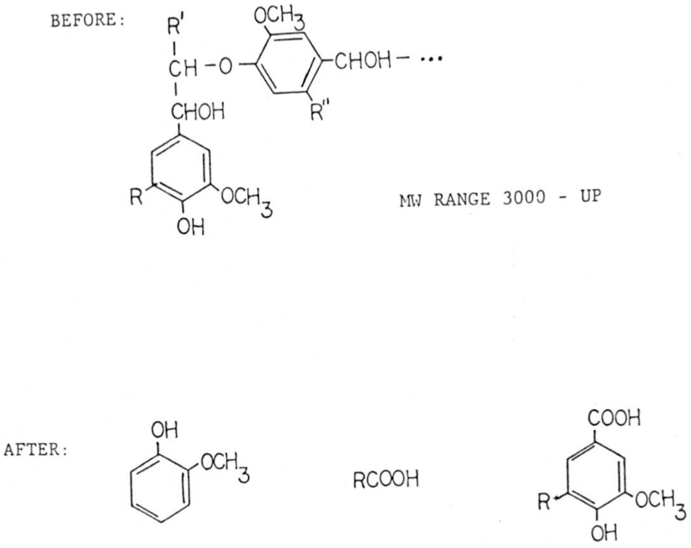

BEFORE:

MW RANGE 3000 - UP

AFTER: RCOOH

Fig. 1.3. Solubilization and Oxidation of Lignin

limited alkaline oxidation, the pretreated peat is anaerobically fermented. The fermentation conditions are adjusted to encourage formation of higher organic acids (butyric, valeric, caproic) instead of methane. This may require the addition of an electron acceptor such as sulfate or nitrate. The higher acids produced in the fermenter may be removed by liquid-liquid extraction into a hydrocarbon solvent and then concentrated by back-extraction into aqueous base. The acids are then electrolytically decarboxylated and oxidized to form mixed olefins. Butyric, valeric, and caproic acid mixtures will form a mixture of propylene, 1-butene, 2-butene (cis- and trans-), 1-pentene, and 2-pentene (cis- and trans-). The olefins may then be hydrated to form mixed alcohols. The fermentation, liquid-liquid extraction, and electrolytic oxidation steps of this process have already been demonstrated for polysaccharide substrates (Levy et al. 1981a; 1981b). Work is currently being performed to adapt this technology to processing peat.

Subsequent hydration of these mixed olefins will produce an alcohol fuel comprised of propanol, butanol, and pentanol. Acid catalyzed hydration of olefins is the current method of commercial production of most alcohols (Kirk-Othmer 1980 and Considine 1974). Since this reaction is a well-known technology, it is not addressed further in this report.

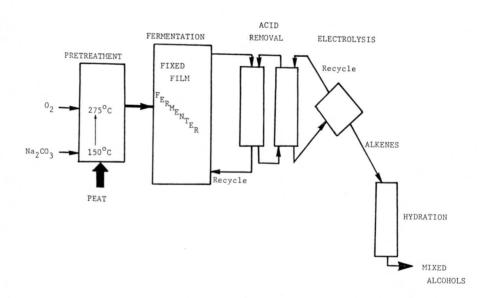

Fig. 1.4. Process Flowsheet

Ethanol has been proven to be a suitable motor fuel in mixtures of up to ten percent with gasoline. Due to ethanol's low energy density and high water solubility, use in higher percentages becomes more difficult. The higher energy densities (Table 1.1) and decreased water solubilities, make the higher alcohols produced by this process better gasoline replacements than ethanol. The projected favorable energy balance for this process also represents an advantage over ethanol processes for liquid fuel production.

Table 1.1

ENERGY DENSITIES OF ALCOHOL FUELS

Alcohol	Heat of Combustion (Btu/lb)
Methanol	9,600
Ethanol	12,820
Propanol	14,420
Butanol	15,530
Pentanol	16,350

2. PROGRAM PLAN AND RECOMMENDATIONS

The program, carried out over a two year period, was intended to examine the proposed process operating parameters establishing optimal loading rates, temperatures, residence times, flow-rates, etc., for each step of the process. The program plan included work on pretreatment of peat, fermentation of pretreated material, electrolytic oxidation of organic acid fermentation products, design of a product removal system, engineering design and economic analysis for a full-scale facility, and preliminary design of a pilot-scale facility.

While the technical feasibility of the process shows great promise, the economic considerations are not encouraging when compared with current liquid fuel market values. It appears that some advantage can be gained in the process by concentrating on the following areas:

- Reduction of operating costs particularly by reducing the alkali requirement, possibly by implementing a recycling scheme.

-   Lessening the severity of the pretreatment, even if the yield of fermentables suffers, to reduce capital costs.

-   Improving on the estimated yield of high weight organic acids in the fermentation step. This is the step with the most room for improvement

As indicated in the sensitivity analysis, the low yield of product is the best area to address to improve the economic projection for this process. Other approaches to the utilization of peat may also be considered. Reduction of the number of process steps may be the easiest way to improve the process yield. Two suggestions toward this end are to recover the pretreatment product as a liquid fuel or to produce methane in the fermentation step as a gaseous fuel or for conversion to electricity.

Even though the economic analysis reflects significant problems with the process, some worthwhile technical achievements include documentation of the effects of changing various pretreatment parameters (temperature, time, oxygen flow rate, solids loading) on yield of low molecular weight material, development of culture adaptation techniques for organic acid production from pretreated peat, careful documentation of methane bioconversion rate from pretreated peat, and the development of a mechanistic understanding of the electrolytic oxidation of aliphatic organic acids.

3. PRETREATMENT - LIMITED ALKALINE OXIDATION OF PEAT

3.1 Background

Peat is a heterogeneous material with a high lignin content. Figure 3.1 shows the principal monomeric precursors of lignin. These compounds are polymerized to form a structural and protective material for the plants that synthesize them. Palmer and Evans (1983) have described the structure of conifer lignin (consisting primarily of coniferyl alcohol monomers) and the types of bonds formed between aromatic nuclei. The most common bond, accounting for approximately 50% of the polymerizing bonds in lignin, is the coupling between the 4-OH and B-Carbon of the alkyl side chain as depicted in Figure 3.2. Other bonds described are between the 4-OH and α-carbon, the 4-OH and an aromatic carbon, direct aryl-alkyl bonds, and biphenyl bonds between aromatic nuclei. Our approach for increasing biodegradability of the lignin in peat is to degrade the lignin first by alkaline hydrolysis and then by oxidative destruction of the polymeric bonds. The material produced

p-coumaryl alcohol   coniferyl alcohol   sinapyl alcohol

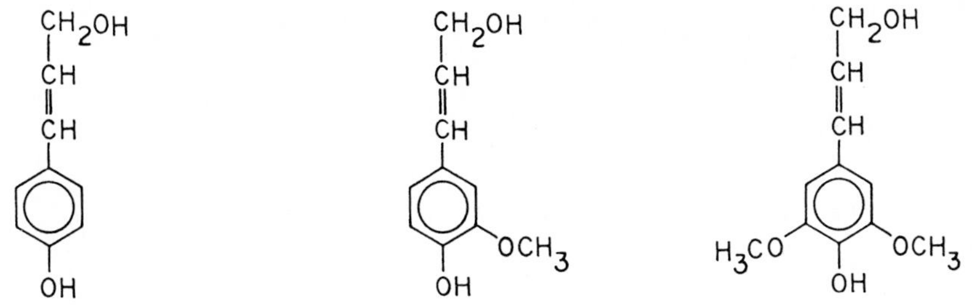

Fig. 3.1. Monomeric Precursors of Lignin

will be aromatic monomers and dimers, somewhat more oxidized than their lignin precursors.

In its natural state, peat is not biodegradable by anaerobic microbes (McCarty et al. 1979, 1983). Hydrolysis of the lignin matrix releases lower molecular weight, water-soluble compounds and improves the biodegradability of peat. While cellulose is fairly stable in the presence of alkali, even at elevated temperatures, lignin is extensively solubilized into a variety of large and mostly refractory polymers as well as a number of lignin monomers. Lignin monomers which have been produced by alkaline hydrolysis include hydroxybenzoic acid, vanillic acid, vanillin, syringaldehyde, acetaldehyde, and catechol (McCarty et al. 1983). These monomers are largely biodegradable by anaerobic bacteria (Healy and Young 1979), but yields after alkaline hydrolysis are quite low (5 - 10%).

Biodegradability can be further increased when limited oxidation is allowed to occur after solubilization (Dynatech Report No. 2115), 1981; Wise et al. 1983; Ruoff et al. 1980). McCarty and co-workers (1979, 1983) noted that the presence of oxygen decreased the yield of solubilized organics, but an optimization of conversion to biodegradable products as conditions of the oxidation are varied was not reported.

The proposed limited alkaline oxidation to produce water-soluble monomers is based on the "wet oxidation" process developed for application to waste disposal in the paper industry (Flynn 1979; Zimmerman and Diddams 1960). Any substance capable of burning that remains dissolved in water, can be oxidized at elevated temperatures. To obtain high conversions of organic matter to CO_2, temperatures in the vicinity of 250 - 300°C may be necessary. Oxidation of soluble sugar and paper pulp waste is shown as a function of reactor temperature in Figure 3.3.

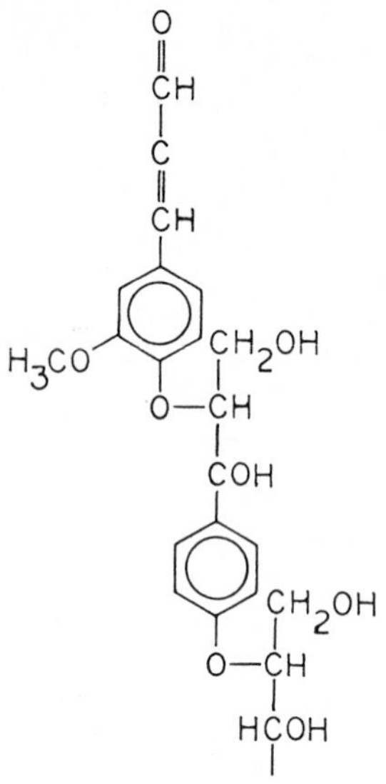

Fig. 3.2. Primary Lignin Polymerizing Bond

Suitable materials for wet oxidation processes are those which contain substantial amounts of water and cannot easily be concentrated to sustain combustion under conventional burning conditions. Although elevated temper- atures are necessary, at sufficient solids concentrations, the wet oxidation reaction will be spontaneous above some threshold temperature. Wet oxidation technology may be readily applied to peat processing. This type of process- ing avoids the difficult and energy intensive dewatering of peat necessary for direct combustion or gasification. Complete wet oxidation of peat and vaporization of the associated water is self-sustaining at peat concentra- tions above 12%.

Wet oxidation of peat proceeds when the solubilized organic portion is exposed to oxygen at elevated temperatures.   If the peat is not solubilized prior to oxygen addition, wet carbonization will occur forming "peat coal". Since the wet carbonized peat is easier to mechanically dewater and has a higher heat of combustion than raw peat, wet carbonization is being devel- oped as a process to obtain fuel from peat (Mensinger 1980).

Peat may be solubilized by addition of alkali at moderate temperatures (75 - 150°C). The products of this alkaline hydrolysis of peat consist of many types of substituted phenols, such as condensed and uncondensed guai-

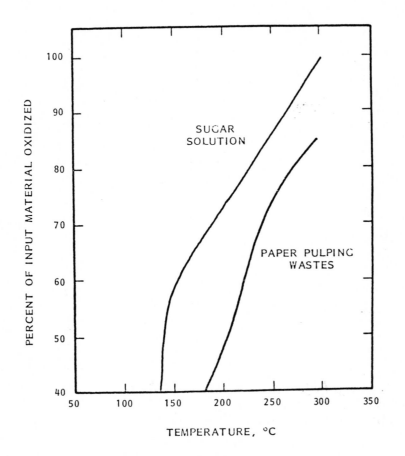

Fig. 3.3.* Wet Oxidation of Dissolved Materials
*From Teletzke (1964)

acyl as well as higher molecular weight entities. The substituent groups
are carbon side chains containing 3 to 5 carbon atoms. Oxidation of the
dissolved peat initially yields single ring aromatics (aldehydes, alcohols,
and carboxylic acids) with methoxy side-chains. Low molecular weight alkyl
carboxylic acids are also formed. Continued oxidation leads to the
degradation of the aromatic nuclei (Oki et al. 1978). The oxidation of
solubilized peat is postulated to follow a general reaction scheme comprised
of three types of reactions (Gallo and Sheppard 1981).

Reaction 1: Peat lignin + O_2 → CO_2 + H_2O + Heat
Reaction 2: Peat lignin + O_2 → Organic acids + Heat
Reaction 3: Organic acids + O_2 → CO_2 + H_2O + Heat

The organic acids formed in Reaction 2 may be aliphatic or aromatic. While Reactions 1 and 2 are believed to proceed rapidly, Reaction 3 is relatively slow. Gallo and Sheppard (1981) reported first-order rate constants for Reaction 2 approximately 10 times the rate constant measured for Reaction 3. This implies that reaction conditions may be chosen so that organic acids are accumulated.

The methoxy content of the product also decreases with time, ostensibly due to the elimination of side chains from the aromatic nuclei. Complete oxidation will eventually result in the conversion of the dissolved peat products to carbon dioxide and heat evolved as steam. A summary of the oxidation pathways of peat is shown in Figure 3.4. Upon addition of alkali and heat to the peat slurry, some organic material is dissolved and some remains suspended. At elevated temperatures, even in the presence of oxygen, the suspended material is "carbonized"; i.e. converted to "peat coal". The peat coal produced has a lower oxygen content and higher heat of combustion than the raw peat feed (Mensinger 1980).

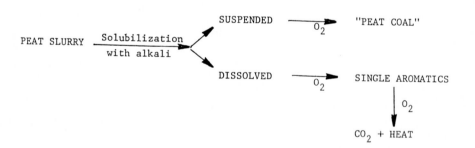

Fig. 3.4. Peat Oxidation Pathways

When solubilized peat organics are subjected to a limited oxidation, single ring aromatics and linear carboxylic acids are believed to be formed from the lignin portion of the peat. Methanol and ethanol may be formed from the carbohydrate portion as degradation products of the hemi-cellulose.

In the presence of excess oxygen at temperatures between 275 - 300°C, the solubilized peat is completely oxidized to CO_2 and H_2O.

3.1. Program Plan for Pretreatment of Peat

The objective of peat pretreatment is to render the peat susceptible to microbial attack. Experiments in this phase of the project were designed to determine conditions which would result in the highest yield of fermentable pretreatment products. The assumption that increasing the yield of low molecular weight products would result in improved biodegradability was made early in the pretreatment work. This assumption was based on recent work at Dynatech (Wise et al. 1983) and work reported in the literature (Colberg and Young 1982; Healy and Young 1979; Healy et al. 1980). Later in the project, fermentation experiments were performed to verify this assumption. These experiments indicated that the fermentable fraction of the pretreated peat consisted of primarily low molecular weight species (<500 MW).

The solubilization of raw peat was found to occur readily in the presence of alkali. A linear correlation between alkali concentration and solubilization (with constant peat concentration) was seen. The yield of low molecular weight products was found to be most sensitive to conditions of the oxidation step. Experiments were designed and performed to investigate the effect of oxygen tension (flow-rate), temperature, and time on the yield of fermentable material.

3.2.1 Task Description

The experiments planned for pretreatment of peat were in the following four areas:

1. Adjustment of solubilization parameters
2. Adjustment of oxidation parameters
3. Characterization of pretreated peat products
4. Documentation of biodegradability of pretreated peat by fermentation

The detailed experiments planned to investigate these areas and results of those experiments are reported in the following sections (3.3 - 3.5). The goals of each task are outlined below.

Task 1 - Adjustment of Solubilization Parameters

The degree of solubilization is dependent on the concentration of available alkali. Experiments will be performed to document this dependence using sodium carbonate as alkali. Data will also be obtained for determining the rate constant for peat solubiliza- tion. In addition, other alkali including sodium hydroxide and ammonia (as ammonium hydroxide) will be tested. Ammonia has advantages of ease of recovery and, at residual concentrations, non-toxicity to anaerobic bacteria.

Task 2 - Adjustment of Oxidation Parameters

The oxidation of solubilized peat appears to be the most critical part of the pretreatment in obtaining high yields of fermentable material. Experiments will be performed to optimize the yield of low molecular weight organics by varying oxygen availability, temperature, and time. Data will be collected to allow engineer- ing design of a full-scale oxidation reactor.

Task 3 - Characterization of Pretreated Peat Products

Products will be characterized by size distribution using ultra- filtration through membranes with successively smaller pore sizes and by gel permeation chromatography. Product liquor from the oxi- dation will be compared at various stages of oxygen consumption and after fermentation to aid in choosing optimum oxidation conditions. Analysis of oxidation products by gas chromatography and high pressure liquid chromatography is also planned to further characterize the low molecular weight products.

Task 4 - Documentation of Biodegradability of Pretreated Peat by Fermentation

This task is described in more detail in Section 4 of this chapt- er. Information on the ultimate conversion of pretreated material is essential in determining optimum conditions. Two primary points are addressed in this task. First the measured biodegradability of pretreated peat is to be determined. Also, the size distribution of the biodegradable fraction is sought to strengthen the initial assumption that maximizing yield of low molecular weight materials will provide the best substrate for microbial conversion.

3.2.2 Experimental Approach

All previous work on the anaerobic digestion of lignin-type materials indicates that it is hydrolyzed lignin that is susceptible to microbial digestion. Ideally, optimization of the pretreatment process should incorporate bioconversion measurements of the pretreated material produced under a variety of pretreatment conditions. It is recognized, however, that cultures would have to be adapted to each sample formed under different conditions to obtain reliable information from biodegradation studies. This procedure would involve long analysis times (30 days) and thus severely limit the number of variables that could practically be investigated.

Our approach was to make the assumption that an increase in yield of low molecular weight water soluble products would indicate an increase in biodegradability. The yield of water soluble low molecular weight material could be monitored very accurately by separation using ultrafiltration and volatile solids measurement by weight.

Once near-optimal conditions had been determined, more attention could be focused on justification of the initial assumptions. Some effort also could be directed at characterization of the pretreated peat product. This product characterization would be limited to determination of size distribution as a function of pretreatment time, comparison of size distribution following fermentation, further separation of fermentable peaks and matching of those peaks with model compounds.

The series of experiments proposed will define conditions of peat pretreatment which yield the highest fraction of fermentable material. Information for engineering design of the reactors will be obtained and a limited characterization of the product performed.

3.3 Determination of Peat Composition

Peat was donated by the Minnesota Gas Company for use in our experimental program. It is of the reed-sedge variety. An elemental analysis of the peat was performed by Galbraith Labs (Knoxville, Tennessee). Further characterization of the raw peat was performed as described by Levy et al. (1981a). The procedure is briefly described below.

Total ash content was determined by burning an oven-dried sample at 600°C in a muffle furnace and comparing weights before and after burning. A more detailed analysis was performed to determine the nature of the non-ash fraction. An oven-dried sample was extracted with 2:1::toluene:ethanol until constant weight was achieved (approx. 24 hours). Then, the non-extracted portion was hydrolyzed with 10 ml/g material of 72% H_2SO_4 at 30°C for 1 hour, diluted with water to 6.5% H_2SO_4 and heated at 110°C for 3 hours. The non-hydrolyzed material was then filtered off, dried and weighed. This material was identified as lignin and non-dissolved ash. The lignin portion was determined by burning in a 600°C muffle furnace. The acid solubilized material was carbohydrate and soluble ash.

Elemental Analysis (by weight)

```
%C :  48.47
%H :   4.91
%N :   2.80
%S :   0.32
%O :  31.75
```

Ethanol/Toluene Extractable :   8.5%

Acid Solubilized Material   :   30.9%

Lignin   :   57.5%   **Average of 3 samples**

Non Acid-Detergent Ash   :   3.1%

Heat of Combustion   : 8,077 Btu/lb total solids
                               (average of 2 samples)

Total Ash   :   11.1%

Fig. 3.5.   Peat Composition

The heat of combustion of the raw peat was measured in a Parr Bomb Calorimeter® on oven-dried samples. A summary of the composition of the peat used in our experiments is given in Figure 3.5. The total ash content of the raw peat is approximately 11%. A portion of the ash (about 8%) is contained in the acid solubilized material, the remainder of which (almost 25% of the raw dried peat) is carbohydrate. This high carbohydrate content is reflected in the relatively low (8000 Btu/lb) heat of combustion.

3.4   Pretreatment of Peat-Experimental Procedures
3.4.1 Preparation of Peat

Reed-sedge peat is hand-picked free of twigs, chunks of logs, and sandy clay inclusions. These materials, although present in slight amounts, are removed to make the peat more homogenous for sampling purposes, and to remove any large objects which might clog reactor values. The peat is weighed (the usual reactor charge is 600g wet peat), diluted to ≈1l with distilled water, mixed with the alkali charge (usually 22.4g granular Na_2CO_3) and the slurry homogenized in a 4-l Waring Commercial Blender in three 1-minute bursts at the high setting, with rinsing down of the mixer chamber between bursts. This step results in better solubilization yields during pretreatment, possibly as a result of particle comminution. The slurry is diluted to 1,400 ml with distilled water, and a raw slurry sample taken; it is then sealed into the reactor.

3.4.2 Solubilization and Oxidation

Solubilization, or breakdown of the peat to suspended and dissolved materials (see Section 3.4.3 for definition), is accomplished by heating the slurry past approximately 150°C, preliminary to oxidation.

Two modes of oxidation have been used: batch and continuous. In batch oxidation, the slurry is heated to 150°C with 500 RPM stirring. Stirring is stopped, a sample taken, and a known amount of oxygen charged, as determined by headspace, less steam, pressure. The static reactor is heated to the desired oxidation temperature and 900 RPM stirring started, giving (at 150°C or above) a noticeable exotherm, for example about 20°C in one minute at 275°C. After one hour, the reactor is cooled. This mode of oxidation was used to model very high oxygen flow rate reaction conditions.

For continuous oxidation, the reactor is heated with 500 RPM stirring directly to oxidation temperature, and a liquid sample taken. Stirring is increased to 900 RPM, and oxidation started using air or oxygen. The gas input is measured through a calibrated Rotameter R7640 (Matheson Gas Products). All reactor pressures above 200 psig were run with air (hydrocarbon-free air, Matheson). The air (or oxygen) input control system is arranged as follows.

From the tank regulator, the air passes through the Rotameter, and is pressure-gauged before passing through a needle valve, the main flow control. Passing through the valve, the air is drawn into and pressurized by the gas booster pump, whose stroke rate is regulated to 0.2-1 Hz by throttling its drive air. The high pressure air then is injected into the reactor, near the bottom of the chamber. Reactor exit gas is led slightly upwards to a vertical-mounted water condenser, cooled therein to room temperature, cracked through a valve (or through a back pressure regulator; however this tends to clog and is time-consuming to clean), passed by a septum port and through an ice-packed 500-ml flask condenser, and exhausted through a wet-test meter which measures (ambient temperature) gas volumes.

Exit gas samples are analyzed for O_2, CO_2 and N_2 by gas partition chromatography as described elsewhere; slurry samples (each consisting of the contents of one or two 75-ml sampling cylinders charged from a liquid sampling tube (<10 ml dead volume), sealed, removed, cooled, and emptied) are analyzed as below.

3.4.3 Sample Analysis

Essentially the slurry samples are incrementally cleared of various size ranges of materials, with analyses at each stage.

Thus, a portion of total slurry is taken for volatile and total solids determination (optionally, volatile liquids are determined by GC as described elsewhere), for which it is weighed into a tared 70 - 80 mm diameter porcelain evaporating dish (Coors), acidified to pH 2.0, dried at 110°C, weighed, ashed at 600°C (requiring 2 - 6 hours to ash completely), and reweighed. Total solids (dry weight), ash, volatile solids (total solids less ash) and percent volatile solids of wet weight are calculated, and also percent ash/wet weight, which indicates the degree of evaporative concentration of the slurry.

The total slurry sample is next centrifuged at 12,000 RPM ($\approx$8800 x g) for 15 minutes in 28 x 100 mm polypropylene centrifuge tubes to spin out larger and heavier particles, and the supernatent collected. Percent of volatile solids is determined on it, and this datum (corrected for evaporative concentration), versus the raw slurry percent volatile solids, is termed "percent solubilization", or "yield of solubilization".

The supernatent is then neutralized to pH $\approx$7 (to protect the membranes), and successively ultrafiltered through Amicon XM 300, PM 10, and UM 05* membranes in a Model 52 Ultrafiltration cell (Amicon, Danvers, MA) at pressures of 10, 10, and 40 psig respectively. Samples are taken at each stage and analyzed for volatile solids. Thus a profile of products by size ranges is obtained for each sample drawn from the reactor.

Additional analytical procedures are applied to the UM05 ultrafiltrate as desired, in particular fermentation tests, separation by solvent-gradient HPLC, and Sephadex gel chromatography. The former two procedures are described elsewhere, work on Sephadex gels will be described here.

Sephadex gel columns 1.5 x 45 - 55 cm were poured in 100 ml burettes with buffer-preswollen G-10-120, G-50-150, and G-100-120 beads. The buffer initially consisted of a 5 wt% $NaHCO_3$ solution, brought to pH 9.00 with NaOH, and preserved with 0.02% NaN_3; later work incorporated 0.5% Tween-20, recommended (but not observed) to give decreased hydrophobic attachment to the Sephadex. (We wished to fractionate by molecular size).

Flow-rates of 0.6 - 1.0 ml/min were used; eluent was collected in a Buchler LC-100 Fraction Collector and the UV absorbance of each fraction in the range of 250 - 360 nm checked. With some exceptions, 250 nm absorbances sufficed to indicate UV-active species. (See Section 3.5.3 for detailed discussion and figures).

---

*This type of ultrafiltration membrane is now discontinued and replaced by YC05, which differs by being surface uncharged instead of negatively charged and having a molecular weight passage cut-off at nominally 300 (vs. 500) daltons.

3.4.4 <u>Equipment</u>

The reactor is a 2l chamber volume PMD Pressure Reactor (Pressure Products Industries, div. Duriron Co., Inc., Warminster, PA). The chamber is cylindrical, $\simeq 30$ x 9 cm, with 8-bolt "Flexatalic" closure of a head onto an asbestos/stainless steel spiral-wound gasket. A magnetically-externally coupled "Gaspersator" turbine stirs 4 cm from the well bottom. The chamber has a gas inlet tube at 1 cm above the well bottom, a sample withdrawing tube at 7 cm above bottom, and a gas outlet tube at 1 cm from the chamber top. An internal cooling coil surrounds the stirrer shaft, and an external mantle, driven by two powerstat transformers (Superior Electric Co., Bristol, CN), supplies heat. Temperature is sensed by a thermocouple in a well at 13 cm from chamber bottom, and actuates a Model 2157 Dynasense Electronic Temperature Controller (Cole-Parmer, Skokie, IL) to power one Powerstat for fine temperature tuning. (The other Powerstat is hand-set to heat or to maintain approximate temperature). A temperature record is kept on a Model SR-204 Heath/Schlumberger Strip Chart Recorder.

Exit gas is controlled through a Model 5-91XW (100 - 3000 psig rating) Back Pressure Regulator (Grove Valve and Regulator Co., Oakland, CA).

3.5 <u>Results and Discussion of Pretreatment Experiments</u>

Results from the alkaline oxidation of peat experiments, presented in the following tables and graphs, show a decrease in high molecular weight material and concommitant production of lower weight products over time. The general trends observed are:

1. Linear correlation between concentration of alkali and degree of peat solubilized up to 0.004 mole alkali per gram peat volatile solids.
2. First-order rate constant for peat solubilization with excess alkali measured at 80° is 0.33 hr^{-1}.
3. Little change in the yield of low molecular weight material seen when oxidation is carried out below 225°C.
4. Greater than 90% combustion of peat can be accomplished with excess oxygen at 300°C.
5. Only half the oxygen is absorbed at 300°C when oxidation is not preceded by solubilization.
6. At 250°C, the lower molecular weight portion of product first increases and then decreases over time indicating the initial formation reactions are faster than decomposition reactions.

7. There is an optimal air flow rate (coinciding to an optimal oxygen partial pressure) for production of low weight product.

8. At higher substrate loadings, similar results are observed but yields of low weight products are lower.

9. Gel Filtration Chromatography studies show the low weight material eluting between standards with molecular weights from 122 to 709, the middle molecular weight material eluting at less than 10,000, and the high molecular weight material between 10,000 and 40,000.

10. Biodegradation is limited to the lower molecular weight portion of the pretreated lignin in peat (see Section 4).

3.5.1 Solubilization Studies

Results from these studies were presented in our earlier report on this project (Levy et al. 1982). The results are reiterated here. High solubilization of peat organics (up to 90%) was achieved at moderate temperatures

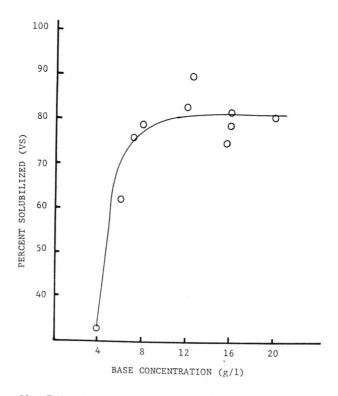

Fig. 3.6. Peat Solubilization with Sodium Carbonate

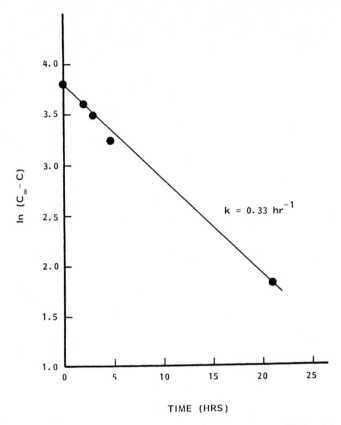

Fig. 3.7. Rate Constant of Peat Solubilization at 80°C with Sodium
Carbonate (Run No. 014183)

(up to 150°C). At 150°C, solubilization of peat was completed after 30 minutes at temperature. A series of experiments was performed to demonstrate
the effect of base concentration on solubilization. All experiments were on
4% peat slurries at 150°C with sodium carbonate used as the base. The
results (Figure 3.6) indicate that the degree of solubilization is strongly
dependent on base concentration of up to .004 moles base per gram of peat.
Similar results were obtained with sodium hydroxide used as base. These
results agree with earlier work performed at Dynatech with peat of a
different composition (Dynatech Report 2115, 1981).

Solubilization experiments using ammonia as alkali were also attempted.
In all cases, low solubilization of peat was achieved with ammonia, probably
due to its high volatility at elevated temperatures. A first-order rate
constant of 0.33 hr^{-1} was determined for solubilization of peat in excess
sodium carbonate at 80°C (Figure 3.7). At 80°C, 68% of the peat organics

were solubilized. At higher temperatures, somewhat higher solubilization was achieved at faster rates. For design purposes, the rate of solubilization is assumed to double for every 10°C increase in reaction temperature. To achieve 80% solubilization in a reactor modelled as a CSTR at 100°C, a 3-hour residence time is required.

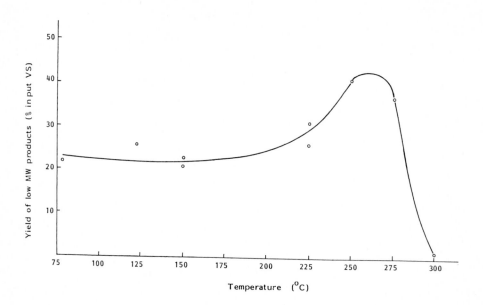

Fig. 3.8. Effects of Temperature on Ultimate Yield of Low Molecular Weight (MW) Products

3.5.2 Oxidative Treatment of Solubilized Peat

The effects of temperature, oxygen concentration or flow rate, and time on the yield of low molecular weight products were studied. Figure 3.8 shows the effect of varying the oxidation temperature on yield of low weight product. All experiments are performed on peat solubilized with sodium carbonate prior to introduction of oxygen. Yields shown are the maximum observed from samples taken at timed increments during oxidations at each temperature. Note that no increase in yield is observed at temperatures below 200°C. Reduction in high weight material is observed at all temperatures, even though there is no build-up of the low weight intermediates at lower temperatures.

It is anticipated that if the reaction of lignin polymer to lower weight organic acids is more rapid than degradation of the organic acids to

(Run #014118; 250°C; 72.5g VS loaded;
0.37 l/min. O_2 as air)

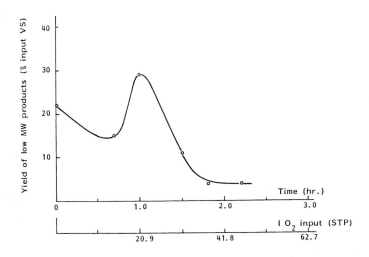

Fig. 3.9. Effect of Time on Low MW Product Yield

CO_2 + H_2O, the concentration of the organic acids will build up to a maximum before decreasing.   This is demonstrated in Figure 3.9 where the maximum yield is seen after 1.0 hour.   The O_2 partial pressure, represented by the O_2 flow rate in these oxidations, also affects the kinetics of the reactions involved. An optimal flow-rate corresponding to 0.002 mole O_2/min·g peat VS, was observed (Figure 3.10).

Results from experiments which have the O_2 loaded into the reactor in a batch mode are shown in Table 3.1.   The same trend is seen from these experiments as from the continuous-flow oxidations. With no oxidation, most of the material is in the high molecular weight range with very limited production of low weight carboxylic acids by alkaline hydrolysis.   As the

96

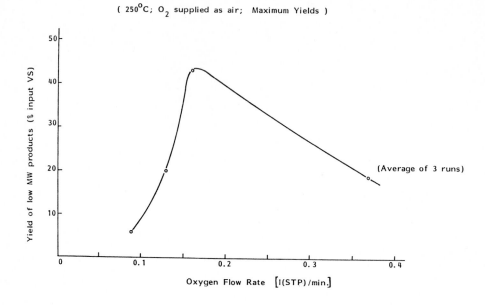

Fig. 3.10. Effect of Oxygen Flow Rate on Low Molecular Weight Product Yield

Table 3.1

YIELDS FROM BATCH PEAT OXIDATIONS

RUN NUMBER	TEMP., °C	OXYGEN TO VOLATILE SOLIDS*	PERCENT OF INPUT VS			
			HIGH MW	MID MW	LOW MW	VOLATILE LIQUIDS
014687	275	no O_2	50.1	17.5	6.7	2.7
014675	275	4	31.2	10.1	9.2	9.0
014680	275	8	12.7	9.8	13.4	10.6
014684	275	12	4.9	4.0	12.4	11.4
014661	285	16	1.0	1.8	10.6	15.4
014668	295	5	2.3	1.3	1.1	12.3
014672	300	25	0.2	0.6	0.9	7.6

* Ratio of moles of O_2 loaded into reactor to moles of peat (196 g volatile solids) loaded into reactor.

oxygen supplied is increased, the high weight material is oxidized and is reduced in yield while CO_2 and low weight products are produced.

Tables 3.2 and 3.3 present data used in drawing Figures 3.9 and 3.10. Table 3.4 shows the effect of alkali on high temperature oxidation. Twice the oxygen was consumed when alkaline solubilization preceded oxidation. In the run without alkali, a carbonaceous residue was recovered which was not present in the other run. This, once again, demonstrates the difference between wet carbonization in which little or no solubilization occurs and wet oxidation in which solubilization is achieved before oxidation is attempted.

Table 3.2

EFFECT OF OXYGEN FLOW RATE

FLOW-RATE (ℓ O_2 (STP)/MIN)	TIME (HR)	YIELD OF LOW MW PRODUCT (%)
0.08	4.7	6
0.13	2.4	20
0.16	0.9	43
0.37	1.0	19 (average of 3)

Table 3.3

LOW MW PRODUCT YIELD VS TIME (Run #014118)*

TIME (HR)	CUMULATIVE O_2, INPUT (ℓ)	YIELD OF LOW MW PRODUCT (%)
0	0	22
0.7	13.1	15
1.0	20.9	29
1.5	28.9	11
1.8	34.8	4
2.2	40.0	4

* Carried out at 250°C; 72.5 g VS initial loading.

Table 3.4

EFFECT OF ALKALI ON OXYGEN ABSORPTION

RUN NUMBER	ALKALI	TEMPERATURE	O_2 ABSORBED
014785	None	300°C	5.25 moles/196 g Peat
014672	.004 moles/ 196 g peat	300°C	8.55 moles/196 g Peat

* Theoretical Maximum O_2 Absorption: 9.8 moles/196 g Peat.

Since high peat concentrations are required in the process, a run containing 12% volatile solids in the initial charge was carried out. Following solubilization, the nearly solid mass initially loaded into the reactor was transformed into a liquid slurry with dissolved and finely suspended solids. This solubilized material did not appear to present clogging or pumping problems. The molecular weight distribution of dissolved and suspended solids, however, was much higher than for comparably pre-treated peat at 4% solids loading. Figure 3.11 shows the molecular weight distributions in the three categories as a function of oxidation time. One of the major differences in conditions is the limit on oxygen flow rate (per volatile solids loaded) in the high solids run imposed by our equipment. The same trend of reduction of high weight material and increase in low molecular weight product is observed in the high solids run. At comparably higher O_2 flow rates, we would anticipate results similar to those observed in other experiments.

The results from the pretreatment experiments have allowed us to project process operating conditions and yields. The projected optimal conditions for peat pretreatment are:

- 15% total solids loading;
- O_2 flow rate of 0.002 mole O_2/min·g peat VS;
- oxidation of 40% of input peat VS;
- conversion of 25.7% of input peat VS to soluble, fermentable material;
- conversion of 22.3% of input peat VS to soluble, but non-fermentable material; and
- conversion of 20.0% of input peat VS to insoluble material, 40% of which is ultimately oxidized.

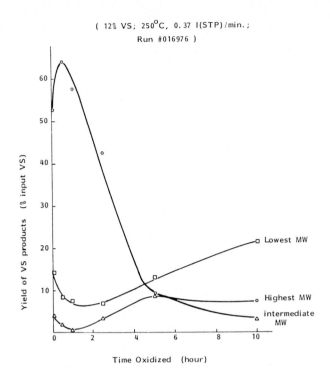

(12% VS; 250°C, 0.37 l(STP)/min.;
Run #016976)

Fig. 3.11. High Solids Peat Pretreatment

3.5.3 Gel Filtration Studies

The analytical procedure of successive ultrafiltrations of the soluble products through 300,000, 10,000, and 500 nominal molecular weight cut-off membranes has been used as a technique of separating the product into high, middle, and low molecular weight size ranges. It is recognized that this does not reflect the molecular weight distributions of the products. In order to gain some knowledge of actual molecular weights, product liquors were passed through Sephadex gels (G-100, G-50, and G-10) and eluate examined by UV absorbance to determine retention time on the gels. By comparing retention times with those of known molecular weight standards, the product molecular weight distribution could be estimated.

For these studies, only the lignin fraction of the peat was used. The "lignin peat" is prepared by slurrying whole peat in 72% $H_2SO_4 \cdot 2H_2O$ and incubating at 30°C for one hour. The slurry is then diluted with water to 6.5% sulfuric acid and incubated at 100°C for three hours. The solids remaining undissolved after this treatment are the lignin portion of the peat which is recovered by filtration and washed with distilled water. Gel filtration studies are carried out on samples at various stages of oxidation.

The lignin peat (56g) was slurried with Na_2CO_3 (22.5g) and diluted to 1,450 ml with water. Solubilization was carried out at up to 250°C and oxidation was accomplished with air at 250°C and an oxygen flow rate of 0.37 l/minute. Liquid samples were collected at various times before and during the oxidation. All samples were centrifuged to remove undissolved material.

Figure 3.12 is an oxidized sample eluted on a Sephadex G-50. Fractions were collected and absorbance at 256 nm measured. Dextran, 10,500 MW and benzoic acid elution volumes are indicated by A and B. This figure indicates that the high molecular weight peak is greater than 10,000 MW and the low molecular weight material is less than 100 MW, but not resolved on this column.

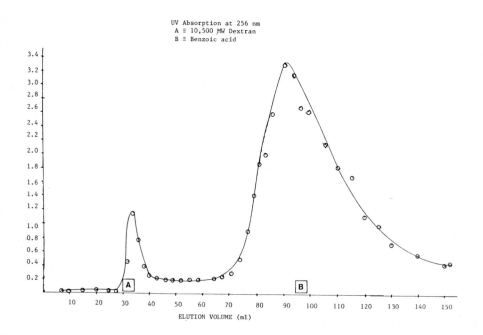

Fig. 3.12. Oxidized Peat Lignin Eluted from Sephadex G-50

The same oxidized peat lignin sample was then eluted from a Sephadex G-100 column. Here it can be seen (Figure 3.13) that the high molecular weight peaks are spread out between the 40,000 and 10,500 MW Dextran Standards. The large low molecular weight peak is not resolved on this column.

The liquid samples from the lignin peat cook were ultrafiltered through UMO5 membranes (<500 MW) before elution on a Sephadex G-10 column for separation of the low molecular weight products. Retention volumes for a series of standard compounds is shown in Figure 3.14. Elution of successively oxidized samples on this column is shown in Figure 3.15. Increased heights are seen for 256 nm absorbing peaks from the solubilized only sample to the 20 minute and 31 minute oxidized samples. More oxidized samples are not shown here, but no further increase in peak size is observed through 65 minutes of oxidation. Most of the UV absorbing material appears to be eluted between standards for benzoic acid (122 MW) and NADH (709 MW). Quantitative statements cannot, however, be made concerning the weight or yield of any material without positive identification of the products.

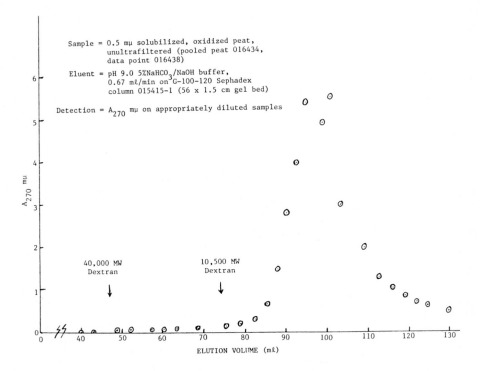

Fig. 3.13. G-100 Gel Filtration of Oxidized Peat Lignin

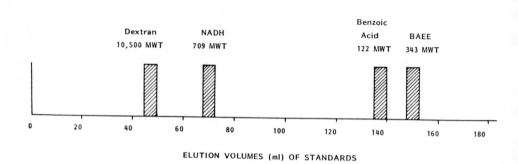

50% NaHCO$_3$/NaOH pH 9 with 0.5% Tween-20
+ 0.05% NaN$_3$

Dextran
10,500 MWT

NADH
709 MWT

Benzoic
Acid
122 MWT

BAEE
343 MWT

ELUTION VOLUMES (ml) OF STANDARDS

Fig. 3.14.   Elution Volumes for Standards on Sephadex G-10 Column

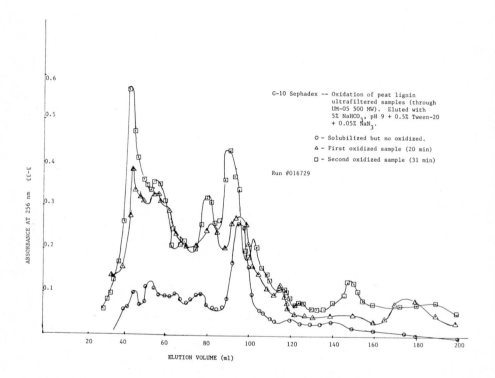

3-33

ABSORBANCE AT 256 nm

G-10 Sephadex -- Oxidation of peat lignin
    ultrafiltered samples (through
    UM-05 500 MW). Eluted with
    5% NaHCO$_3$, pH 9 + 0.5% Tween-20
    + 0.05% NaN$_3$.

O - Solubilized but no oxidized.
△ - First oxidized sample (20 min)
□ - Second oxidized sample (31 min)

Run #016729

ELUTION VOLUME (ml)

Fig. 3.15.   Successive Oxidation of Peat Lignin

3.5.4 HPLC Studies

Some pretreated peat samples were separated by High Pressure Liquid Chromatography (HPLC) to provide further characterization. Two points were documented in the HPLC studies. First, changes in peak size and retention times (indicating change in composition) were observed as the oxidative pretreatment progressed. Also, discrete peaks disappeared following ferment- ation of the pretreated peat indicating utilization of specific components of the material by anaerobic bacteria.

The HPLC characterization of oxidized peat was performed on samples taken at various times from a continuous peat oxidation at 250°C. Prior to HPLC analysis, samples were ultrafiltered through 5-OPS (1000 MW cutoff) membranes (Amicon). The samples were injected onto a C-18 reversed phase column and eluted with a water-methanol gradient (O - 99% methanol over 20 minutes). Both water and methanol contained 1% H_3PO_4. Figures 3.16a and 3.16b show successively more oxidized samples. Early eluted peaks are observed to increase in size in the later samples while other peaks decrease. Detection is by UV absorption at 254 nm.

A separate study was performed to observe differences in chromatograms before and after fermentation. In this study, peaks were separated on

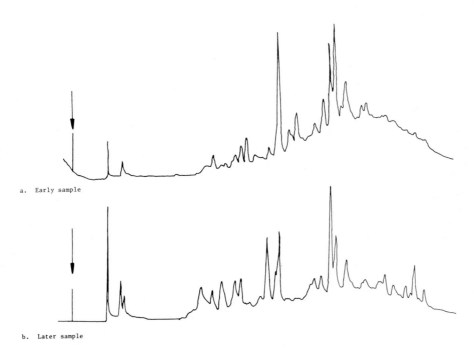

a. Early sample

b. Later sample

Fig. 3.16. HPLC of Successively Oxidized Samples

Sephadex G-10 columns (Section 3.5.3) prior to elution on the C-18 HPLC column. The fraction containing the major UV-absorbing peak (254 nm) was collected and concentrated, and injected onto the HPLC. The eluent contained water with 1% acetic acid mixed with 30% methanol containing 1% acetic acid. The methanol content is linearly increased to 70% over 15 minutes (flow rate at 1.5 ml/min) and then held constant. Eluent is detected at 254 nm. Figures 3.17a and 3.17b show appearance and disappearance of peaks after fermentation of pretreated peat. Both chromatograms show peaks after subtraction of solvent-only background.

The HPLC studies are inconclusive concerning the identification of specific peaks. Future work can concentrate on the identification of peaks used by the anaerobic microorganisms and optimization of pretreatment to maximize the yield of these compounds. The methods developed in this work can be used to separate products in future characterization and identification work.

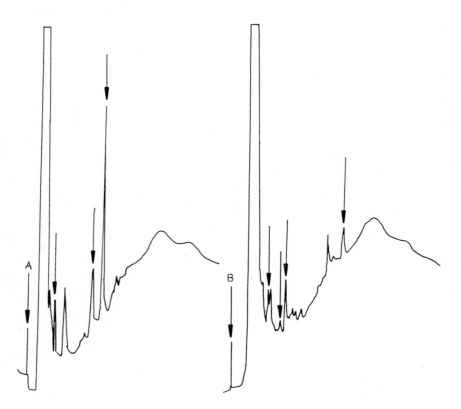

Anaerobic Digestion. Arrows show peaks which disappear or appear after digester.

Fig. 3.17. HPLC of Pretreated Peat Before (A) and After (B)

4. MICROBIAL CONVERSION OF PRETREATED PEAT AND MODEL COMPOUNDS TO ORGANIC ACIDS

4.1 Program Objectives and Task Description

Microbial digestion of pretreated (partially oxidized) peat will be accomplished by methane-suppressed, mixed culture, anaerobic fermentation. The desired products of the microbial digestion are linear organic acids (acetic through hexanoic). The pretreated peat substrate is believed to be primarily composed of single-ring aromatic compounds with methoxyl, hydroxyl, and carboxyl groups attached to the ring. Previous work has documented the conversion of aromatic model compounds, pretreated peat, and wood lignin to methane by mixed-culture anaerobes. Production of organic acids from these aromatic substrates has not been investigated.

Parallel experiments are planned for selection and characterization of cultures capable of digesting model compounds and pretreated peat. After culture acclimatization to the selected substrates beginning with inoculum from sewage digester contents, the rate and conversion efficiency (percent of organic matter ultimately converted to product) of fermentables to methane is measured. Experiments are then designed and executed in serum vials to determine the best conditions for suppression of methane formation and accumulation of organic acids. Once the method is chosen, the rate and efficiency of these conversions will be measured. All this work is planned for bench-scale systems (up to 3.0 l liquid volumes).

High performance digesters are appropriate for this fermentation because all of the substrate will be soluble. While a fairly extensive amount of work has been reported on fixed-film methane digesters, suppressed-methane fixed-film digestions have not been investigated. The goal in this task is to demonstrate acid formation and accumulation with this type of digester using a simple substrate (glucose) and then test conversion of peat to organic acids in a high-performance digester.

In summary, the goals of the fermentation work on this project are:

1. Adapt cultures capable of fermenting model aromatic substrates to methane.
2. Measure rates of substrate utilization and methane formation and conversion efficiencies of these cultures.
3. Determine conditions and adapt cultures for organic acid formation from model compounds.
4. Adapt cultures capable of fermenting pretreated (partially oxidized) peat to methane.

5. Measure rate of peat digestion and conversion efficiency to methane.

6. Determine conditions for organic acid formation from peat and demonstrate feasibility, measure rate, and determine efficiency of this conversion.

7. Demonstrate organic acid formation from glucose in a high performance (anaerobic packed-bed) fermenter.

8. Demonstrate organic acid formation from peat in a high performance fermenter.

4.2 Background

The fermentation employed in the process may be described as a mixed culture, methane-suppressed, anaerobic fermentation whose major products are the organic acids, acetic through caproic (hexanoic) and carbon dioxide (Sanderson et al. 1979a). The fermentation is similar to methane fermentations used in sewage treatment plants. These fermentations proceed in two discrete biochemical stages. Initially, a class of microorganisms, called acetogens, metabolize the material, secreting acetic and other aliphatic organic acids as well as carbon dioxide and hydrogen. These products (acetic acid, CO_2, H_2) are metabolized by a second class of microorganisms, called methanogens, which produce methane and carbon dioxide as end products. Digestion of polysaccharides (cellulose and hemicellulose) to methane has been extensively researched and is fairly well understood. Degradative bacteria, containing extra-cellular cellulolytic enzymes, are capable of dissolving and metabolizing polysaccharides to produce acetic acid, CO_2, and H_2. Sufficient energy is derived from this conversion to enable the bacteria to maintain growth. Methanogenic bacteria may then use the energy remaining in the acetic acid, CO_2, and H_2 by converting these intermediate products to methane and carbon dioxide.

The fermentation may be altered by inhibiting methane formation. This has been accomplished by using the specific inhibitor 2-bromoethane sulfonic acid (Sanderson et al. 1979b; Balch and Wolfe 1976). Figure 4.1 shows the effect of 2-bromoethane sulfonic acid (BES) on organic acid production from the fresh water plant Hydrilla. In the absence of BES, the organic acids produced by the acetogens are degraded to methane and carbon dioxide. The BES eliminates methane production and allows a stable build-up of the organic acid product.

Experience with fermentation of polysaccharides has led to the observation that as the organic acid level in the fermenter increases and the pH drops, higher molecular weight acids are formed (Levy et al. 1981a). The

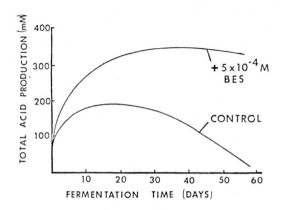

Fig. 4.1. Effect of Bromoethane Sulfonic Acid (BES) on Organic Acid Production from Hydrilla

CO_2 and H_2 produced initially may be used to form acetic acid (Ljundahl and Wood 1969) or the H_2 may be incorporated into the formation of higher acids. The higher acids appear to be formed by addition of 2-carbon units to existing acids. The formation of butyric acid from the condensation and reduction of two molecules of acetyl-CoA has been hypothesized (Keenan 1979). An increase in the production of butyric acid in the presence of high acetic acid levels has been demonstrated experimentally (Levy et al. 1981b). Similar experiments were also reported (Levy et al. 1981a) in which valeric or caproic acid production was increased by elevating propionic or butyric acid levels.

The details of the actual biosynthetic route have yet to be proven, but production of higher acids seems to proceed via 2-carbon additions to propionic or acetic acid primers. Higher acid production is carried out by bacteria classified as fermentative which are likely to be capable of secreting higher or lower weight organic acids depending on conditions in the fermenter.

In this research program, the technology of suppressed-methane fermentation is to be applied to the fermentation of pretreated (solubilized and partially oxidized) peat. The limited alkaline oxidation of peat results in an array of substituted single-ring aromatics, carboxylic acids, and methanol. These are all potential substrates for digestion by anaerobic microorganisms (Healy and Young 1979; Healy et al. 1980; Ferry and Wolfe 1976; Keith et al. 1978; Ruoff et al. 1980; Levy et al. 1981a; Colberg and Young 1982). In model compound experiments, using eleven substituted aromatic compounds, Healy and Young (1979) demonstrated high conversions to

Table 4.1

ANAEROBIC DEGRADATION OF AROMATIC COMPOUNDS TO METHANE*

SUBSTRATE	PERCENT CONVERTED
Vanillin	72 ± 1.4
Vanillic acid	86 ± 2.8
Ferulic acid	86 ± 2.8
Cinnamic acid	87 ± 8.1
Benzoic acid	91 ± 7.8
Catechol	67 ± 1.6
Protocatechuic acid	63 ± 1.8
Phenol	70 ± 3.2
p-Hydroxybenzoic acid	80 ± 2.7
Syringic acid	80 ± 1.6
Syringaldehyde	102 ± 13.3

* Healy and Young, Appl. Environ. Microbiol.
38:84-89 (1979).

methane as indicated in Table 4.1. Mixed microbial inoculum was used and lag periods of a few days to a few weeks were experienced while the bacteria became acclimized to the aromatic substrates. In some instances, cross-acclimization (immediate acceptance of one substrate after adaptation to another) occurred, indicating the existence of similar metabolic pathways.

Cultures isolated from freshwater mud and sewage sludge samples and identified as Acetobacterium woodii have been found capable of anaerobic fermentation of methoxylated benzoic acid derivatives to acetic acid (Bache and Pfennig 1981). However, this fermentation is believed to occur by de-methoxylation rather than ring cleavage. Substantial documentation of other anaerobic cultures which are capable of cleaving the aromatic ring of benzoic acid and its derivatives has been provided (Keith et al. 1978; Aftring and Taylor 1981; Healy and Young 1979; Healy et al. 1980; Ferry and Wolfe 1976; Williams and Evans 1975; Clark and Fina 1952; Taylor et al. 1970; Evans 1977). The biochemical pathways of the anaerobic catabolism of aromatic substrates have been postulated by a number of researchers. Recent work with wood lignin (the most similar to peat of any substrates examined) has demonstrated that high temperature alkaline treatment renders the material degradable by anaerobic bacteria (Colberg and Young 1982). Their work indicated that solubilized compounds of approximately 700 and 300 MW were metabolized. Work by Zeikus and co-workers (1982) indicated that pre-treated lignin of molecular weight less than 300 is microbially converted to CO_2 and CH_4 while material of greater than 850 MW was not decomposed by the same mixed bacterial inoculum. These workers also demonstrated the bio-degradability of a "dimer" model lignin compound, that is, a compound containing two aromatic rings linked by a β-aryl ether bond (Figure 4.2).

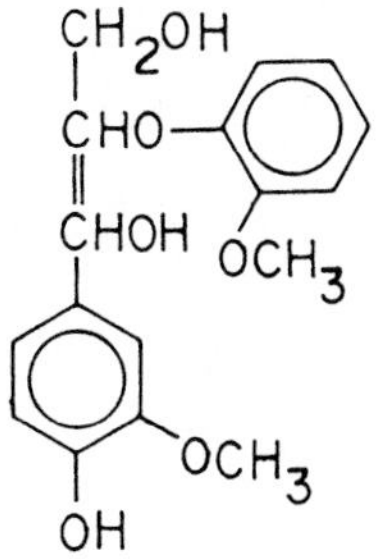

Fig. 4.2. Guaiacylglycerol-β-(o-methoxyphenyl) ether

The pathway for degradation of the aromatic compound ferulic acid to methane has been described in detail (Healy et al. 1980) and is presented in Figure 4.3. These workers added 10^{-4} molar 2-bromoethane sulfonic acid (BES) to the fermentation medium to specifically inhibit methanogenesis. In the presence of BES, the organic acid intermediates began to accumulate. The ability to produce organic acids by anaerobic fermentation when methano-genesis is suppressed has been demonstrated using polysaccharide substrates (Levy et al. 1981a, 1981c; Sanderson et al. 1979c). As these fermentations proceed, organic acids build-up by two-carbon additions onto acetic or propionic acid primers. Linear organic acids up to caproic (hexanoic) are formed. Inhibition of methane formation should promote accumulation of the organic acids in a manner analogous to that demonstrated for polysaccharide substrates.

In anaerobic respiration, bacteria use inorganic electron acceptors to metabolize substrates instead of O_2. Thus, NO_3^- is reduced to N_2; SO_4^{-2} to S^{-2}; and CO_2 to CH_4. These reactions are particularly important when thermo-dynamic considerations discussed in Section 4.3 are taken into account. In nitrate and carbon dioxide reducing cultures, intermediates following ring opening include formic, acetic, propionic, and other aliphatic carboxylic acids (Taylor et al. 1970; Evans 1977; Keith et al. 1978; Ferry and Wolfe 1976). Reduction of nitrate continues to N_2, though nitrite does appear as an intermediate (Williams and Evans 1975).

Recent research into the biochemistry of sulfate reducers has mostly been related to their effects on methane fermentations. Sulfate utilizing

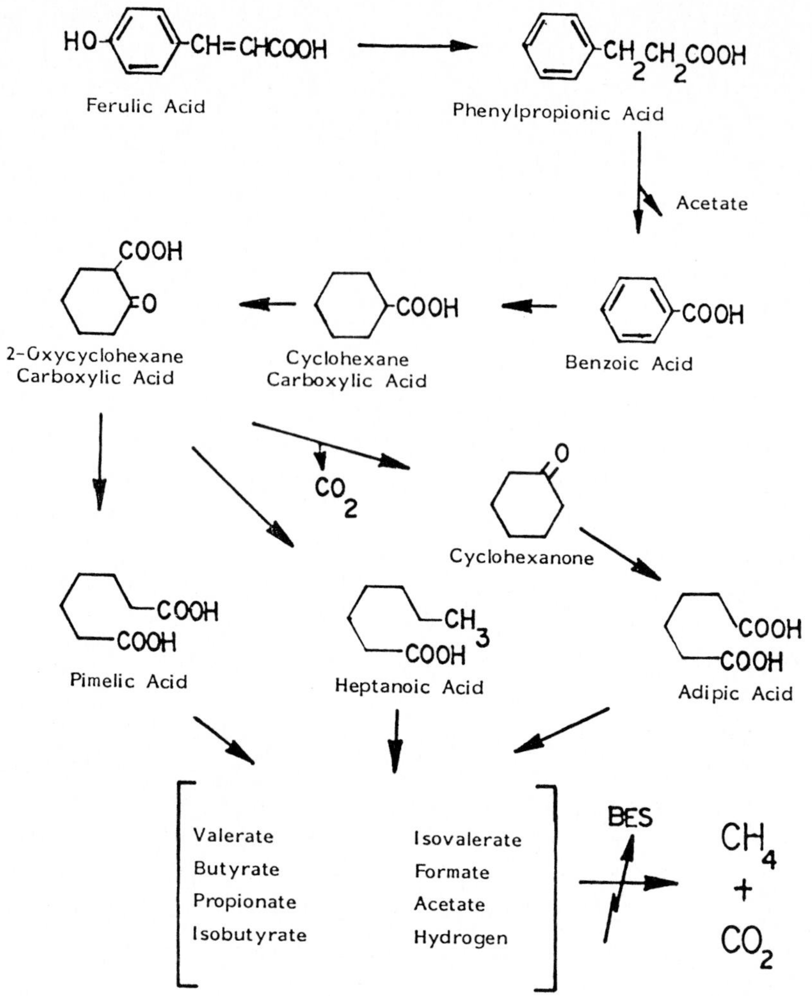
HO
CH=CHCOOH
Ferulic Acid
CH2CH2COOH
Phenylpropionic Acid
Acetate
COOH
O
2-Oxycyclohexane
Carboxylic Acid
COOH
Cyclohexane
Carboxylic Acid
COOH
Benzoic Acid
CO2
O
Cyclohexanone
COOH
COOH
Pimelic Acid
CH3
COOH
Heptanoic Acid
COOH
COOH
Adipic Acid
Valerate
Butyrate
Propionate
Isobutyrate
Isovalerate
Formate
Acetate
Hydrogen
BES
CH4
+
CO2

* Healy, Young, and Reinhard, Appl. Environ. Micro., 39:436-444 (1980).

Fig. 4.3. Model for Decomposition of Ferulic Acid to Methane*

bacteria are capable of outcompeting methanogens for available H_2 and acetate (Winfrey and Zeikus 1977; Schonbert 1982), though evidence suggests that not all sulfate reducers rely on acetate as a substrate (Banat et al. 1981). The advantage of sulfate utilizing anaerobic bacteria may be explained by their thermodynamic advantage over CO_2 reducers and by their greater affinity for uptake of available acetate (Schönheit 1982).

The fermentation of pretreated peat to methane has been demonstrated in the laboratory (Buivid et al. 1980; Ruoff et al. 1980; Dynatech Report No. 2115, 1981). Procedures similar to those used by Healy and co-workers on model compounds were used. Sewage sludge was used as inoculum and cultures were slowly acclimated to the new substrate which was introduced at low concentrations. Conversion of over 30% of the pretreated peat to methane and carbon dioxide was consistently observed in these experiments.

The objectives of the current experimental program is to combine the anaerobic fermentation of peat with the technique of inhibiting methano-genesis to produce organic acids. This is to be achieved by developing mixed cultures which produce methane from pretreated peat and then altering the conditions of the fermentation so only the acetogens can thrive. It is anticipated that besides the addition of BES, an electron acceptor such as nitrate or sulfate must be supplied. Model compound experiments, using benzoic acid as the substrate, are also being run to help gain a more basic understanding of the anaerobic digestion of the aromatic substrates. The experimental program is designed to demonstrate two points. First, that conditions can be found for the anaerobic degradation of model compounds and pretreated peat consisting of low molecular weight water soluble compounds without the production of methane (i.e. without reduction of CO_2). The second point to be shown is that in these fermentations, organic acids can be produced and accumulated as the final product. Because the presence of inorganic oxidizing agents (sulfate or nitrate) will favor complete oxidation of organic carbon, organisms will have to be isolated or developed that do not possess a pathway for metabolism of the desired products.

4.3 Experimental Program Rationale
4.3.1 Thermodynamic Considerations

Among the unique attributes of living organisms is the ability for self-replication. This ability implies the necessity for organisms to extract the energy required for reproduction from their environment. Chemo-trophic anaerobic bacteria, to which this discussion is limited, extract the energy required for processes such as biosyntheses, active transport, and active movement from chemical reactions. The energy producing reactions are catabolic processes in which the organism degrades the substrate to products of a lower energy state (Figure 4.4). Energy derived from such reactions can then be used in the energy requiring functions (such as biosynthesis) of the cell. The energy is stored and transferred by high energy compounds such as adenosine 5'-triphosphate (ATP) which is capable of releasing considerable energy of hydrolysis (7-10 Kcal/mole) from its conversion to

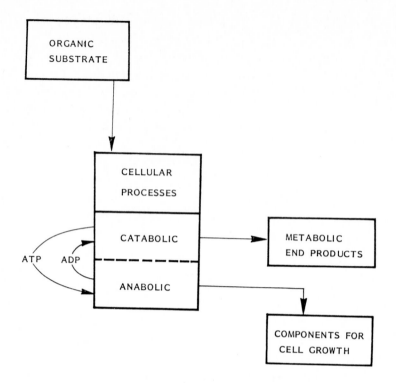

Fig. 4.4. Energy Flows in Anaerobic Microorganisms

adenosine 5'-diphosphate (ADP).   Under physiological conditions, approximately 10 Kcal/mole are required to regenerate ATP from ADP. This energy
must be provided by the available growth substrate. Since fermentation
(anaerobic chemotrophy) yields only a small portion of usable energy from
the substrates compared to respiration (aerobic chemotrophy), the availability of an energy source is often the growth limiting factor for anaerobic
organisms (Decker et al. 1970).

    For a microorganism to accomplish the conversion of a particular sub-
strate to a desired metabolic end-product, two primary conditions must be
met:

1.   The organism must be capable of extracting sufficient energy from
      the conversion to produce ATP or other high energy intermediates.

2.   The organism must possess metabolic pathways capable of accepting
      the substrate and secreting the product.

The first condition is easily examined a priori. That is, we can determine if the desired conversion is thermodynamically favorable under conditions suitable for the organism's growth. Since the identity of the materials contained in the pretreated peat substrate is uncertain, the following discussion is limited to model aromatic compounds for which free energies of formation are available in the literature. Some analogous conclusions may be made for the pretreated peat.

Sugar substrates may be converted by acid-forming anaerobic bacteria to acetate, bicarbonate, and hydrogen gas according to the stoichiometry

$$C_6H_{12}O_6 + 4H_2O \rightarrow 2CH_3COO^- + 2HCO_3^- + 4H^+ + 4H_2.$$

This conversion is accompanied by a favorable free energy change of 49.3 Kcal/mole glucose, providing sufficient energy for ATP formation (Thauer et al. 1977). The products of this metabolism are also capable of providing growth energy (31.3 Kcal/reaction) to methanogenic bacteria from the formation of methane

$$CO_2 + 4H_2 \rightarrow CH_4 + 2H_2O.$$

The formation of methane subsequent to acetic acid production also helps to drive the first reaction by reducing the concentration of its products.

Providing substrates other than glucose alters the free energy yield of production of acetate and methane. Two model compounds, benzoic acid and ferulic acid (Figure 4.5), are considered. The free energy change which accompanies the conversion of these model compounds to methane or acetate

$C_7H_6O_2$

Benzoic Acid

$C_{10}H_{10}O_4$

Ferulic Acid

Fig. 4.5. Model Fermentation Substrates

has been calculated for physiologic conditions. Benzoic acid conversions
are listed in Table 4.2. Conversion to methane is sufficiently exergonic to
be spontaneous, while conversion to acetate is unfavorable. The conversion
to acetate can be made favorable enough to support ATP formation with the
addition of an electron acceptor such as sulfate or nitrate. In the presence
of oxidizing agents such as nitrate, the greatest free energy yield can be
attained by complete oxidation of benzoate to carbon dioxide:

$$C_6H_5COO^- + H_2O + 6NO_3^- \rightarrow 7CO_2 + 3N_2 + 7OH^-.$$

This conversion yields over 600 Kcal/mole of benzoate.

Examination of other model compounds reveals that some can be converted
with favorable free energy changes to organic acids (Table 4.3). Still, the
conversion to methane is thermodynamically more favorable. Part of the
challenge of selecting cultures which produce organic acids from aromatic
compounds is suppressing the more favored conversions to methane or, when
inorganic electron acceptors are present, to carbon dioxide.

4.3.2 Experimental Approach

Microbial cultures capable of metabolizing the single-ring aromatic
compounds produced by alkaline pretreatment of peat are to be developed from
sewage digester inoculum. As the preceding discussion points out, two
conditions must be met for successful cultures to result. First, organisms
must be present which are capable of metabolizing the substrate material and
producing organic acids. Work cited in Section 4.2 leads us to believe that
such organisms are present in sewage digesters.

The second condition is that the desired conversion to organic acids is
thermodynamically favorable enough to support cell maintenance and replic-
ation. Sample calculations in Section 4.3.1 indicate that this may not be
the case. To compensate for this unfavorable situation, the fermentation
medium may be adjusted to make the free energy yield of these conversions
capable of producing ATP for cell growth. This will be achieved by the
addition of inorganic electron acceptors (nitrate, sulfate) which enable the
substrate to be more completely oxidized, yielding more energy per mole of
substrate metabolized.

In all cases, the production of organic acids is not the most thermodynamically favorable conversion. To allow organic acids to accumulate as
the final product, production of some other products must be blocked. In
the fermentations where no oxidizing agent is added (no inorganic electron
acceptor), the production of methane must be inhibited. This is achieved by

Table 4.2
FREE ENERGY OF REACTION FOR CONVERSION OF BENZOIC ACID*

1. Benzoate to methane:

$$C_6H_5COO^- + H^+ + 4.5H_2O \rightarrow 3.75CH_4 + 3.25CO_2$$
$$\Delta G = -29.0 \text{ Kcal/mole}$$

2. Benzoate to acetate:

$$C_6H_5COO^- + 4.5H_2O + 0.5CO_2 \rightarrow 3.75CH_3COO^- + 2.75H^+$$
$$\Delta G = +23.47 \text{ Kcal/mole}$$

3. Benzoate to acetate with sulfate reduction:

$$C_6H_5COO^- + 0.42H^+ + 4H_2O + 2.67SO_4^{2-} \rightarrow 1.08CH_3COO^- + 2.67H_2S + 4.83HCO_3^-$$
$$\Delta G = -10.59 \text{ Kcal/mole}$$

4. Benzoate to acetate with nitrate reduction:

$$C_6H_5COO^- + 3.67H_2O + 0.67NO_3^- \rightarrow 3.33CH_3COO^- + 2H^+ + 0.33HCO_3^- + 0.33N_2$$
$$\Delta G = -55.4 \text{ Kcal/mole}$$

---

*Free energy changes are calculated assuming pH 7.0, 25°C, and aqueous concentrations of 1.0 M for all reactants and products except water (55 M), methane (1.7×10^{-4}M), hydrogen sulfide (7.75×10^{-2}M), and nitrogen (6.46×10^{-4}M).

Values for solubilities and free energies of compounds are from Thauer et al. (1977), Perry and Chilton (1973), and CRC Handbook of Chemistry and Physics (1969).

Table 4.3
FREE ENERGY OF REACTION FOR CONVERSION OF FERULIC ACID*

1. Ferulic acid to methane:

$$C_9H_9O_2COO^- + H^+ + 5.5H_2O \rightarrow 5.25CH_4 + 4.75CO_2$$
$$\Delta G = -147.1 \text{ Kcal/mole}$$

2. Ferulic acid to acetate:

$$C_9H_9O_2COO^- + 5.5H_2O + 0.5CO_2 \rightarrow 5.25CH_3COO^- + 4.25H^+$$
$$\Delta G = -75.2 \text{ Kcal/mole}$$

3. Ferulic acid to caproate:

$$C_9H_9O_2COO^- + 2.88H_2O \rightarrow 1.31C_5H_{11}COO^- + 2.13CO_2 + 0.31H^+$$
$$\Delta G = -69.6 \text{ Kcal/mole}$$

---

* Same concentrations and references as Table 4.2.

the addition of 2-bromoethane sulfonic acid (BES). This compound, a chemical analogue of coenzyme M, specifically inhibits methane formation without affecting organic acid production.

In the presence of an oxidizing agent, organic carbon will be converted to carbon dioxide, producing the greatest free energy yield. Organisms capable of metabolizing acetate and other organic acids are selectively eliminated from the culture by treatment with antibiotics. After adapting the culture to metabolizing the aromatic substrate, the fermentation is continued until all substrate is utilized. After addition of acetate, the acetate metabolizing organisms begin to grow while those producing acetate from the aromatic substrate remain dormant. Penicillin is then added to destroy the growing organisms, leaving the acetate producing microbes as the only viable bacteria. Fresh substrate is then supplied and acetate and other organic acids allowed to accumulate. These microorganism selection procedures are shown schematically in Figure 4.6.

The fermentation experiments are performed concurrently with benzoic acid as a model compound and pretreated peat prepared by high temperature alkaline oxidation. Experiments are also carried out to measure conversion rates (particularly to methane), ultimate conversion of the pretreated peat, and suitability of anaerobic packed bed fermenters for these suppressed-methane fermentations.

4.4 Experimental Procedures
4.4.1 MIT and NBF Fermenters

In order to investigate the fermentation of pretreated peat and the model compound benzoic acid, two chemostats were set up. One chemostat was a New Brunswick Scientific Microferm® Model MF-105 and the other was assembled according to the MF-105 design.

The MF-105 chemostat is a bench-scale fermenter equipped with a 5 l glass vessel. In addition, the fermenter is equipped with instruments to monitor and maintain temperature and agitation. The actual volume of liquid in the fermenter was 3.0 L. Temperature was maintained at 37°C and the stir rate at 200 RPM. An inlet/outlet line was provided for wasting and feeding, and a gas sampling port was provided for monitoring gas composition. In addition, a wet-test meter was connected to the chemostat to measure the volume of gas produced.

The second chemostat consisted of similar components. Heat tape was wrapped around the 5 l glass vessel and connected to a Thermistemp® temperature controller to maintain a temperature of 37°C. A magnetic stir bar provided agitation and a gas sampling port allowed monitoring of gas distribution.   In addition, an inlet/outlet line was provided for wasting and feeding, and a wet-test meter was connected to the chemostat to measure the volume of gas produced.

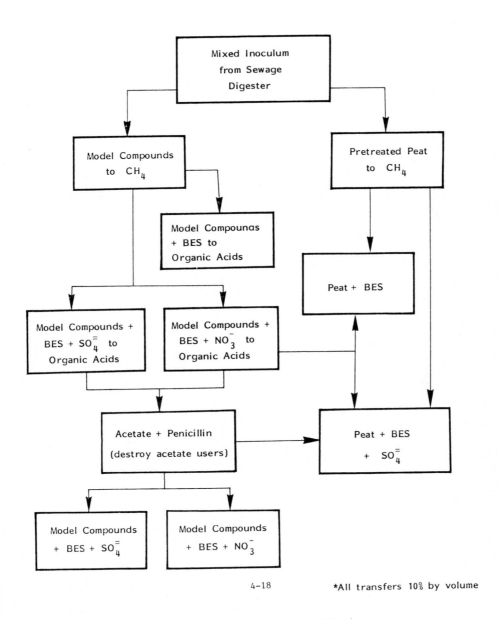

4-18   *All transfers 10% by volume

Fig. 4.6. Selection of Organic Acid Producers

118

Initially, the two chemostats were treated identically. Each received 3.0 ℓ of sewage sludge mixed with medium as shown in Table 4.4. Wasting and feeding were not done for several days. Then, a glucose feed solution was prepared (Table 4.6) and each day 100 ml of this solution were fed to each chemostat.

After two weeks, the two chemostats were given different feed solutions. One chemostat (NBF) was fed a benzoic acid feed solution (Table 4.7) and the other (MIT) was fed a pretreated peat feed solution (Table 4.8). For several weeks, both feed solutions were supplemented with glucose, in decreasing concentration, until the cultures were acclimated to the substrates. Then the glucose was no longer added to the feed solutions and the cultures were maintained on the pretreated peat or benzoic acid as their substrates.

Over a period of about a year, each chemostat was fed daily. Gas composition was also determined daily, on a Fisher-Hamilton Gas Partitioner Model 29, and the pH of the effluent was determined as well. Organic acid analysis was performed intermittently. As the cultures adapted to the substrates and successfully converted them to products, CO_2 and CH_4 production stabilized. When sufficient data had been collected for a given concentration of substrate, the concentration of substrate in the feed

Table 4.4

START-UP INOCULUM FOR CHEMOSTATS

A. Preparation of Medium

COMPONENT	CONCENTRATION
NaHCO$_3$	2.6 g/ℓ
S-3 Stock Solution*	1.6 mℓ/ℓ
S-4 Stock Solution*	15.2 mℓ/ℓ
S-5 Stock Solution*	1.5 mℓ/ℓ
S-8 Stock Solution*	1.0 mℓ/ℓ

Dissolve components in ~900 mℓ distilled H$_2$0.
Adjust pH to 7.2 and bring volume to 1 ℓ.

B. Preparation of Inoculum

Mix 5 ℓ sewage sludge (total solids = 2.4%) with 1 ℓ medium and add 20 g glucose.

* See Table 4.5 for composition of stock solutions.

Table 4.5

STOCK SOLUTIONS FOR NUTRIENT MEDIA

STOCK SOLUTION	COMPOUND	CONCENTRATION
S-3	$(NH_4)_2HPO_4$	26.70 g/ℓ
S-4	NH_4Cl	13.10 g/ℓ
	$MgCl_2 \cdot 6H_2O$	120.00
	KCl	40.00
	$MnCl_2 \cdot 4H_2O$	1.33
	$CoCl_2 \cdot 6H_2O$	2.00
	H_3BO_3	0.38
	$CaCl_2 \cdot 2H_2O$	0.18
	$Na_2MoO_4 \cdot 2H_2O$	0.17
	$ZnCl_2$	0.14
S-5	$FeCl_2 \cdot 4H_2O$	44.94 g/ℓ
S-8	Folic acid	2 mg/ℓ
	Pyridoxine HCl	10
	Riboflavin	5
	Thiamin	5
	Nicotinic acid	5
	Pantothenic acid	5
	Cobalamin	0.1
	para-aminobenzoic acid	5
	Thioctic acid	5

Table 4.6

GLUCOSE FEED SOLUTION

COMPONENT	CONCENTRATION
Glucose	Variable*
$NaHCO_3$	2.6 g/ℓ
S-3 Stock Solution**	1.6 mℓ/ℓ
S-4 Stock Solution**	15.2 mℓ/ℓ
S-5 Stock Solution**	1.5 mℓ/ℓ
S-8 Stock Solution**	1.0 mℓ/ℓ

Dissolve components in ~900 mℓ distilled H_2O and bring volume to 1 ℓ.

* Concentration of glucose was decreased from 100 g/ℓ to 14 g/ℓ over a period of 2 weeks.
** See Table 4.5 for composition of stock solutions.

solutions was increased. The concentration of benzoic acid in the NBF feed
solution was increased from 5 mM to 100 mM over a period of three months.
However, due to limited CH_4 production by the MIT fermenter during the first
two weeks, the concentration of pretreated peat in the MIT feed solution was
decreased to 160 ml/l. Over the next five months, it was increased to 500
ml/l.

Table 4.7

BENZOIC ACID START-UP FEED SOLUTION

Component	Concentration
Benzoic Acid (5mM)	0.6 g/l
$NaHCO_3$	2.6 g/l
S-3 Stock Solution*	1.6 ml/l
S-4 Stock Solution*	15.2 ml/l
S-5 Stock Solution*	1.5 ml/l
S-8 Stock Solution*	1.0 ml/l

Benzoic acid is added to ~800 ml distilled H_2O
and dissolved by adding concentrated NH_4OH to
neutralize the acid. Then pH is adjusted to
7.6 with concentrated HCl. Other components
are added to the solution and volume is brought
to 1.0 l with distilled H_2O.

* See Table 4.5 for composition of stock
solutions.

Table 4.8

PRETREATED PEAT START-UP FEED SOLUTION

COMPONENT	CONCENTRATION
Pretreated peat #014951	330 ml/l
S-4 Stock Solution*	15.2 ml/l
S-8 Stock Solution*	1.0 ml/l

Dilute components to ~900 ml with distilled
H_2O. Adjust pH to 7.50 with concentrated HCl
and bring volume to 1.0 l.

* See Table 4.5 for composition of stock
solutions.

In addition to increasing the concentration of substrate in the feed solution, decreasing the retention time for the MIT was also investigated. Initially at 500 ml/l pretreated peat, the retention time for the MIT was thirty days and, over a period of about seven months, it was decreased to four days. Steady-state gas production data at each retention time was taken for periods coinciding to feeding of at least one reactor volume.

The stoichiometry of the fermentation reaction of benzoic acid is:

$$4C_6H_5COOH + 18H_2O \rightarrow 15CH_4 + 13CO_2$$

Therefore, for 100 ml of 25 mM benzoic acid feed solution, 225 ml CH_4 is formed at 100% conversion (20°C).

However some assumptions must be made for the fermentation of pretreated peat. Assuming 1% solids in the feed and that ferulic acid is representative of the solubilized and oxidized peat, then each 100 ml pretreated peat feed solution contains 0.005 gmoles substrate. The stoichiometry of the fermentation reaction for ferulic acid is:

$$4C_{10}H_{10}O_4 + 22H_2O \rightarrow 21CH_4 + 19CO_2$$

Therefore, for each 100 ml pretreated peat feed solution (with 1% solids), 651 ml CH_4 is formed at 100% conversion (20°C).

4.4.2 Serum Via Experiments with Benzoic Acid

To determine the effects of various compounds on the fermentation of benzoic acid, several serum vials were set up. The inoculum for the vials consisted of combined effluents from the benzoic acid continuous digester (NBF) and from the two packed-bed reactors (PBC No. 4 and No. 5) using a benzoic acid feed solution.

The inoculum was prepared as follows: 400 ml of the effluent from the NBF and 250 ml of the effluent from each of PBC No. 4 and No. 5 were mixed thoroughly; 30 ml of this solution were removed and replaced with 30 ml of fresh 25 mM benzoic acid feed solution (Table 4.9).

After thorough mixing, 80 ml of the final solution were placed in each of ten serum vials (160 ml capacity). The vials were then sealed with butyl rubber stoppers and crimped caps. Finally, each sealed vial was purged with 80% N_2/20% CO_2 and gas distribution in the head space was determined to be sure that all the O_2 had been removed. Gas composition was determined on a Fisher-Hamilton Gas Partitioner Model 29. Vials were then incubated at 37°C.

Gas composition was checked daily to determine CO_2 and CH_4 production. Each vial was wasted and fed weekly with 8 ml of the 25 mM benzoic acid feed solution for three weeks, when production of CH_4 was steady. The pH in the vial was checked when the vials were wasted and fed.

Table 4.9

25 mM BENZOIC ACID FEED SOLUTION

Component	Concentration
Benzoic Acid	3.05 g/ℓ
$NaHCO_3$	2.60 g/ℓ
S-3 Stock Solution*	1.6 mℓ/ℓ
S-4 Stock Solution*	15.2 mℓ/ℓ
S-5 Stock Solution*	1.5 mℓ/ℓ
S-8 Stock Solution*	1.0 mℓ/ℓ

Benzoic acid is added to ~800 mℓ to distilled H_2O and dissolved by adding NH_4OH to neutralize the acid. Then pH is adjusted to 7.6 with concentrated HCl. Other components are added to the solution and volume is brought to 1.0 ℓ with distilled H_2O.

* See Table 4.5 for composition of stock solutions.

Table 4.10

FEED SOLUTIONS FOR SERUM VIAL EXPERIMENT

Feed Solution A: 25 mM Benzoic Acid Feed Solution Prepared as shown in Table 4.9.

Feed Solution B: 25 mM Benzoic Acid Feed Solution with 5×10^{-4}M 2-bromoethane sulfonic acid (BES).

Feed Solution C: 25 mM Benzoic Acid Feed Solution with 5×10^{-4}M BES and 2.4 g/ℓ Na_2SO_4.

Feed Solution D: 25 mM Benzoic Acid Feed Solution with 5×10^{-4}M BES and 1.7 g/ℓ KNO_3.

Feed Solution E: 25 mM Benzoic Acid Feed Solution with 5×10^{-4}M BES and 0.17 g/ℓ KNO_3.

Table 4.11

FEED SOLUTIONS CONTAINING ACETATE AND PENICILLIN

Stock Feed Solution

COMPOUND	Concentration
$NaHCO_3$	2.6 g/ℓ
CH_3COONa	4.1 g/ℓ
Penicillin	0.1 g/ℓ
(Sodium salt)	
S-3 Stock Solution*	1.6 mℓ/ℓ
S-4 Stock Solution*	15.2 mℓ/ℓ
S-5 Stock Solution*	1.5 mℓ/ℓ
S-8 Stock Solution*	1.0 mℓ/ℓ

pH of this solution = 7.81

* See Table 4.5 for composition of stock
solutions.

New feed solutions containing various compounds were then prepared
(Table 4.10). (See note on Table 4.11). Each feed solution was fed to two
serum vials, to provide duplicate data. Vials fed with Feed Solution A
served as controls.

Both vials receiving each feed solution were treated the same for four
weeks. Gas composition was checked daily and vials were wasted and feed
weekly with 8 ml of the appropriate feed solution. Organic acids analysis
was done with a Perkin-Elmer Model F-30 Gas Chromatograph, the columns
packed with Chromosorb 101 with no liquid phase.

At this point, another experiment was begun. One vial of each pair was
wasted and fed as usual, checking gas composition each week, prior to
wasting and feeding. The other vial was left alone for six weeks to allow
complete substrate utilization.

Then different feed solutions were prepared for the vials which had
been left alone (Table 4.11). Each vial was wasted and fed 10 ml of the
stock feed solution to which was added penicillin, acetate, and the
appropriate salts as shown in Table 4.12. This feed solution is used to
eliminate acetate utilizing organisms. The vials were purged with 80%
N_2/20% CO_2 after wasting and feeding.

Table 4.12

FEED SOLUTIONS FOR VIALS AFTER PENICILLIN TREATMENT

ORIGINAL FEED SOLUTION	COMPOUNDS ADDED TO 10 ml STOCK FEED
Feed Solution A	Nothing added
Feed Solution B	1.0 mg BES
Feed Solution C	1.0 mg BES and 24 mg Na_2SO_4
Feed Solution D	1.0 mg BES and 17 mg KNO_3
Feed Solution E	1.0 mg BES and 1.7 mg KNO_3

* See Table 4.5 for composition of stock solutions.

After one week, the residual penicillin was destroyed by adding to each vial 0.1 ml of a penicillin solution (1000 units dissolved in 10 ml water). Two days later, the weekly wasting and feeding schedule was resumed, using 8 ml of the appropriate original feed solution.

The stoichiometry of the fermentation reaction is

$$4C_6H_5COOH + 18H_2O \rightarrow 15CH_4 + 13CO_2.$$

Therefore, for each 8 ml of 25 mM benzoic acid feed solution, we provide 0.2 mM benzoic acid. Finally, the maximum yield of methane would be 0.75 mM or approximately 16.8 ml per vial per feeding. The extent of benzoic acid utilization is determined by methane or acetate production compared to theoretical and by measurement of residual benzoic acid by HPLC.

4.4.3 Serum Vial Experiments with Pretreated Peat and Ultrafiltered Pretreated Peat

To investigate the fermentation of pretreated peat and the effects of various compounds on that fermentation, several serum vials were set up. The inoculum for these vials consisted of effluent from the pretreated peat continuous digester (MIT).

The inoculum was prepared as follows: effluent from the MIT digester was collected and stored frozen, until 500 ml were collected. After thawing and thorough mixing, 60 ml were placed in each of eight serum vials, then 20 fresh pretreated peat feed solutions (Table 4.13) were added to each vial. Each feed solution was added to two vials to provide duplicates.

Table 4.13

PRETREATED PEAT FEED SOLUTIONS FOR SERUM VIALS

Feed Solution A

Component	Concentration
Pretreated Peat	250 mℓ/ℓ
S-4 Stock Solution*	15.2 mℓ/ℓ
S-5 Stock Solution*	1.0 mℓ/ℓ

Components are mixed together with ~600 mℓ distilled H_2O. The pH is adjusted to 7.0 – 7.2 with concentrated HCl and the volume is brought to 1.0 ℓ with distilled H_2O.

Feed Solution B

Feed Solution A + 0.12 mg/mℓ bromoethane sulfonic acid (BES)

Feed Solution C

Feed Solution B + 0.8 mg/mℓ Na_2SO_4

Feed Solution D

Feed Solution B + 2.4 mg/mℓ Na_2SO_4

* See Table 4.5 for composition of stock solutions.

The vials were sealed with butyl rubber stoppers and crimped caps, and the headspace was purged with 80% N_2/20% CO_2. Gas composition was determined with a Fisher-Hamilton Gas Partitioner Model 29.

Each vial was wasted and fed once a week with 10 ml of the appropriate feed solution. The pH of the effluent was determined and gas composition was checked at that time. In addition, organic acids analysis was conducted once every other week with a Perkin-Elmer Model F-30 Gas Chromatograph, the columns packed with Chromosorb 101 (no liquid phase).

In addition to the pretreated peat feed solutions described above, ultrafiltered pretreated peat feed solutions were also used as substrates for fermentation. The ultrafiltered pretreated peat was prepared by pressurized filtration through a succession of membranes. The membranes have a specified MW cut-off which allows solutes smaller than the cut-off to pass through the filter. In this case, the ultrafiltered pretreated peat solution was passed through membranes with nominal cut-offs of 300,000, then 10,000 and finally 500 MW.

126

Table 4.14

ULTRAFILTRATE PRETREATED PEAT FEED SOLUTION

COMPONENT	CONCENTRATION
Ultrafiltered pretreated peat	160 mℓ/ℓ
S-4 Stock Solution*	15.2 mℓ/ℓ
S-8 Stock Solution*	1.0 mℓ/ℓ

Prepare using distilled water. Adjust pH to 7.0
with concentrated HCl.

* See Table 4.5 for composition of stock solutions.

The inoculum for the serum vials in this experiment was prepared as
follows: 200 ml effluent from the MIT fermenter were collected and centrifuged at 12,000 rpm for ten minutes. The pellet, containing the microbes, was
washed once with 0.9% NaCl and resuspended with a volume of water (containing the appropriate salts and vitamins) to a total volume of 200 ml. After
thorough mixing, 80 ml of the final suspension were placed in each of two
serum vials (160 ml capacity).

The vials were then fed the ultrafiltered pretreated peat feed solution
(Table 4.14). The vials were sealed and the headspace purged with N_2/CO_2 as
previously described. The gas composition was determined and the vials
incubated at 37°C.

In order to acclimate the microbes to the ultrafiltered pretreated peat
feed solution, the concentration of ultrafiltered pretreated peat in the
feed solution was decreased to 10%, then increased to and maintained at 20%
for the remainder of the experiment.

Gas composition was determined daily for three months to monitor CH_4
production. When the CH_4 production rate was stable, the vials were wasted
and fed once a week, the pH of the effluent was determined, and gas composition was also determined once a week. Organic acids analysis was performed
occasionally.

Using ferulic acid as a model compound for both pretreated peat and the
ultrafiltered pretreated peat, the assumed stoichiometry of the fermentation
reaction is:

$$4C_{10}H_{10}O_4 + 22H_2O \rightarrow 21CH_4 + 19CO_2$$

Following this stoichiometry, at 1% volatile solids in the feed, 100 ml feed solution would contain 0.005 gmoles substrate. Therefore, quantitative conversion to products would yield 0.027 gmoles CH_4, equivalent to 0.69 l CH_4 (37°C).

4.4.4 Other Serum Vial Experiments

In order to establish the appropriate concentration of the various compounds used in the benzoic acid and pretreated peat serum vial experiments, several preliminary serum vial experiments were conducted. In addition, an experiment on the fermentation of lignin was also conducted.

The first experiment established the concentration of bromoethane sulfonic acid (BES) required to inhibit CH_4 production. The inoculum used for this experiment consisted of effluent from the benzoic acid continuous digester (NBF) and was prepared as follows: 100 ml of effluent were thoroughly mixed and 3.3 ml were removed. To the remainder, 3.3 ml 75 mM benzoic acid feed solution (Table 4.15) were added. After thorough mixing, 80 ml of the final solution were placed in a serum vial (150 ml capacity). The vial was sealed and the headspace purged with N_2/CO_2 as previously described.

A total of six vials were prepared in this way. The two control vials were fed the 75 mM benzoic acid feed solution described in Table 4.15. Two other vials were fed the same solution with 10^{-3} M BES added. The last two vials were fed the benzoic acid feed with 10^{-4} M BES added. In all vials, the volume wasted and fed weekly was 10 ml.

In order to monitor the CH_4 production, gas compositions were determined daily as previously described. In addition, vials were wasted and fed every week or every other week and the pH of the effluent determined. Organic acids analysis was performed occasionally, as previously described.

Another experiment was designed to determine the appropriate concentration of Na_2SO_4 for maximum formation of organic acids. The substrate for this experiment was glucose and the inoculum was sewage sludge. In this case, all serum vials contained BES to inhibit CH_4 production. The vials were set up as follows: 70 ml medium (Table 4.16) and 10 ml sewage sludge were added to each of six vials, then Na_2SO_4 was added to the appropriate vials. The vials were sealed and purged as previously described.

The vials were wasted and fed 8 ml of the appropriate feed solution just once, on the third day. Gas production was monitored daily for two weeks. The pH of the effluent was determined and organic acids analysis was done twice, as previously described.

Table 4.15

75 mM BENZOIC ACID FEED SOLUTION

COMPONENT	CONCENTRATION
Benzoic acid	9.15 g/ℓ
NaHCO$_3$	2.60 g/ℓ
S-3 Stock Solution*	1.5 mℓ/ℓ
S-4 Stock Solution*	15.2 mℓ/ℓ
S-5 Stock Solution*	1.5 mℓ/ℓ
S-8 Stock Solution*	1.0 mℓ/ℓ

Benzoic acid is added to ~800 mℓ distilled H$_2$O and dissolved by adding concentrated NH$_4$OH to neutralize the acid. Then the pH is adjusted to 7.5 with concentrated HCl. Other components are added to the solution and volume is brought to 1.0 ℓ with distilled H$_2$O.

* See Table 4.5 for composition of stock solutions.

Table 4.16

GLUCOSE FEED SOLUTIONS

Feed Solution A

COMPONENT	CONCENTRATION
Glucose	3 g/ℓ
NaHCO$_3$	2.6 g/ℓ
S-3 Stock Solution*	1.6 mℓ/ℓ
S-4 Stock Solution*	15.2 mℓ/ℓ
S-5 Stock Solution*	1.5 mℓ/ℓ
S-8 Stock Solution*	1.0 mℓ/ℓ
Bromoethane sulfonic acid	106 mg/ℓ

Feed Solution B

Feed Solution A + 0.24 mg/mℓ Na$_2$SO$_4$

Feed Solution C

Feed Solution A + 2.4 mg/mℓ Na$_2$SO$_4$

* See Table 4.5 for composition of stock solutions.

Table 4.17

LIGNIN FEED SOLUTIONS

COMPONENT	CONCENTRATION
Solubilized lignin or ultrafiltered solubilized lignin	160 ml/ℓ
S-4 Stock Solution*	15.2 ml/ℓ
S-8 Stock Solution*	1.0 ml/ℓ

Add components to ~800 mℓ distilled H_2O. Adjust pH to 7.0 with concentrated HCl and bring volume to 1 ℓ.

* See Table 4.5 for composition of stock solutions.

The final significant serum vial experiment was conducted using lignin as a substrate. The lignin was isolated from the peat by acid hydrolysis with H_2SO_4. The peat was dried at 110°C for 18 hours, ball-milled to less than 25 mesh, then redried at 110°C for 3 hours. A portion of the prepared peat was stirred into 75% H_2SO_4**. After cooling, the resulting slurry was filtered through a sintered glass filter. The filter cake was washed several times with water and then dried at 110°C. A portion of the lignin was then suspended in distilled H_2O and sodium carbonate was added. The usual peat solubilization and oxidation (at 250°C) procedures were then followed.

After the pretreatment, a portion of the final solution was ultra-filtered, as previously described in Section 4.4.3. The unfiltered and ultrafiltered solutions were used as components in the feed solutions for this experiment (Table 4.17). Two serum vials that contained cultures acclimated to pretreated peat were used. The unfiltered lignin feed solution was fed to one of the vials and the ultrafiltered lignin feed solution was fed to the other. Wasting and feeding with 10 ml feed solution were done once a week, and pH and gas composition determined at that time, according to previously described procedures. Organic acids analysis was not done. The vials were used to determine ultimate conversion of lignin and ultrafiltered lignin to products.

***

**The sulfuric acid was at ≈30°C. This slurry was left for two hours then diluted to 6.5% H_2SO_4 and boiled for three hours.

130

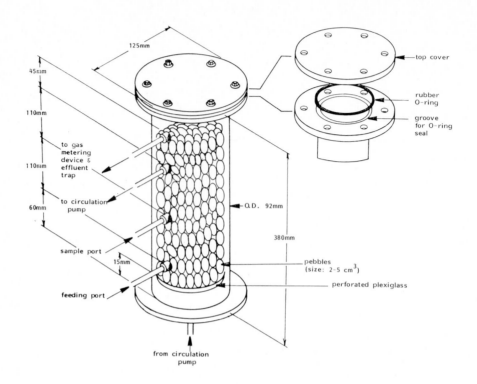

Fig. 4.7. Packed Bed Column (PBC) Fermenter

4.4.5 Start-Up and Operation of Packed-Bed Fermenters

In addition to chemostat and serum vial fermenters, a third type of fermenter was designed (Figure 4.7). The stones packed in the chamber provide a surface to which the cultures may attach, enhancing the cell mass in the digester.

The chamber consists of a plexiglas cylinder, sealed at each end. The bottom disk was fused to the cylinder. The top was sealed by placing a rubber O-ring between two disks and bolting them together. A perforated plexiglas plate was placed 15 mm from the bottom of the cylinder, acting as a support for the stones and a dispersion plate for inflowing liquid. Sampling ports were provided as indicated in Figure 4.7. In addition, a feed inlet line, an effluent line, a gas sampling port, and a vessel for collecting waste effluent were provided. Due to the volume of the stones used to pack the column, the void volumes vary somewhat from column to column (Table 4.18).

Columns nos. 1, 2, and 3 were each inoculated with 500 ml sewage sludge. Columns 4 and 5 were each inoculated with 500 ml effluent from the NBF chemostat, which was being fed a benzoic acid feed solution. Column 6 was inoculated with 500 ml effluent from the MIT chemostat, which was being fed a pretreated peat feed solution.

In order to maintain the pH between 7 and 8, the concentration of $NaHCO_3$ in the feed solution was varied when necessary (Table 4.19). The general operating procedure is presented below.

For the first week, the reactors were not wasted and fed, although gas samples were taken every other day to determine CH_4 production. The second through fourth weeks, the reactors were wasted and fed 250 ml of the appropriate feed solution twice each week, giving a retention time of ≈7 days. Analyses were performed as well. The pH of the effluent from each feeding was determined. Gas analysis was done every other day, using a Fisher-Hamilton Gas Partitioner Model 29. Organic acid analysis was done on a Perkin-Elmer Model F-30 Gas Chromatograph, the columns packed with Chromosorb 101 with no liquid phase.

Starting with the fifth week, retention time was decreased to ≈4 days by wasting and feeding 250 ml every other day. Organic acid analysis was performed periodically. This schedule was maintained until the reactors were converted to continuous systems.

Table 4.18

VOID VOLUMES OF PACKED BED FERMENTERS

COLUMN NUMBER	VOID VOLUME WITHOUT STONES (mℓ)	VOID VOLUME WITH STONES (mℓ)	RATIO	FEED SOLUTION
1	890	490	0.55	glucose
2	970	455	0.47	glucose
3	975	445	0.46	pretreated peat
4	965	445	0.46	benzoic acid
5	985	445	0.45	benzoic acid
6	970	445	0.46	pretreated peat
7	8,260	3,550	0.43	glucose

Table 4.19

PACKED BED COLUMN REACTOR FEED SOLUTIONS

I. PBC #1 and #2: Glucose Reactors

COMPOUNDS	CONCENTRATION
Glucose	4 g/ℓ
$NaHCO_3$	5 g/ℓ
S-3 Stock Solution*	1.6 mℓ/ℓ
S-4 Stock Solution*	15.2 mℓ/ℓ
S-5 Stock Solution*	1.5 mℓ/ℓ

Prepare using tap water; no adjustment of pH.

II. PBC #3 and #6: Pretreated Peat Reactors

COMPOUNDS	CONCENTRATION
Pretreated Peat	160 mℓ/ℓ
S-4 Stock Solution*	15.2 mℓ/ℓ
S-8 Stock Solution*	1.0 mℓ/ℓ

Prepare using distilled water. Adjust pH to
7.0 with concentrated HCl.

III. PBC #4 and #5: Benzoic Acid Reactors

COMPOUNDS	CONCENTRATION
Benzoic Acid	3.05 g/ℓ
$NaHCO_3$	2.60 g/ℓ
S-3 Stock Solution*	1.6 mℓ/ℓ
S-4 Stock Solution*	15.2 mℓ/ℓ
S-5 Stock Solution*	1.5 mℓ/ℓ
S-8 Stock Solution*	1.0 mℓ/ℓ

Prepare using distilled water. Dissolve benzoic acid by addition of NH_4OH. Adjust pH to
7.6, then add $NaHCO_3$ and stock solutions.
Bring to volume with distilled water.

* See Table 4.5 for composition of stock solutions.

After 6 - 8 months on this schedule, several of the reactors were converted to continuous flow systems. The recirculation through the reactor was maintained by Masterflex pumps (Cole-Parmer Instrument Co.). The flow rate for reactors no. 1 and 2 was ≈40 ml/min and for reactors no. 3 and 4 was ≈43 ml/min.

Each reactor was fed fresh substrate once daily (see Section 4.5 for results) and pH was determined every other day. Organic acid analysis and COD determinations were performed occasionally.

4.4.6 Analytical Procedures

In order to evaluate progress and compare results of these experiments, several standard analytical procedures had to be chosen. The methods were needed for measurement of pH, volume of gas produced, identification of gases produced, quantitation of organic acids produced, and determination of benzoic acid utilized for the various experiments conducted.

The pH of the broth was monitored in all experiments as an indication of the health of the cultures. The pH was determined with an Orion Model 501 digital readout pH meter equipped with a combination polymer body gel-filled electrode (Fisher Scientific Co., No. 13-639-252).

The volume of gas produced was measured for the chemostats (NBF and MIT) and for the packed-bed fermenter, after they were converted to continuous reactors. There were two types of devices used to measure volume of gas produced. The first was a standard wet test meter (GCA/Precision Scientific). The second device was a liquid displacement apparatus, shown in Figure 4.8.

In order to estimate the extent of conversion of substrate to products, it is necessary to determine the proportion of the various gases in the head space. This is determined on a Fisher-Hamilton Gas Partitioner Model 29. Two standard mixtures are used, one containing O_2 and N_2 (breathing air) and the other containing known amounts of CO_2 and CH_4. The standards are run each time the gas partitioner is started up. Using a 1.0 ml gas-tight syringe (Precision Scientific, Series A-2, No. 050033), 0.5 ml of each standard or sample is injected into the gas partitioner. All four gases can be resolved, each producing a peak. Peak height of the sample is compared to the peak height of the standard, allowing calculation of the proportion of each gas in the sample.

Another method used to determine conversion of substrate to products is organic acids analysis done by gas chromatography. Since some of the components of the substrates may interfere with this analysis (e.g. benzoic acid), an extraction is performed to remove the interfering compounds. The

Gas Inlet,

From Fermenter

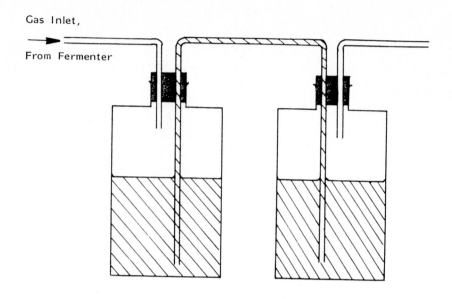

Fig. 4.8. Liquid Displacement Apparatus

extraction procedure is as follows: three drops of concentrated H_2SO_4 are added to the 1.0 ml of the sample (standard, blank). The solution is mixed and 4.0 ml diethyl ether are added to the vial. Each solution is mixed for one minute on a Vortex mixer and then centrifuged at ≈ 2000 rpm for several minutes to separate the aqueous and organic phases. To a vial containing 1.0 ml in 1 N NaOH, 3.0 ml of the diethyl ether are added. Each solution is mixed and centrifuged as before. Then, 0.6 ml of the aqueous layer is transferred to a clean vial. This solution is acidified with three drops of concentrated H_2SO_4. Finally, three drops of concentrated $CuSO_4$ are added and the sample is mixed and run on the GC after removing any precipitate.

The gas chromatograph used is a Perkin-Elmer Model F-30, with columns packed with Chromosorb 101 with no liquid phase. The program used is shown in Table 4.20. The concentration of each organic acid present in the sample is calculated by comparing peak height of the sample to the peak height of standards run under the same conditions.

The final method used to determine the extent of conversion of sub-strate to products is high performance liquid chromatography (HPLC). In this case, the disappearance of substrate is determined rather than the appearance of products. The sample was prepared as follows: 2.0 ml effluent was removed from the fermenter (serum vial or NBF) and then filtered through

a Millipore HA 0.45 μm filter. When dilutions were necessary, the sample
was diluted with distilled water and filtered a second time.

The HPLC system used is a Waters System equipped with an M-45 solvent
delivery system, a U6K injector, a Model 440 absorbance detector with a 254
nm wavelength, and a 10 mV strip-chart recorder. The column used was a Regis
Octadecyl Workhorse (30 cm x 4.6 mm). The isocratic elution system used was
45% MeOH/1% H_3PO_4. Spectrograde glass-distilled methanol was obtained from
Burdick and Jackson Laboratories, Inc. Phosphoric acid is Mallinckrodt
reagent grade. Distilled water is filtered, deionized, and passed through a
carbon bed in a system from Sybron/Barnstead. After the components of the
elution system have been mixed thoroughly, the eluent is filtered through a
Millipore HA 0.45 μm filter and degassed under vacuum.

Table 4.20

PROGRAM FOR ORGANIC ACIDS ANALYSIS
ON PERKIN-ELMER MODEL F-30 GAS CHROMATOGRAPH

PARAMETER	SETTING
N_2 (Carrier Gas)	50 psi
Flow A	27
Flow B	28
Injection Temperature	200°C
Detection Temperature	250°C
Initial Oven Temperature	160°C
Time at Initial Temperature	1 minute
Rate of Temperature Increase	10°C/minute
Final Oven Temperature	220°C
Time at Final Temperature	3 minutes
Air Pressure at Inlet	20 psi
H_2 Pressure at Inlet	20 psi

4.5 Results and Discussion

Fermentation experiments were performed to investigate four types of
conversions:

    (a) Benzoic acid to methane

    (b) Benzoic acid to aliphatic acids

    (c) Peat to methane

    (d) Peat to aliphatic acids

While progress was made in all four areas, not all fermentations attempted were successful. The data presented in this report is limited to successful experiments which allow some interpretation of the results. All cultures were developed from sewage digester inoculum using non-sterile techniques. All fermentations were carried out at 37°C.

Table 4.21

METHANE PRODUCTION FROM BENZOIC ACID
(CSTR, 75 mM Feed, 30 Day Retention Time, 3.0 ℓ Reactor)

DAY	$\%CH_4$	pH	ℓCH_4/DAY (STP)
6/19	59.2	7.19	0.42
6/20	59.6	7.18	0.32
6/21	59.8	7.09	0.38
6/22	62.7	7.18	0.52
6/23	58.8	7.10	0.46
6/24	60.1	7.12	0.45
6/25	63.2	7.18	0.55
6/26	61.7	7.21	0.51
6/27	– – – – – – N O D A T A – – – – – – – –		
6/28	55.6	7.16	0.39
6/29	55.2	7.17	0.35
6/30	60.3	7.20	0.36
7/1	60.8	7.12	0.48
7/2	62.4	7.22	0.52
7/3	– – – – – – N O D A T A – – – – – – – –		
7/4	– – – – – – N O D A T A – – – – – – – –		
7/5	– – – – – – N O D A T A – – – – – – – –		
7/6	60.7	7.25	0.16
7/7	57.7	7.17	0.33
7/8	60.7	7.15	0.55
7/9	59.6	7.24	0.53
7/10	60.8	7.27	0.56
7/11	62.1	7.32	0.60
7/12	63.3	7.18	0.50
7/13	60.4	7.30	0.54
7/14	60.1	7.18	0.55
7/15	60.9	7.25	0.56
7/16	60.0	7.22	0.56
7/17	61.5	7.30	0.57
7/18	– – – – – – N O D A T A – – – – – – – –		
7/19	62.9	7.25	0.38
7/20	59.6	7.22	0.43
7/21	58.6	7.20	0.50
7/22	56.4	7.14	0.51

4.5.1 Conversion of Benzoic Acid to Methane

A culture from sewage inoculum was adapted to convert benzoic acid to methane. While this conversion is well-documented in the literature, rate data was not available. The chemostat fermenter was fed a solution containing increasing concentrations of benzoic acid up to 75 mM (9.15 g/l) at which concentration the rate data was obtained. The reactor was operated at a thirty day retention time. The daily methane production is shown in Table 4.21. A calculated first-order rate constant of approximately 0.5 day^{-1} (Table 4.22) is reasonable for a dissolved substrate. The low steady-state benzoate concentration (2.8 mM) is reflective of the favorable thermodynamics of conversion of the substrate to methane.

Table 4.22

BENZOATE TO METHANE FERMENTATION

$$4C_7H_6O_2 + 18H_2O \rightarrow 15CH_4 + 13CO_2$$

CSTR Digester:

    75 mM Benzoate Loading
    30 day Retention Time

Rate Constant:

    Based on Benzoate Utilization: 0.86 day^{-1}
    Based on Methane Formation: 0.3 day^{-1}

Steady-State Benzoate Concentration in Reactor:

    2.8 mM

4.5.2 Conversion of Benzoic to Acetic Acid

The suppressed-methane fermentations of benzoic acid were carried out in 160 ml serum vials. In all these experiments 5×10^{-4} M 2-bromoethane sulfonic acid (BES) was used to inhibit methane formation. Some vials also contained KNO_3 (0.17 mg/ml) or Na_2SO_4 (0.24 or 2.4 mg/ml). Acetic acid production from 25 mM benzoic acid feed is shown in Figure 4.9. This vial fermenter was operated with an eight week retention time and no added electron acceptor. Although our thermodynamic considerations led us to believe this conversion was unfavorable, significant acetic acid build-up

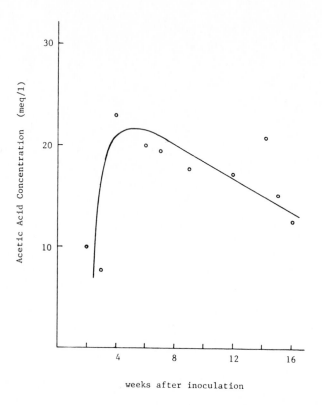

Fig. 4.9. Acetic Acid Production from Benzoate
(Vial No. 15574-C) 5×10^{-4} BES

was observed. Methane was also produced in this fermentation, possibly providing the energy for microbial growth. In the serum vial fermentation containing sodium sulfate (Figure 4.10), higher acetic acid levels were achieved and conversion of benzoic acid varied from almost 100% to about 50%. Methane production was observed towards the end of this fermentation, contributing to the diminished acetic acid yield.

In order to eliminate the methanogens from the microbial culture, penicillin and acetate were added without benzoic acid in the feed medium. By inducing only the acetate utilizers (methanogens and other organisms capable of oxidizing acetate in the presence of nitrate) to grow, the penicillin could selectively destroy them. In Figure 4.11, two cultures metabolizing benzoic acid in the presence of nitrate were monitored for acetic acid production. Only vial "J" had the penicillin treatment, resulting in a marked increase in acetic acid concentration. No vials without this treatment had significant acetic acid build-up in the presence of nitrate.

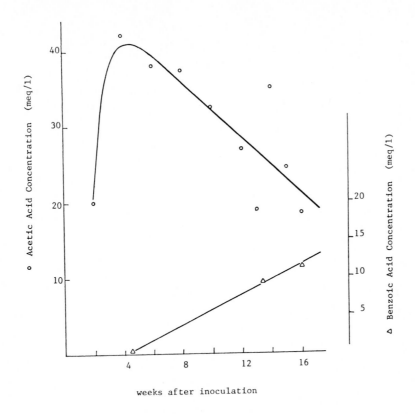

Fig. 4.10. Acetic Acid Production From Benzoate
(Vial No. 15574-E) 5 x 10^{-4}M BES & 2.4 mg/ml Na$_2$SO$_4$

4.5.3 Conversion of Pretreated Peat to Methane

    A microbial culture was developed from sewage digester inoculum which
was capable of converting pretreated peat to methane and carbon dioxide.
This culture was maintained in a chemostat (CSTR) fermenter with 4.0 l
liquid volume. Rate data was measured at various retention times with peat
(pretreated at 250°C) pooled to ensure constant composition of the feed. The
feed was diluted to about 1.1% volatile solids and had a chemical oxygen
demand (COD) of 9,180 mg O$_2$/l. The elemental composition (performed by
Galbraith Labs, Knoxville, TN) and stoichiometries of oxidation and methane
conversion are shown in Table 4.23. The theoretical maximum methane production, based on these stoichiometries is 327 ml (STP) methane per 1.0 g
peat volatile solids converted or 352 ml methane per g COD reduced.

    A second type of continuous fermenter was also operated on the same
pool of pretreated peat feed which was diluted to 7,290 mg/l COD. This was
a Packed Bed Column Fermenter (PBC No. 3) which has a high microbial density

Table 4.23

COMPOSITION AND STOICHIOMETRIC CONVERSIONS OF
POOLED PRETREATED PEAT

ELEMENT	ANALYSIS	NORMALIZED VALUE
C	30.45%	43.03%
H	2.75%	3.89%
O	37.56%	53.08%

<u>Stoichiometry of Pretreated Peat Oxidation:</u>

$$C_{1.081} H_{1.173} O_{1.000} + 0.87425 O_2 \rightarrow 1.081 CO_2 + 0.5865 H_2O$$
COD: 0.928 g O_2/1.000 g peat VS

<u>Stoichiometry of Conversion to CH_4/CO_2:</u>

$$C_{1.081} H_{1.173} O_{1.000} + 0.29 H_2O \rightarrow 0.44 CH_4 + 0.65 CO_2$$
CH_4 potential: 327 mℓ/1.000 g peat VS

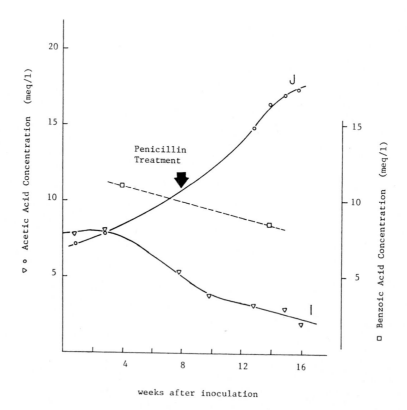

Fig. 4.11. Acetic Acid Production Following Penicillin Treatment
(Vial No. 15574 I & J) 5 x 10^{-4}M BES + 0.17 mg/ml KNO_3

and therefore a high methane production per volume of reactor. The void volume in this reactor is 0.45 l. The ultimate conversion to methane of the pooled pretreated peat was determined by batch fermentations in serum vials. Total methane production and COD reduction were determined.

Methane production from the continuous reactors is summarized in Table 4.24. Reactor performance at various retention times is correlated in Table 4.25. At 16-day retention time in the CSTR fermenter, ultimate conversion to methane is achieved. At a 1.5 day retention time in the PBC fermenter ultimate conversion is achieved. (The measured COD reduction actually slightly exceeds that achieved in the batch experiments, but this may be attributed to experimental error.) A first-order rate plot for the CSTR fermenter is shown in Figure 4.12 and a rate constant of 0.25 day^{-1} was calculated.

Table 4.24

SUMMARY OF DAILY METHANE PRODUCTION FROM CONTINUOUS PEAT FERMENTERS

FERMENTER	RETENTION TIME	MEAN CH$_4$ PRODUCTION (m l/day)	NUMBER OF DAYS AT RETENTION TIME	REACTOR LIQUID VOLUME	FEED COD (mg/ l)
CSTR (MIT)	16 days	395 + 34 (14)	19	4.0 l	9,180
CSTR (MIT)	11.4 days	498 + 48 (18)	21	4.0 l	9,180
CSTR (MIT)	4 days	572 + 110 (11)	18	4.0 l	9,180
PBC #3	1.8 days	377 + 68 (14)	14	0.45 l	7,290
PBC #3	1.5 days	414 + 54 (17)	22	0.45 l	7,290

Table 4.25

PERFORMANCE OF CONTINUOUS PEAT TO METHANE FERMENTERS

FERMENTER TYPE	FEED COD (mg/ l)	EFFLUENT COD (mg/ l)	RETENTION TIME	PERCENT COD REDUCTION	m l CH$_4$ (STP) PRODUCED PER g COD REDUCED
CSTR	9,180	4,512	16 days	50.8	327
CSTR	9,180	4,940	11.4 days	46.2	323
CSTR	9,180	6,280	4.0 days	31.6	325
PBC	7,290	3,263	1.8 days	55.2	348
PBC	7,290	3,350	1.5 days	54.0	342
BATCH	9,180	4,714	62 days	48.6	349
BATCH	9,180	4,634	62 days	49.5	354

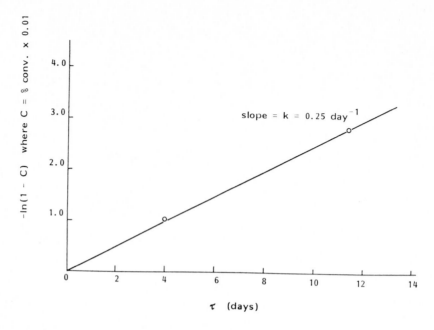

Fig. 4.12. Chemostat CSTR Fermenter Pretreated Peat to Methane

4.5.4 Conversion of Pretreated Peat to Organic Acids

A number of approaches to adapting microbial cultures to convert pre-treated peat to acetic and higher molecular weight acids were attempted. These included:

(a) First developing cultures for conversion of peat to methane, then adding BES.

(b) Adding BES at the start of adapting the cultures.

(c) Taking cultures adapted as in (a) and subjecting them to the penicillin treatment described for benzoic acid cultures.

(d) Taking cultures adapted to convert benzoic acid to acetic acid and switching to pretreated peat feed.

The only approach which resulted in a sustained production of acetic acid was approach (d), using cultures already producing acetic acid from benzoic acid. The other approaches produced cultures that were inactive or converted the peat to methane and/or carbon dioxide. Even the successful cultures had a limited amount of methane production (i.e. the methanogens were not entirely inhibited).

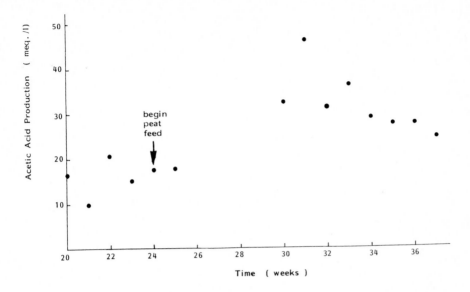

Fig. 4.13. Acetic Acid Production From Peat (Vial No. 015574-C)
(10 week retention time; 5×10^{-4} M BES)

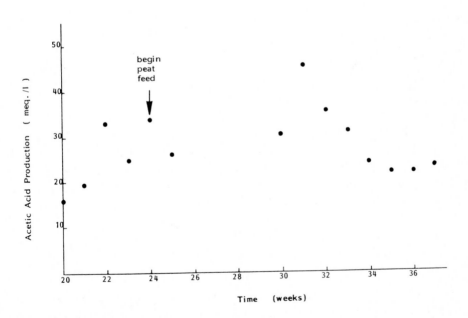

Fig. 4.14. Acetic Acid Production From Peat (Vial No. 015774-F)
(10 week retention time; 5×10^{-4} M BES + 2.4 mg/l Na_2SO_4)

144

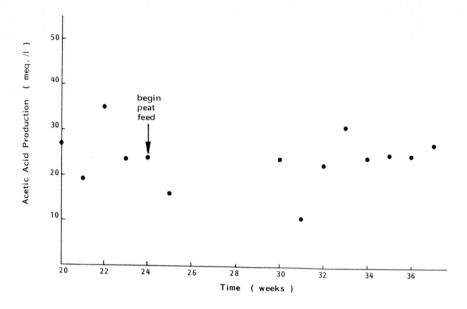

Fig. 4.15. Acetic Acid Production From Peat (Vial No. 015574-E)
(10 weeks retention time; 5×10^{-4} M BES + 2.4 mg/l Na_2SO_4)

Acetic acid production from three vial fermenters, initially adapted to benzoic acid production, then inhibited with BES, and then treated with pencillin is shown in Table 4.26 and Figures 4.13 through 4.15. Vial C had BES added and Vials E and F had BES and 2.4 mg/l Na_2SO_4 added. The acetic acid production in Vials E and F was slightly higher than in Vial C.

A continuous chemostat (CSTR) fermenter was also operated converting pretreated peat to acetic and butyric acids. The organic acid production by this fermenter is shown in Table 4.27. This is the only instance where constant levels of butyric acid were observed in the fermenter. No higher acids (valeric or caproic) were detected, but conditions favoring their formation (low pH and higher substrate concentrations) were not checked.

4.5.5 Fermentation of Whole and Ultrafiltered Peat Lignin

A portion of peat was acid washed to remove the cellulosic portion as described earlier. The remaining peat, primarily a material that can be considered lignin, was pretreated (solubilized and oxidized at 250°C). The product of the pretreatment was centrifuged to remove undissolved solids and the liquid portion divided into two portions. One portion was diluted 3.2/10 into nutrient medium and the second portion was ultrafiltered through

145

Table 4.26

ACETIC ACID PRODUCTION IN CONTINUOUSLY FED VIALS
AFTER INITIATING PEAT FEED
(8 Week Retention Time)

WEEK	ACETIC ACID CONC. (meq/ℓ)		
	VIAL C	VIAL E	VIAL F
24	17.6	24.0	33.6
25	17.8	16.0	26.0
30	32.3	23.9	30.3
31	46.2	10.8	45.4
32	31.3	22.6	35.5
33	36.1	31.0	31.0
34	28.8	24.1	24.1
35	27.7	24.9	22.0
36	27.7	24.9	22.0
37	24.3	27.3	23.0

Table 4.27

ORGANIC ACID PRODUCTION FROM PEAT IN CONTINUOUS FERMENTER

Liquid Volume: 3.0 ℓ
Retention Time: 30 days
Feed COD 9,180 mg/ℓ
Methane Suppressant: 5×10^{-4} M BES

DAY	ACETIC meq/ℓ	BUTYRIC (meq/ℓ)
1	0.0	0.0
14	11.8	1.6
24	19.1	2.8
25	18.9	2.8
28	12.6	2.5
29	16.7	3.2
30	18.3	3.7
31	17.7	4.1
32	18.7	3.8
37	17.3	3.5
38	19.3	3.7
42	18.5	3.7
43	18.0	4.3
44	16.5	3.1
45	18.0	4.0
46	21.0	4.4
49	17.0	4.0

146

Table 4.28

FERMENTATION OF PEAT LIGNIN (WHOLE AND ULTRAFILTERED)

SUBSTRATE	RETENTION TIME	% VS IN FEED	ml CH_4 FORMED PER 10 ml FEED	% CONVERSION OF THEORETICAL
Whole Peat Lignin	18.7 days	1.01	3.06 ± 1.01 (7)	46.9
Ultrafiltered Peat Lignin	18.7 days	0.55	3.00 ± 1.25(7)	84.5

Theoretical Stoichiometry:

$$1.00 \ C_{1.03} \ H_{0.85} \ O_{1.00} + 0.32 \ H_2O \rightarrow 0.37 \ CH_4 + 0.66 \ CO_2$$

a 500 MW cut-off membrane prior to diluting into a fermentation medium. Each of these feed mixtures was used in separate serum vial fermentations and converted to methane. The retention time for each fermentation was 18.7 days and vials were wasted and fed three times per week.

The pretreated peat lignin was sent to Galbraith Labs for elemental analysis and stoichiometry of conversion to methane determined. The results of these fermentations are shown in Table 4.28. The yield (per ml of feed) is the same for the whole and ultrafiltered feeds. The conversion of only the ultrafiltered part of the peat feed (less than 500 MW material) can account for all the methane production in both fermentations, implying that only the low molecular weight material is bioconverted. This result confirms the assumption made at the outset of the pretreatment experiments: to optimize fermentation yield, the production of low molecular weight material in the pretreatment must be maximized.

4.6 Conclusions

The ability to convert model aromatic compounds to methane has been conclusively demonstrated by other researchers (see references.)   Rate of conversion for these compounds are not available in the literature. The measured rate constant of between 0.3 and 0.8 day^{-1} for benzoic acid conversion to methane is in the range of expected values for soluble substrates. The suppressed-methane fermentations of benzoic acid performed in this work have demonstrated the technical feasibility of converting aromatic compounds to acetic acid by anaerobic digestion. Particularly promising are the results of fermentations following penicillin treatment in which methane formation and oxidation of acetate in the presence of nitrate were eliminated. This technique offers a useful tool for future work in this area. The lack of formation of higher molecular weight acids than acetic is a shortfall in this work. The process calls for maximized production of caproic acid which is essential in obtaining reasonable product yields. The next step in the development of these microbial cultures is to find conditions under which higher acid formation occurs.

Past work on the fermentation of pretreated peat and plant lignin is fairly nebulous. This is the result of variations in the nature of the substrate after pretreatment. The present work demonstrates higher product yields in the methane fermentations (25% of the organic fraction of peat starting material is ultimately converted) than has been reported by other investigators. Rates of conversion to methane (0.25 day^{-1}) and the ability to operate packed bed digesters on pretreated peat feed are useful data from these experiments. The ability to produce acetic acid from peat by anaerobic digestion is the first step in adapting cultures for this process. Though this has been demonstrated, the results are not totally positive since methane formation could not be completely inhibited. Also, as with benzoic acid, formation of caproic acid from peat was not demonstrated. This, of course, is essential in the development of the process.

Future work in this area should focus on the suppressed-methane fermentation of pretreated peat. A number of points are worth pursuing with the culture acclimation techniques developed on this project. These points include:

1. Demonstration of higher acid formation (up to caproic) from peat.

2. Rate measurements of product formation in suppressed-methane fermentations.

3. Operation of packed-bed column digesters on peat in the suppressed-methane mode.

4. Further optimization of the severity of pretreatment vs. fermentability.

5. Investigation of the effect of addition of electron acceptors to peat fermentations.

5. ELECTROLYTIC OXIDATION OF ALIPHATIC CARBOXYLIC ACIDS

Recovery of organic acids from relatively dilute aqueous solutions (1M or less) is a formidable separation problem with substantial capital and energy requirements. To circumvent this difficulty, organic acids in aqueous solution can be converted (by electrochemical means) to hydrocarbons which are subsequently recovered by phase separation. Conversion of organic acid salts to either paraffins or olefins are most applicable to the development of processes for synthetic fuel production. These reactions may be described as:

$$R-COO^-Na^+ + R'-COO^-Na^+ + 2H_2O \rightarrow R-R' + 2CO_2 + H_2 + 2NaOH$$

and

$$RHCH_2COO^-Na^+ + H_2O \rightarrow R = CH_2 + CO_2 + H_2 + NaOH,$$

reactions which also produce hydrogen and carbon dioxide and regenerate the alkali necessary to form salts of the organic acids. While energy must be supplied as electricity to accomplish these conversions, negligible energy is required to separate the products from the aqueous stream.

Two points must be addressed in the development of this step of the proposed process. First, given a particular mixture of organic acids (primarily butyric, valeric, and caproic) how can the product distribution (paraffins, olefins, alcohols, esters) be controlled? As stated above, conversion to paraffins or olefins has the advantage of ease of separation of products from reactants. The second point is to determine the minimum applied potential necessary for the chosen conversion reducing the electrical requirement. This is equivalent to operating the electrolytic cell at the lowest possible voltage.

Our work to date has focused primarily on the first point. Experiments have been performed aimed at developing an understanding of the mechanisms of the electrochemical oxidation of organic acids and, from this understanding, determining operating conditions favoring production of the chosen products. An experimental system was designed and built that enabled examination of these reactions by varying parameters such as current density, electrode material, temperature, pressure, reacting species, supporting electrolytes, concentration of reacting species, flow rate, and pH. This section gives background information on the reactions being studied, describes the experimental methods, presents results, and draws a conclusion concerning our understanding of the mechanism of these reactions and the conditions necessary to use this step in a process for production of fuels.

5.1 Background

The electrochemical oxidation of aliphatic carboxylic acids is among the oldest of organic reactions. Kolbe (1849) described the decarboxylation and dimerization of organic acids to produce alkanes. Hofer and Moest (1902) described the decarboxylation of organic acids to produce alcohols. In addition, a variety of other products have been observed under both sets of reaction conditions particularly when acids other than acetic are electrolyzed including olefins, esters, non-dimerized alkanes and cyclopropane derivatives (Koehl 1964; Utley 1974).

The stoichiometry for the Kolbe and Hofer-Moest reactions are, respectively:

$$RCOOH + R'COOH \rightarrow R-R' + 2CO_2 + H_2$$
$$RCOOH + H_2O \rightarrow R*OH + CO_2 + H_2$$

where $*R$ refers to R and isomers of R collectively.

The former (Kolbe) reaction is predominant at smooth platinum anodes with acetic acid and the longer chain fatty acids (C_6 and up) (Weedon 1952). The second (Hofer-Moest) reaction is observed to predominate at carbon anodes (Eberson 1973). In both cases under most conditions the main side product is olefin:

$$RCOOH \rightarrow R'HC = CHR'' + CO_2 + H_2$$

where R' and R'' are either hydrogen or organic radicals. With most other electrode materials and at low current densities the primary side product (in aqueous solution to which this discussion is limited) is oxygen.

Speculation as to the mechanism of the Kolbe and Hofer-Moest reactions has been going on almost since the reactions were discovered. Brown and Walker (1891) presented a rudimentary mechanism involving the oxidation of organic acid anions, e.g. acetate to radicals as the primary anode process. Other theories, focusing on the fact that in the absence of organic acids water electrolysis will occur under the same conditions, assumed that oxidation of organic acid was not the primary electrode process. One of these, which was an elaboration of an early hypothesis presented by Schall (1896), suggested that active (Singlet) oxygen is produced at the anode by water electrolysis, which subsequently oxidizes the organic acid to a peroxy acid. Observed products are formed from subsequent decomposition of the peroxy acid (Fichter 1939). A similar mechanism proposed by Glasstone and Hickling (1934) suggests that the primary anode process is the oxidation of water to

150

hydroxyl radicals which then dimerize to form hydrogen peroxide or react
directly with the organic acid to produce observed products.   A supporting
argument to this theory is that anode materials which catalyze the decomposition of hydrogen peroxide produce oxygen not Kolbe or Hofer-Moest
products.

Subsequently, two series of papers were published which form the basis
of the current view of the mechanism of the Kolbe reaction (Conway and
Dzieciuch 1963a,b,c; Dickinson and Wynne-Jones 1962a,b,c).   These papers
present convincing evidence that oxygen evolution is prevented under Kolbe
conditions by the formation of an adsorbed layer of organic acid or Kolbe
intermediate which prevents water from reaching active sites on the electrode surface. This insight coupled with the notion that the Hofer-Moest
product and many of the other products observed are formed via a carbonium
ion intermediate, which was suggested by Walling (1957) and confirmed
elegantly by Corey et al. (1960), allowed Utley (1974) to construct a
diagram similar to the one presented in Figure 5.1

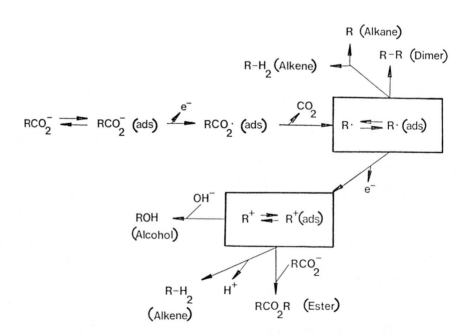

Fig. 5.1.   Electrolytic Oxidation Reaction Mechanisms

Whereas this diagram fairly represents the current understanding of the mechanism of electrochemical oxidation of aliphatic organic acids, it does not explain many of the experimental observations made by many researchers and confirmed in this work. Some of these observations are:

(a) the ratio of alkane dimer to olefin in platinum increases with the length of the linear aliphatic acid (Petersen 1900, 1906);

(b) alkane dimer is rarely observed using a carbon anode;

(c) the Hofer-Moest reaction (alcohol formation) occurs at a lower potential than alkane dimer formation (Wynne-Jones 1962a);

(d) olefin formation is favored at higher temperatures on carbon anodes; and

(e) oxygen formation is suppressed at lower current densities on carbon than on platinum.

Subsequently an elaboration of this basic mechanism has been developed which is capable of explaining these observations. The basic hypothesis along with supporting data are presented in Sections 5.3 and 5.4.

5.2   Experimental
5.2.1 Materials and Apparatus

The electrolysis reactions under investigation take place in a flow-through electrolytic cell manufactured by ECO, Inc., Cambridge, Massachus-etts. A schematic diagram of the cell is shown in Figure 5.2. The cell has been modified and reinforced to allow it to be operated at pressures up to 200 psig. The electrodes are separated by cardboard gaskets giving an electrode spacing of 0.15 mm. The exposed internal surface area of the electrodes is 27 or 7.8 cm^2. The inlet and outlet ports are both located on the cathode side of the cell.

The cell is integrated into a total system, shown in Figure 5.3, which is capable of measuring and regulating liquid temperature and pH, as well as current, potential difference, gas flow rate, and pressure. A Welker diaphragm pump draws the aqueous acid solution from a 1 l holding vessel and circulates it through the electrolytic cell at a rate of ≃10 ml/sec. The liquid and generated gas return to the holding vessel, and the gaseous products exit through a back pressure regulator.

The liquid in the holding vessel is stirred continuously and monitored for temperature and pH. The system pressure is measured by a pressure gauge

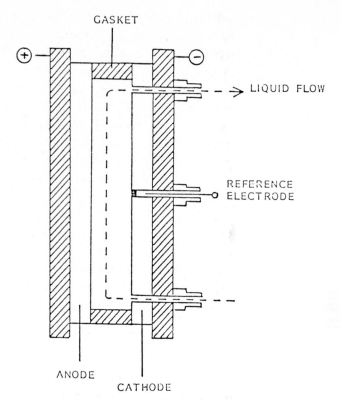

Fig. 5.2.   Schematic Diagram of Electroprep^{tm} Electrolysis Cell

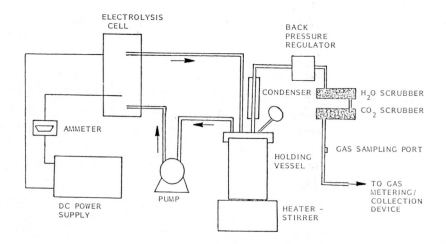

Fig. 5.3.   Schematic Diagram of Experimental Apparatus

connected directly to the holding vessel, as shown in Figure 5.3. The electrolytic cell is connected to a Sorenson SRL10-100 DC power supply, capable
of delivering up to 100 amps at between 0 and 10 V. The current is precisely
monitored by a Sargent-Welsh ammeter connected in series between the anode
and the positive terminal of the power supply.

Water vapor is removed from the exit gas steam by using a "U" tube
filled with indicating drierite. Once a satisfactory balance between decarboxylated products and carbon dioxide production was demonstrated, another
"U" tube containing indicating Ascarite was added to the system to remove
CO_2 from the exit gas to simplify the analysis of the gas composition. Gas
production is measured by water displacement.

Experiments have been run up to 200 psi and up to 150°C. For experiments over 50°C, a condenser is added prior to the back pressure regulator to
minimize water loss in the gas stream.

All reagents are reagent-grade materials obtained from major supply
houses and are used without purification. In many experiments, sodium sulfate is used as a supporting electrolyte to reduce the electrical resistance
of the electrolyte solution.

5.2.2 Product Recovery and Analysis

The headspace of the holding vessel is purged with nitrogen both before
and after the electrolysis experiment. This allows detection of water electrolysis and recovery of volatile products dissolved in the water immiscible
layer. The volume of gas produced is measured as described earlier. The composition of these gases is determined by injecting one sample into a Fisher
Model 25V Gas Partitioner using argon carrier gas to determine the hydrogen
concentration and another sample into a Fisher Model 29 Gas Partitioner
using helium as the carrier gas to measure the concentrations of the other
sample gases, primarily O_2, N_2, and CO_2.

Other gaseous products (olefins, etc.) are separated and quantified
using a Gow-Mac 750 gas chromatograph equipped with a flame ionization
detector. These products are separated on a 1/8" x 20' stainless steel
column packed with n-octane on Durapak using nitrogen carrier gas at
48°C.

After the gas flow rate and composition are determined, the hydrogen
flow rate is compared to the electron flow rate to ensure against gas leaks.
The electron flow rate may be calculated as follows:

$$e = \frac{60 \cdot I}{F}$$

where: e = Electron Flow Rate (electrons/minute);
           I = Current (amperes); and
           F = Faraday's Constant (96500 coul./mole).

Since two electrons must be exchanged to produce a molecule of hydrogen at the cathode, the hydrogen flow rate should be half that of the electron flow rate.

The reaction mixture is removed from the holding vessel and allowed to phase separate. The pump line and the electrolysis cell are flushed with distilled water to ensure all products are recovered quantitatively. The water immiscible products are analyzed on a programmable Perkin-Elmer F30 gas chromatograph equipped with a flame ionization detector. Glass columns $\frac{1}{4}$" dia x 6' packed with Chromasorb 101 (no liquid phase) are used with nitrogen carrier gas to separate these products at 130° to 260°C. The water soluble products, which are mainly alcohols, are quantified using the same instrument by injection of the acidified aqueous solution using the same temperature program.

The aqueous solution is often diluted five-fold before acidification so that a homogeneous solution will result.

For experiments with lower acids, e.g. propionic acid and butyric acid, dimeric products such as butane, hexane and heptane may be present as gases due to their high vapor pressure. These products may be quantified by injecting the gas sample into the Perkin-Elmer gas chromatograph and comparing to known standards as discussed above.

Concentrations of organic acids before and after electrolysis are determined by potentiometric titration. A sample of the solution is acidified with hydrochloric acid and back titrated with standard sodium hydroxide. The pH as a function of base added is plotted, and the organic acid concentration is determined from the difference between the inflection points above and below pH 4.75. At the same time the amount of dissolved carbon dioxide may be estimated by measuring the difference in moles of base required to bring the sample and blank back to the starting pH.

5.3 Results

Experiments were run in which the following reaction conditions were varied:

1. Reactor pressure with either butyric, valeric, or caproic acid as the reacting species.
2. Temperature with butyric or valeric acid as the reacting species.

3. Liquid flow rate through the cell with valeric acid as the reacting species.
4. Concentration of valeric acid.
5. Current density with valeric acid as the reacting species.
6. Concentration of caproic acid in mixtures with propionic or butyric acid.

In the tabulation of the results of these experiments, the product distributions are reported as percent of organic acid reactant converted to each product based on the total amount of that organic acid reacted in the experiment. In each experiment reported, hydrogen is recovered accounting for the total current supplied. Material balances indicated in the results show the percentage of anode product recovered compared to the total current supplied.

Running at elevated total pressures increases the olefin partial pressure in the gas phase which should result in a decrease in olefin yield. Tables 5.1 through 5.3 present the effect of pressure on the yields of the various products for butyric, valeric, and caproic acids, respectively.

Table 5.1

EFFECT OF PRESSURE ON THE PRODUCT YIELDS FROM THE ELECTROLYSIS
OF BUTYRIC ACID ON PLATINUM IN AQUEOUS SOLUTION

PRODUCT	RUN 82 PRESSURE: 100 KPa % YIELD	RUN 81 PRESSURE: 1.1 MPa % YIELD
Hexane	0	2.7
Propyl Butyrates	53.0	67.0
Isopropanol	3.1	2.3
n-Propanol	1.0	0.9
Propylene	42.0	27.0
pH Range	5.0 - 5.2	5.0 - 5.3
Concentration	2.0 M	2.0 M

156

Table 5.2

EFFECT OF PRESSURE ON THE PRODUCT YIELDS*
FROM THE ELECTROLYSIS OF VALERIC ACID ON PLATINUM
IN AQUEOUS SOLUTION

PRODUCT	RUN 82 PRESSURE: 100 KPa % YIELD	RUN 81 PRESSURE: 1.1 MPa % YIELD
n-octane	42.9	63.3
sec-butyl valerate	2.7	3.4
n-butyl valerate	0.9	0
sec-butanol	4.1	1.7
n-butanol	1.4	1.3
other aqueous	10.5	2.1
1-butene	15.5	8.9
trans-2-butene	5.5	2.1
cis-2-butene	4.1	1.7
n-butane	2.3	0.8
other gaseous	10.0	4.2
uncollected	0	10.5
pH range	7.4 - 9.2	7.4 - 8.2

* Mole percent of acid consumed to produce the specified
products.

Note that in each case a substantial reduction in olefin formation is
achieved. Even more interesting is the observation that it is the alkane
dimer production which increases the pressure. Table 5.4 gives the results
of additional experiments on the effect of pressure on linear alkane dimer
yield under a variety of experimental conditions. The reactions were all
run in pairs with pressure the only intentional variable except Run 75,
which is included because it gave the higher linear alkane yield from
valeric acid obtained to date.

A number of experiments were run using carbon, instead of platinum, as
the anode material. In one series of experiments, the olefin produced from
electrolyzing valeric acid as a function of reactant temperature was

Table 5.3

EFFECT OF PRESSURE ON THE PRODUCT YIELDS
FROM THE ELECTROLYSIS OF CAPROIC ACID ON PLATINUM

PRODUCT	RUN 82 PRESSURE: 100 KPa % YIELD*	RUN 81 PRESSURE: 1.1 MPa % YIELD
Decane	76.0	88.0
Pentyl Caproates	0	0
2-Pentanol	3.3	1.0
1-Pentanol	0	0
Pentenes	20.0	11.0
pH Range	7.4 - 8.9	7.4 - 8.0

* Average of two runs.

Table 5.4

EFFECT OF PRESSURE ON LINEAR ALKANE YIELD

RUN NUMBER	ACID	pH RANGE	PRESSURE	CONCENTRATION	%ALKANE*
71	Valeric	7.5 - 8.6	1.1 MPa	0.7 M	67%
72	Valeric	7.5 - 9.1	100 KPa	2.0 M	51%
73	Valeric	7.4 - 9.2	100 KPa	2.0 M	45%
74	Valeric	7.4 - 8.2	1.1 MPa	2.0 M	59%
75	Valeric	7.0 - 7.5	1.1 MPa	2.0 M	71%
76	Caproic	7.7 - 8.3	1.1 MPa	1.0 M	80%
77	Caproic	7.4 - 8.8	100 KPa	1.0 M	68%
78	Caproic	7.4 - 8.0	1.1 MPa	1.0 M	88%
79	Caproic	7.4 - 8.9	100 KPa	1.0 M	76%

* Mole percent of acid consumed to form alkane dimer product.

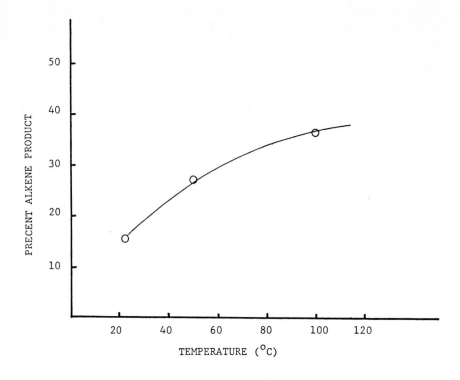

Fig. 5.4. Effect of Temperature on Olefin Formation
(Valeric Acid on Carbon Anode)

examined. Figure 5.4 is a plot of olefin yield vs. temperature showing increased olefin production at higher temperatures. The plateau observed above 100°C may be due to the increase in pressure necessitated by operating temperatures above the boiling point of the solvent.

Another experiment designed to demonstrate the reversability of olefin formation was run using valeric acid in an electrolysis vessel initially pressurized with propylene. Figure 5.5 shows the distribution of liquid products from this experiment. The significant point is that n-propanol and i-propanol were identified in the product mix, indicating that olefins may be converted to alcohols under the reaction conditions. Similar experiments were run using a platinum anode in an attempt to demonstrate the reversibility of the radical to olefin reaction. In these experiments, valeric acid was electrolyzed in a reaction vessel pre-pressurized with propylene. A reaction of this type would be expected to produce normal and isomeric heptanes, decanes, and undecanes along with the usual n-octane product if

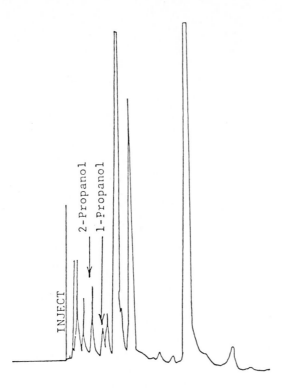

Valeric acid on Carbon anode with
propene in headspace

Fig. 5.5. Gas Chromatogram (Liquid Products)

the radical to olefin reaction is reversible. In contrast to the reaction
in which alcohols were formed from olefins, no heptanes, decanes, or
undecanes were identified.

Mixtures of butyric and caproic acids were electrolytically oxidized on
a platinum anode. The initial pH in all these experiments is 7 and final pH
is 9 (±0.3). The yield of butyric acid products is shown in Table 5.5. The
yield of dimeric products formed from butyric acid (n-hexane and n-octane)
increases as the initial concentration of caproic acid increases, up to
0.35 M, as seen in Figure 5.6. The ratio of butyric to caproic acid reacted
in these experiments is observed to decrease as caproic acid concentration
is increased (Figure 5.7). This ratio can be normalized as indicated in the
figure, and the normalized ratio of acids reacted is approximately 1:3
(Butyric:Caproic) at all caproic acid concentrations above 0.35 M.   The

160

Table 5.5

PERCENT YIELDS OF BUTANOIC ACID PRODUCTS FROM ELECTROLYTIC OXIDATION
OF MIXTURES OF BUTANOIC AND HEXANOIC ACIDS ON PLATINUM ANODE*
(Current Density .28 amp/cm^2; pH 7-9; Atmospheric Pressure)

RUN NUMBER	REACTING ORGANIC ACIDS	PROPANOLS	ESTERS	OLEFINS	DIMERIC PARAFFINS	H$_2$ BALANCE	MATERIAL BALANCE	BUTANOIC HEXANOIC
118	1.4 M Butyric	45.7	4.0	31.1	17.1	100%	93.0%	---
130	1.4 M Butyric	42.6	1.7	33.6	20.5	100%	89.0%	---
134	0.7 M Butanoic/0.1 M Hexanoic	40.0	4.3	35.6	17.5	100%	82.3%	5.14
133	0.7 M Butanoic/0.2 M Hexanoic	25.6	11.5	28.7	31.2	98%	84.7%	1.90
136	0.7 M Butanoic/0.3 M Hexanoic	19.5	14.7	16.5	46.7	100%	82.2%	0.75
132	0.7 M Butanoic/0.35 M Hexanoic	13.9	12.5	14.1	56.4	100%	93.6%	0.66
122	1.4 M Butanoic/0.7 M Hexanoic	2.9	3.7	10.7	76.8	100%	88.7%	0.59
137	0.7 M Butanoic/0.5 M Hexanoic	7.8	14.5	13.1	62.1	100%	81.5%	0.47
119	0.7 M Butanoic/0.7 M Hexanoic	15.6	6.3	15.6	53.2	98%	86.0%	0.30
131	0.7 M Butanoic/0.7 M Hexanoic	1.7	20.6	12.3	62.2	96%	84.2%	0.32
135	0.7 M Butanoic/1.0 M Hexanoic	2.0	28.1	10.2	57.1	100%	89.2%	0.23

* Yields based on mole % of total butanoic acid consumed to form measured quantities of each product.

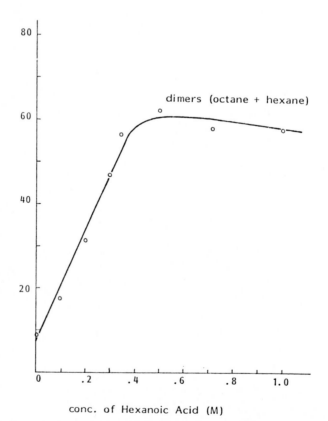

Fig. 5.6. Yield of Dimeric Products from Butanoic Acid as a Function of
Hexanoic Acid Concentration

formation of dimeric products from caproic acid (n-octane and n-decane) follows a similar pattern to that observed for butyric acid when examined as a function of caproic acid concentration (Figure 5.8).

An analogous increase in the formation of dimeric products (butane and heptane) from propionic acid was seen when caproic acid was added to the electrolysis solution (Table 5.6). In these experiments, the normalized ratio of propionic/caproic acid reacted is 1/6.25.

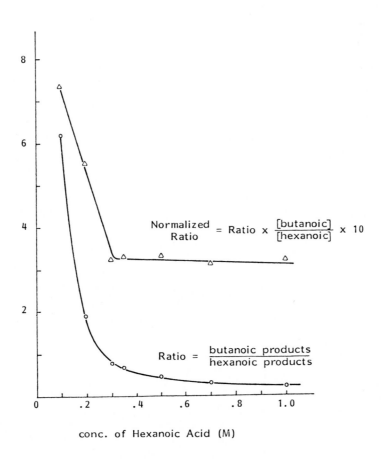

Fig. 5.7.   Ratio of Butanoic Acid Reacted in Mixed Runs

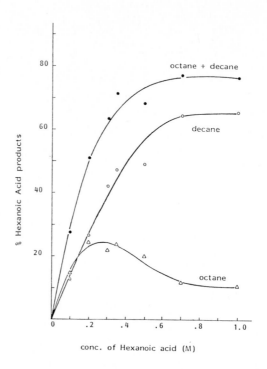

Fig. 5.8. Dimer Yield from Caproic Acid as a Function of Caproic Acid Concentration

The effects of temperature, current density, concentration, and liquid flow rate were observed in electrolysis solutions containing a single reacting organic acid (butyric or valeric). Table 5.7 shows an increase in alcohol formation and decrease in dimer formation at increased temperature. Table 5.8 shows a decrease in dimer formation with decreasing flow rate (Runs 158, 164, and 166). At a relatively low flow rate, an increase in organic acid concentration restores dimer formation to that seen at high flow rates (Runs 166, 167, and 168).

An effect on the product distribution is observed for current densities below 0.27 amp/cm² (Figure 5.9). As the current density decreases, products requiring a two-electron oxidation are favored. At current densities below 0.13 amp/cm² a significant increase in production of O_2 from water is observed.

Table 5.6

PERCENT PRODUCT YIELDS FROM PROPANOIC ACID IN MIXED ACID RUNS*

RUN NUMBER	REACTING ORGANIC ACIDS	% ETHENE	% BUTENE	% HEPTENE	% ETHANOL	% ESTERS	% MATERIAL BALANCE	PROPANOIC HEXANOIC REACTED
147	2.0 M Propanoic	60.9	0	---	30.8	8.6	94.3	---
151	2.0 M Propanoic/1.0 M Hexanoic	43.1	33.5	15.1	3.6	4.6	90.0	0.32
152	1.4 M Propanoic/1.0 M Hexanoic	28.8	14.8	44.0	2.2	10.3	101.1	0.24
153	1.4 M Propanoic/0.7 M Hexanoic	25.7	9.2	42.0	5.6	5.3	81.0	0.30

* % yields of each product based on amount of propanoic acid reacted to form that product compared to total amount reacted.

Table 5.7

EFFECT OF TEMPERATURE ON YIELD OF PRODUCTS*

RUN NUMBER	REACTING ORGANIC ACIDS	TEMP, °C	% ALCOHOLS	% ESTERS	% DIMER	% OLEFINS	% DISPROPORTIONATION PRODUCTS	% OF CURRENT TO O_2 PRODUCTION	% MATERIAL BALANCE
130	1.4 M Butanoic	22	42.6	1.7	20.5	33.6	1.6	---	89.0
141	0.7 M Butanoic	22	65.6	0	8.9	24.4	1.0	3.4	100.0
138	0.7 M Butanoic	80	73.1	<1	4.9	21.3	0.7	13.3	98.3
140	0.7 M Pentanoic	22	25.3	9.7	36.1	25.4	3.4	3.4	100.0
139	0.7 M Pentanoic	80	44.7	6.1	29.5	16.1	3.6	9.1	100.0

* Yield calculated as mole % acid reacted to form product compared to total moles acid reacted in run.

Table 5.8

EFFECT OF LIQUID FLOW RATE AND ORGANIC ACID CONCENTRATION ON PRODUCT DISTRIBUTION*
(Current Density 0.27 amp/cm^2; cell cross-sectional area 0.016 cm^2; exposed anode area 7.33 cm^2)

RUN NUMBER	REACTING ORGANIC ACID	% BUTENES	% OCTANE	% ALCOHOLS	% ESTERS	% CURRENT TO O_2	% MATERIAL BALANCE	FLOW RATE (mℓ/sec)
158	1.0 M Pentanoic	35.2	41.1	15.5	8.2	12.7	98.4	8.0
164	1.0 M Pentanoic	32.5	38.3	20.3	8.9	10.1	100.8	2.8
166	1.0 M Pentanoic	32.3	25.8	36.6	5.3	17.4	104.6	1.4
167	1.5 M Pentanoic	32.0	42.7	18.3	7.0	12.7	98.7	1.8
168	0.6 M Pentanoic	34.4	21.1	37.5	6.9	14.3	105.4	1.8

* Yields presented as percent of reacting organic acid forming each product.

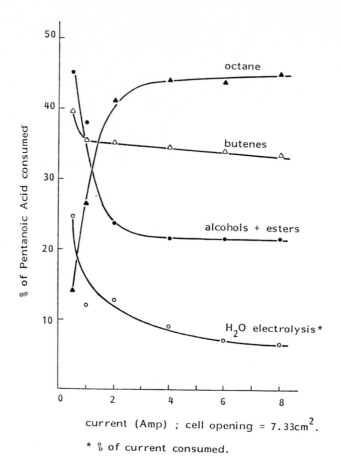

Fig. 5.9. Effect of Current Density on Product Distribution

5.4 Discussion and Conclusions

The electrochemical oxidation of organic aliphatic acids in aqueous solution involves these competing reactions:

1. One-electron oxidation to produce radical products (alkane dimer) E = +2.2V.

2. Two-electron oxidation to produce carbonium ion products (olefin, alcohol, ester) E = +1.9V.

3. Oxidation of water to produce oxygen E = +1.2V.

The primary goal is to understand why each of these reactions pre-dominates under different conditions. Based on the open cell potentials of each reaction vs. the nominal hydrogen electrode, one would expect reaction 3, the oxidation of water, to predominate in all cases. This is in fact observed on all anode materials except smooth platinum, carbon and a few others. As mentioned earlier, current understanding suggests that on these latter materials organic acids are normally adsorbed on the electrode surface to a degree which prevents water from reaching it to be oxidized. However, at low current densities or in dilute solutions of acid, water electrolysis is observed on platinum.

Figure 5.10 represents the electrode pair with details of the suggested concentration densities of ionic and molecular species, which is the key to the current hypothesis. Adjacent to the anode surface is a lipophilic layer formed by the adsorbed organic layer plus hydrocarbon products which have not yet escaped from the electrode surface. The integrity of this layer is a function of current density which affects the production rate of these products, and its thickness is determined by the length of the acid molecules adsorbed on the surface.

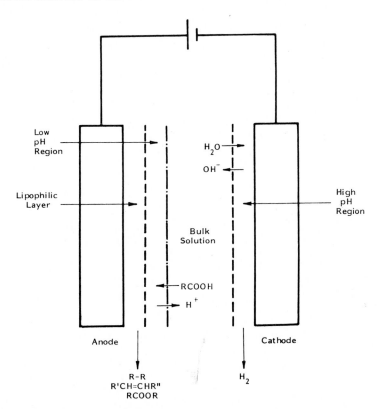

Fig. 5.10. Model of Electrolytic Oxidation of Carboxylates on Platinum

Adjacent to the lipophilic layer on the anode is a low pH region. There is a corresponding high pH region at the cathode surface. Since less is going on at the cathode, formation of the high pH region is easier to comprehend. At the cathode the reaction is:

$$2H_2O + 2e^- \rightarrow H_2\uparrow + 2OH^-$$

That is, for every electron passed through the solution, a hydroxide ion is produced at the cathode. Since the only method available to receive hydroxide ions from the cathode surface is diffusion, a concentration gradient is established such that the diffusion flux is equal to the formation rate under steady state operation for a cell operating at neutral bulk pH and a cathodic current density of $0.25A/cm^2$. Typical for this type of electrolysis, one would anticipate a hydroxide concentration at the cathode surface of 1.6 M in a well stirred solution. Similarly the anode, in order to maintain neutral charge in the solution must produce protons or consume organic acid anions at an equivalent rate. How much of each method occurs depends on the product mix as well as the anion concentration and the current density. Normally one net proton will be produced for each two electrons passed giving a proton concentration adjacent to the electrode solution of 0.8 M under the same assumptions as used for the cathode. This acid concentration is sufficiently high to affect the intraconversion of olefins and alcohols via a carbonium ion intermediate.

To answer the second question, why aren't all of the organic acid molecules oxidized twice to form carbonium ion products if the second oxidation (E = +1.9V) is easier than the first (E = +2.2V), one must consider once again the lipophilic layer on the electrode. The low potential required for the second oxidation is only achieved when water is available near the electrode surface to stabilize the product. In the absence of water the potential required for the second oxidation is hypothesized to be greater than or equal to the 2.2V required for the first oxidation. This hypothesis forms the basis for an explanation why alkane dimer is found with longer chain acids and why olefin is the preferred carbonium ion product with this anode material.

The explanation for the lack of linear alkane formation at the carbon anode is based on the nature of carbon itself, which is known to contain active centers, the density of which is a function of the manner in which the carbon is made. It is assumed that the electrochemical reaction occurs only at these sites, which usually are relatively isolated on the electrode surface. Thus the actual current density at these sites is much higher than the apparent current density based on the overall geometry of the anode. This view of the electrode reaction explains why oxygen formation is suppressed at lower apparent current densities on carbon than on platinum. It also provides two explanations why alkane dimer formation is not normally observed. First that the sites are sufficiently isolated to disfavor a biomolecular reaction if in fact alkane dimer formation involves such a process, and, second, it discourages the formation of a continuous lipophilic layer at the anode surface, which is presumed responsible for keeping the water required to stabilize carbonium ions formed in the second oxidation step away from the electrode surface.

Another point to be addressed is the amount and type of olefin produced. The model being presented assumes that conversion of olefin to carbonium ion is a reversible equilibrium process in the low pH region and that the amount of olefin produced is a function of the escaping tendency of the olefin from the electrode surface. The electrochemical oxidation of the shorter acids produce more olefin (except acetic acid where olefin formation is not possible) than the large ones because of their high vapor pressures. Table 5.9 lists the vapor pressures for various possible olefin products from the electrolysis of butyric, valeric, and caproic acids calculated at 35°C based on the heat of vaporization as near this temperature as is readily available and the boiling point. The pentenes have vapor pressures of ≃1,000 mm Hg, the butenes 2,000, and propylene 10,000 mm Hg, in accord with the typical olefin yields from caproic, valeric, and butyric acids respectively of 20%, 40%, and 55%. The yields of the various olefin isomers from the electrolysis of valeric acid is reported by Koehl (1964) and confirmed in our laboratory of 50% 1-butene, 20% trans-2-butene and 10% cis-2-butene are also in accord with the vapor pressures of these isomers. The mole fraction of each at the electrode surface, which is required along with the vapor pressure of the pure material to make an exact calculation of the isomeric composition, is probably based on the relative stability of each isomer, which would favor the trans-2-butene.

Additional evidence, supporting the hypothesis of an adsorbed lipophilic layer, is seen in the results of the mixed acid runs. The influence of caproic acid on the products formed from butyric (or propionic) acid is interpreted to be caused by the formation of a lipophilic layer by caproic acid and its products. The existence of this layer is most strongly supported by the ratio of butyric/caproic acid reacted in the mixed acid experiments. The normalized ratio of 1:3 (Figure 5.7) suggests a higher partition coefficient for caproic than butyric acid into the lipophilic layer. Observation of a lower normalized ratio between propionic and

Table 5.9

VAPOR PRESSURE OF OLEFINS 35°C

OLEFIN	BP_{760}	ΔH_V Cal/g*	ΔH_V Cal/mole	VAPOR PRESSURE 35°C
Propylene	−47.7°C	104.6 (47.7°C)	4,393	10,600 mm Hg
1-Butene	− 6.3°C	86.8 25°C	4,861	2,600 mm Hg
cis-2-Butene	+ 3.7°C	86.8 25°C	4,861	1,870 mm Hg
trans-2-Butene	0.88°C	91.8 25°C	5,141	2,160 mm Hg
1-Pentene	30°C	85 est.	5,950	890 mm Hg
2-Pentene	37.1°C	85 est.	5,950	710 mm Hg

* Lange, N. A. <u>Handbook of Chemistry</u>, 9th ed. Handbook Publishers, Sandusky, Ohio 1956.

hexanoic acids is consistent with the relative partition coefficients of these organic acids between aqueous and organic solvents.

If it is the influence of the lipophilic layer that primarily controls the product distribution rather than properties of the reacting acids themselves, then the formation of dimers in the mixed acid runs should be governed by a mass action statistical distribution. The formation of dimer products from two reacting organic acids can be represented by the general equation:

$$aR_1COOH + bR_2COOH \rightarrow cR_1-R_1 + dR_1-R_2 + eR_2-R_2 + (a+b)CO_2 + (\tfrac{a+b}{2}) H_2.$$

If "r" is defined as the ratio of the organic acids reacted (a/b), then the distribution of dimer products formed is:

$$R_1 - R_1 = r^2$$
$$R_1 - R_2 = 2r$$
$$\text{and} \quad R_2 - R_2 = 1$$

This interpretation is applied to results from the mixed acid runs in Table 5.10. Expected product distributions are calculated from the ratio of butyric to caproic acid reacted and compared with experimentally observed

Table 5.10

EXPECTED AND OBSERVED RATIOS OF ALKANE DIMER FORMATION
FROM BUTYRIC-CAPROIC MIXED RUNS

RUN NUMBER	REACTING ORGANIC ACIDS	BUTYRIC CAPROIC REACTED (r)	HEXANE/OCTANE		OCTANE/DECANE		HEXANE/DECANE	
			EXPECTED $\frac{r^2}{2r}$	OBSERVED	EXPECTED $\frac{2r}{1}$	OBSERVED	EXPECTED $\frac{r^2}{1}$	OBSERVED
122	1.4 M Butyric/0.7 M Caproic	0.59	0.30	0.35	1.18	1.08	0.35	0.38
131	0.7 M Butyric/0.7 M Caproic	0.32	0.16	0.16	0.64	0.47	0.10	0.07
132	0.7 M Butyric/0.35 M Caproic	0.66	0.33	0.26	1.32	1.04	0.44	0.27
135	0.7 M Butyric/1.0 M Caproic	0.23	0.11	0.10	0.46	0.33	0.05	0.03
136	0.7 M Butyric/0.3 M Caproic	0.75	0.38	0.30	1.50	1.04	0.56	0.31
137	0.7 M Butyric/0.5 M Caproic	0.47	0.24	0.21	0.94	0.83	0.22	0.17

product distributions. In the presence of greater than 0.3 M caproic acid,
good agreement is seen between the expected and observed distribution of
the dimer products. Variations from the expected distribution may be due
to factors not considered by this simple model, such as differences
in diffusivity and competing reactions leading to formation of other
products.

Results from the experiments which varied temperature, current density,
flow rate, and concentration may be interpreted as being supportive of the
hypothesis of control of product distribution by a lipophilic layer. As the
integrity of the layer is disrupted by increased temperature (volatilization
of olefin products) local depletion of reacting organic acid (seen at low
flow rate and low bulk concentration), or lowered current density allowing
impregnation of water, the formation of dimeric (non-carbonium ion) products
is decreased. The effect of pressure on the product distribution is also
consistent with this explanation.

In summary, this hypothesis explains the five points raised at the end
of Section 5.1 in the following manner:

(a) The ratio of alkane dimer to olefin on platinum increases with
the chain length of the acid electrolyzed because the
lipophilic layer thus formed is thicker and is therefore more
efficient at keeping water, which is required for carbonium
ion stabilization, from the anode surface.

(b & c) Alkane dimer is not normally found on carbon because
electrolysis occurs only at active sites on the carbon anode;
thus a continuous lipophilic layer is not formed. In the

absence of a lipophilic layer, double oxidation to a carbonium ion occurs at a lower potential than radical formation.

(d) The amount of olefin product is determined by the escaping tendency of the olefin, which increases exponentially with temperature.

(e) Oxygen formation is suppressed at lower apparent current densities than on platinum because the entire area of the carbon electrode is not electroactive, resulting in higher actual current densities than is apparent.

The ability to improve alkane dimer yields, and in general control the product distribution, has potential applications in conversion of dilute aqueous solutions of organic acids to hydrocarbon products which are easily recovered from the aqueous stream. Particularly in the conversion of mixed organic acids obtained as products of fermentation processes, substantial yields of alkane dimers can be obtained from C_3 to C_5 acids providing a sufficient concentration (>.3M) of caproic acid is maintained or the reaction is carried out under pressure.

The significance of this finding can best be appreciated by quoting Lennart Eberson (1973) in a review article on the electrochemical behavior of carboxylic acids:

Since the possibility of two competing pathways in the Kolbe reaction exists, the immediate problem to answer is whether one can influence the direction of a given process by proper manipulation of experimental variables so that either the radical or carbonium ion mechanism is favored. As we shall see, unidirectional selectivity can easily be achieved, in that one can always find conditions to favor the cationic mechanism; on the other hand, in many cases, because of structural features in the substrate, it may not be possible to realize the radical pathway even under the most favorable experimental conditions.

In short, to date Kolbe electrolysis has not been used commercially for the conversion of mixed and/or dilute organic acids to hydrocarbons because no technique was available whereby the yield of alkane dimer could be increased above those which have been observed for years, as given in Table 5.11 (Allen 1958). By running electrolyses at elevated pressures these values have been substantially exceeded in the current work.

Table 5.11

KOLBE ELECTROLYSIS OF ALIPHATIC MONOBASIC CARBOXYLATES

ACID	STRUCTURE	PERCENTAGE YIELD		
		PARAFFIN	OLEFIN	ESTER
Acetic	CH_3COOH	85	2	2
Propionic	C_2H_5COOH	8	66	5
n-Butyric	C_3H_7COOH	14.5	53	10
iso-Butyric	$(CH_3)_2CHCOOH$	trace	62	10
n-Valeric	C_4H_9COOH	50	18	4
iso-Valeric	$(CH_3)_2CHCH_2COOH$	43	42	5
Methyl Ethyl Acetic	$(CH_3)(C_2H_5)CHCOOH$	10	42	10
Trimethyl acetic	$(CH_3)_3C\ COOH$	13	52	0
Caproic	$C_5H_{11}COOH$	75	7	1.5

Allen (1958), p. 103.

6. PRELIMINARY ENGINEERING DESIGN AND ECONOMIC ANALYSIS

The process for conversion of peat to mixed alcohols is described and major equipment and mass and energy balances provided. A plant processing 2000 tons per day (TPD), dry basis, peat is used for the design basis. Peat is delivered to the plant, following wet harvesting, in a 3 - 5% solids slurry. The slurry is pressed to 15% solids before being fed to a solubilization tank with sodium carbonate which is heated to 100°C (212°F) with an external heat exchanger. Following solubilization, the peat is partially oxidized to produce a soluble substrate for acid-forming anaerobes. After removal of undissolved solids by filtration, linear organic acids ($C_2 - C_6$) are formed by fermentation. These acids are removed by consecutive liquid-liquid extractions, and then electrolytically oxidized to form linear olefins ($C_3 - C_5$). These olefins can then be hydrated forming a mixed alcohol product. Economics for a base case design are presented, including estimates of capital and operating costs. A sensitivity analysis is performed to determine the effects of varying process parameters on unit product cost.

172

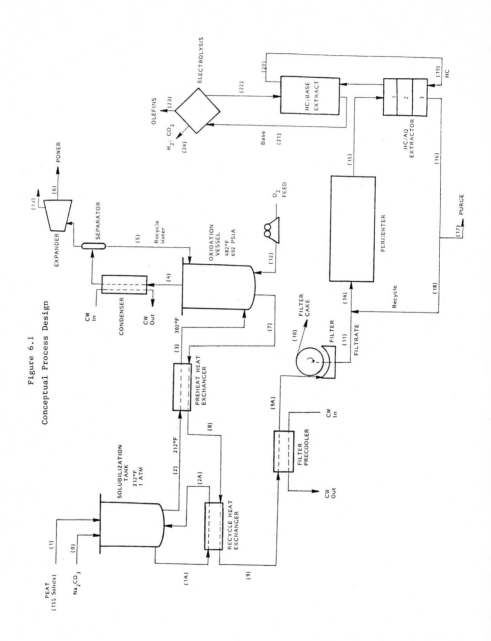

Figure 6.1
Conceptual Process Design

6.1 Base Case Process Design

A conceptual process design for conversion of 2000 TPD peat to mixed olefins is shown in Figure 6.1. The output for this process is 83.6 TPD mixed olefins with an energy content of 3.26×10^9 Btu/day. The process consists of the following major processing steps: water removal by press drying, solubilization, oxidation, fermentation, extraction and electrolytic oxidation. Material flows are shown in Table 6.1. The assumptions for the base case are summarized in Table 6.2.

6.1.1 Peat Dewatering

Peat is delivered to the plant at 3 - 5% solids and is dewatered in two steps. First, a screening operation is used to remove fines and concentrate peat to 8%. Then, the peat is pressed to 15% solids. Figure 6.2 shows the material balance for concentration from 3% to 15% solids. The amount of water removed is 53,330 TPD. A fast rate filtration system is capable of removing 20 gpm/ft² (Perry and Chilton 1973); therefore, a 360 ft² filtration unit would be needed. The efficiency of this process step has been assumed to be 100%.

Mechanical dewatering devices are commercially available from companies such as Ingersoll Rand. These are capable of removing water at the rate of 1,600 TPD for a 3-m belt width unit requiring 150 HP for operation. Eight presses are necessary for the process.

6.1.2 Peat Solubilization

In the solubilization, 80% of the peat volatile solids are solubilized. This step is carried out at 212°F (100°C) and has a residence time in the solubilization reactor of 3 hours. The reactor is designed as a CSTR and required volume is 55,000 ft³ (1.56×10^6 L).

The heat for the solubilization step is provided by external recycle heat exchange with the oxidized peat stream. The peat slurry from the solubilization tank is pumped through the exchanger at a rate dependent on its temperature rise and heat demand. The recycle ratio (RR) is defined as:

$$RR = \frac{lbs \ of \ recycle \ (heat \ exchange \ fluid)}{lbs \ of \ feed \ material \ to \ solubilizer}$$

For a recycle ratio of 7.1, the temperatures of streams for the recycle heat exchanger are shown in Figure 6.4. The heat required is 3.388×10^9 Btu/day and, assuming a heat transfer coefficient of 200 Btu/hr·ft²·°F, the area of the exchanger is 8,000 ft².

174

Table 6.1
PROCESS MATERIAL BALANCE – DAILY BASIS IN TONS PER DAY (TPD)

STREAM	PEAT VOLATILE SOLIDS	PEAT ASH	H_2O	Na_2CO_3	NON-SOLUBLE PEAT	SOLUBLE PEAT	SOLUBLE/FERMENTABLE	SOLUBLE/NON-FERMENTABLE	O_2	CO_2	TOTAL ORGANIC ACIDS	PROPYLENE	BUTENES	PENTENES	H_2	HYDRO-CARBON	TOTAL	TEMP (°F)	PRESSURE (PSIA)
0				360													360	70	
1	1,800	200	11,340														13,340	70	
2		200	11,340	360	360	1,440											13,700	212	14.7
3		200	11,340	360	360	1,440											13,700	302	692
4			2,943							1,386							4,329	482	692
5			2,882														2,882	300	692
6			61							1,386							1,447	300	692
7		200	11,621	360	360		462	402									13,405	482	692
8		200	11,621	360	360		462	402									13,405	393	
9		200	11,621	360	360		462	402									13,405	253	
10		200	1,800	56	360		71	62									2,549	99	99
11			9,821	304			391	340									10,856	99	99
12									1,152								1,152	70	
13			61							1,386							1,447	-88	
14			13,143	407			409.5	455.1			54.3						14,469	99	99
15			13,143	407			74.9	455		9.8	378.87						14,469	99	99
16			13,143	407			75	455.1		9.8	221.5						14,312	99	99
17			9,821	304			56.5	340		9.8	167.2						10,699	99	99
18			3,322	103			18.5	115.1			54.3						3,613	99	
19																37,041	37,041		
20											157.41					37,041	37,198		
21			17,326*								729						18,055		
22			17,326*								570						17,896		
23												35.82	18.45	29.31			83.58		
24										70.49					3.20		73.69		

* Aqueous Base

Table 6.2

PROCESS ASSUMPTIONS

Feedstock:
3 - 5% Peat (Total Solids) Slurry
90% Volatile Solids + 10% Ash (Dry Basis)
Empirical Formula: $C_{4.07}$ $H_{4.90}$ $O_{2.00}$ (Volatile fraction)
15% Peat (Total Solids) feed to solubilization tank
0.2 lbs Na_2CO_3/lb peat

Solubilization:
80% Solubilization of peat volatile solids
Rate constant 0.33 hr^{-1} @ 80°C
80% Solubilization achieved in 3 hr at 100°C (212°F)
Neglect heat of solution of Na_2CO_3

Oxidation:
Carried out at 482°F (250°C) for 2.5 hr
Consumption of O_2 is 0.003 ℓ (STP)/min/g peat VS
40% of input peat VS oxidized
3,000 Btu/lb input peat VS released in oxidation

Filtration:
Product stream from oxidizer heat exchanged to 37°C
  before filtration
All ash and undissolved solids removed by filtration
Filter cake contains 5 parts H_2O per 1 part solids

Fermentation:
Feed to fermenters contains 3.6% soluble fermentables
Fermenter operated with 25% recycle
53.4% of solubilized VS after oxidation are fermentable
85.6% of soluble fermentables converted
Rate constant for conversion is 2.0 day^{-1}

Liquid-Liquid
Extraction:
Designed for 95% removal of valeric acid
No acetic/propionic acid transferred into hydrocarbon
  phase

Electrolysis:
Carried out on platinum electrode
Voltage is 2.8 V
Current density is 0.15 amp/cm^2

Fuel Cell:
60% efficiency of converting CO_2/H_2 to electricity

6.1.3 Peat Oxidation

The limited oxidation of peat is carried out in a plug flow reactor at 482°F and 692 PSIA, which includes an oxygen partial pressure of approximately 100 PSIA. The oxidation of 40% of the input volatile solids is achieved over a 2.5 hour residence time. Oxygen is consumed at the rate of 0.003 ℓ (STP)/min/g peat VS. The stoichiometry of complete oxidation is given by

$$C_{4.07} H_{4.90} O_{2.00} + 4.30 O_2 \rightarrow 4.07 CO_2 + 2.45 H_2O.$$

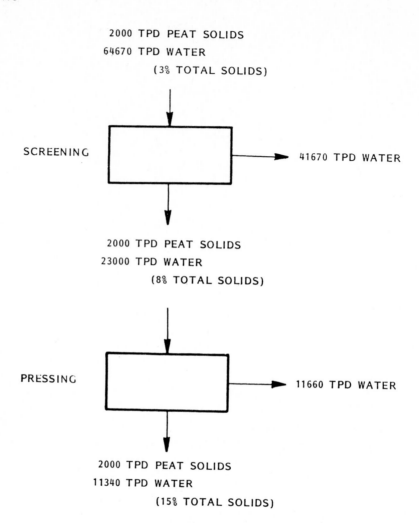

2000 TPD PEAT SOLIDS

64670 TPD WATER

(3% TOTAL SOLIDS)

SCREENING → 41670 TPD WATER

2000 TPD PEAT SOLIDS

23000 TPD WATER

(8% TOTAL SOLIDS)

PRESSING → 11660 TPD WATER

2000 TPD PEAT SOLIDS

11340 TPD WATER

(15% TOTAL SOLIDS)

Fig. 6.2. Peat Dewatering Material Balance

The actual O_2 consumption is 0.64 lbs/lb peat VS. In this step 0.77 lb CO_2 and 0.19 lb H_2O are formed per pound of input peat VS. The solubilized peat remaining after oxidation is 53.4% fermentable. The daily oxygen requirement is 1,152 tons. The O_2 is compressed to reactor pressure using three stages with intercooling to 80°F. The volume of the oxidation vessel is 45,000 ft³ (1.28 x 10^6 l).

Prior to oxidation, the solubilized stream is preheated in a heat exchanger with heat supplied by the oxidized stream. The quantity of peat exchanged is 2,148 x 10^6 Btu/day and temperatures are indicated in Figure 6.5.

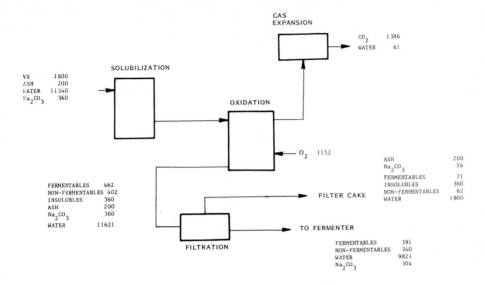

Fig. 6.3. Pretreatment Material Balance (Tons Per Day)

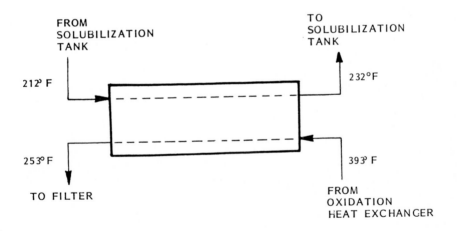

$$q = 3.388 \times 10^9 \text{ BTU/day}$$
$$\Delta T m = 88° \text{ F}$$
$$UA = 1.60 \times 10^6 \text{ BTU/hr}° \text{ F}$$
$$U = 200 \text{ BTU/hr} \cdot \text{ft}^2 \cdot° \text{ F}$$
$$A = 8000 \text{ ft}^2$$

Fig. 6.4. External Recycle Heat Exchanger

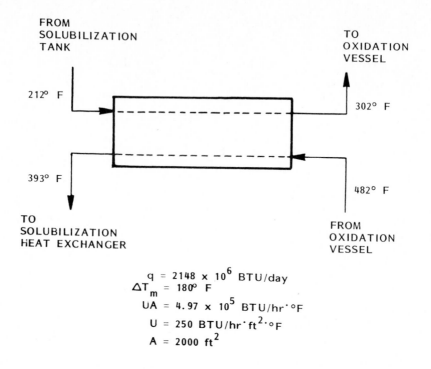

$$q = 2148 \times 10^6 \text{ BTU/day}$$
$$\Delta T_m = 180° \text{ F}$$
$$UA = 4.97 \times 10^5 \text{ BTU/hr} \cdot °F$$
$$U = 250 \text{ BTU/hr} \cdot ft^2 \cdot °F$$
$$A = 2000 \text{ ft}^2$$

Fig. 6.5. Oxidation Preheat Heat Exchanger

The required area, assuming a heat transfer coefficient of 250 Btu/hr-ft²-°F, is 2,000 ft².

Energy can be recovered from the hot gas stream containing primarily CO_2 produced in the oxidation. Water is condensed and recycled back to the oxidation vessel before the hot gas is expanded. The gas stream is cooled from 482°F to 300°F by cooling water which is heated from 80°F to 120°F. A low heat transfer coefficient (65 Btu/hr-ft²-°F) is used since water is condensing in the presence of CO_2. The required area of the condenser is 10,140 ft² for removal of 5.51×10^6 Btu/day. Cooling water flow is 11,485 GPM and is heated from 80°F to 120°F.

The energy recovered by expanding the hot gas stream is calculated in Table 6.3. An efficiency of 50% is assumed and 4.51×10^4 Kwh/day (1.54×10^8 Btu/day) are recovered.

Energy required for compression of oxygen to reactor pressure is calculated assuming 80% efficiency and a 3-stage process with inter-cooling to 80°F. The oxygen consumption is 1,152 TPD (19,701 SCFM at 80°F). This requires 7,629 HP or 1.365×10^5 Kwh/day.

Table 6.3

ENERGY RECOVERED BY EXPANDING HOT EXHAUST GAS FROM OXIDATION

Assumptions:

1. $C_p = 0.25$ Btu/lb-°F

2. $P_{in} = 600$ PSIA; $T_{in} = 300°F = 760°R$

3. $P_{out} = 16$ PSIA

4. $\gamma = C_p/C_v = 1.28$ (for CO_2)

5. $k = \dfrac{\gamma - 1}{\gamma} = 0.2187$

6. Efficiency = $\eta = 0.5$

7. Flow rate ($CO_2 + H_2O$) = 1,447 TPD

$$\Delta H(\text{Btu/lb}) = C_p T_{in} \left[1 - \left(\frac{P_{out}}{P_{in}} \right)^k \right] \eta = 52.00$$

$\Delta H = (52.00)(1,447)(2,000) = 1.50 \times 10^8$ Btu/day or 4.41×10^4 Kwh/day

6.1.4 Filtration of Oxidized Peat

Undissolved ash and volatile solids are removed from the liquid feed before fermentation of the soluble peat. All ash and insoluble peat volatiles are removed by filtration. The filter cake formed is assumed to be five parts water per one part solids. There is about a 15% loss of soluble fermentables in the filter cake as indicated in Table 6.1 The filtered stream fed to the fermenter contains 3.6% soluble fermentables. Assuming a filtration rate of 25 gal/hr-ft² and a flow rate of 108,000 gas/hr, 4,300 ft² of filter area is required.

Prior to filtration, the stream is cooled to 99°F with cooling water, removing $3,717 \times 10^6$ Btu/day. The heat exchanger area, assuming a heat transfer coefficient of 300 Btu/hr-ft²-°F, is 9,000 ft². Cooling water is heated from 80°F to 120°F, requiring a flow of 7,750 GPM.

6.1.5 Fermentation of Oxidized Peat

The pretreated peat stream containing 3.6% fermentable, soluble peat and 3.1% non-fermentables is fed to the fermentation vessels. The fermentation vessels are packed bed columns containing solid supports for microbial attachment. The high microbial densities obtained in this type of reactor

Table 6.4

STOICHIOMETRY OF PEAT CONVERSION TO ORGANIC ACIDS

ORGANIC ACID PRODUCT	STOICHIOMETRY	MAXIMUM % YIELD*
Acetic	$C_{2.44} H_{2.94} O_{1.2} + 1.11 H_2O + 0.14 CO_2 \rightarrow$ $1.29 CH_3COOH$	150.5
Propionic	$C_{2.44} H_{2.94} O_{1.2} + 0.75 H_2O \rightarrow$ $0.74 CH_3CH_2COOH + 0.22 CO_2$	106.5
Butyric	$C_{2.44} H_{2.94} O_{1.2} + 0.61 H_2O \rightarrow$ $0.52 CH_3CH_2CH_2COOH + 0.36 CO_2$	89.0
Valeric	$C_{2.44} H_{2.94} O_{1.2} + 0.53 H_2O \rightarrow$ $0.40 CH_3CH_2CH_2CH_2COOH + 0.44 CO_2$	79.3
Caproic	$C_{2.44} H_{2.94} O_{1.2} + 0.45 H_2O \rightarrow$ $0.32 CH_3CH_2CH_2CH_2CH_2COOH + 0.52 CO_2$	72.2

* % Yield calculated as grams product per gram oxidized peat converted.

Table 6.5

DETERMINATION OF ORGANIC ACID PRODUCT SPECTRUM OF FERMENTATION

ORGANIC ACID	MAXIMUM CONCENTRATION (mM)
Acetic	112
Propionic	61
Butyric	105

Caproic and valeric acids make up remainder in molar
ratio of caproic/valeric = 1.3/1.0.

Acids are formed up to maximum values of lower acids before
higher acids are formed. Conversion, in every case, is 85% of
soluble fermentable fraction.

results in higher conversion rates. The fermentable fraction of peat is 85% converted to a spectrum of organic acids. The concentration of the acids in the broth is partially controlled by recycling and removal in liquid-liquid extraction. The conversion stoichiometries for the fomration of each acid are shown in Table 6.4. The method of determining the concentration of each acid in the effluent stream from the fermenter is shown in Table 6.5. Since lower weight acids are formed prior to build-up of high weight product, maintaining increased concentrations of the low weight acids through recycle will result in improved yield of higher acids. The effect of varying the recycle ratio of the purge stream after higher acids have been extracted is shown in Figure 6.6. At very high recycle ratios the total yield of butyric, valeric, and caproic acids is improved, but the yield of caproic acid is decreased. At low recycle ratios more caproic acid is found, but the total yield of the three higher acids is decreased. By changing the hydrocarbon to aqueous flow rates in the extractors, the efficiency of organic acid removal is affected. Figure 6.7 demonstrates this point. Improved yields are seen at higher hydrocarbon to aqueous flow rates.

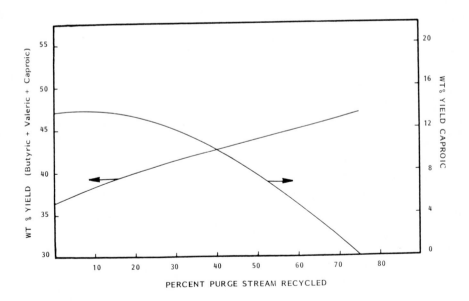

Fig. 6.6. Effect of Recycle Rate on Product Yield
(85% Conversion; HC/AQ = 3.2)

182

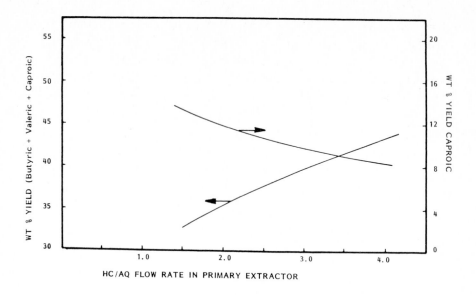

Fig. 6.7. Effect of HC/AQ Flow Rates on Yield of Extracted Acids
(n=3; 40% Recycle)

The packed-bed fermenters are modelled as CSTRs with a first-order rate constant of 2.00 day^{-1}. The fermentable material is 85% converted to organic acids. The required retention time is 2.83 days. In the base case design, 25% of the extracted purge stream is recycled to the fermenters. For this case the fermenter void volume is 1,231,000 ft^3. Assuming 50% void volume in the packed fermenters, the total fermenter volume is 2,462,000 ft^3 (69.7 x 10^6 liters). The fermenters can be constructed as 20 tanks, each 40 feet in diameter and 100 feet high. These tanks feed into five separate trains of extractors.

6.1.6 Liquid-Liquid Extractors

Two sets of liquid-liquid extractors are used to selectively remove the higher acids from the fermenter broth and concentrate these acids as aqueous salts. As demonstrated in Figure 6.7, extractor operation will affect the product yield from the fermenter. In the first step of the extraction (removal of higher acids from the fermenter broth into a hydrocarbon solvent), three stages of mixer-settlers are used with a hydrocarbon to aqueous flow-rate of 3.2/1.0. With these conditions, the removal efficiencies for butyric, valeric, and caproic acids are 56, 95, and 100%, respectively. (The measured partition coefficients are 0.2, 0.7, and 2.9). The material flows

for fermentation and extraction into hydrocarbon solvent are shown in Figure
6.8. The sizes of the mizers and settlers are calculated using the
assmptions in Table 6.6. The required volumes for eachs tage are 4,058 ft³
for the mixers and 24,348 ft³ for the settlers.

In the second set of extractors, the organic acids are removed from the
hydrocarbon in aqueous base. Since the acids become ionized in the aqueous
base, and the partition coefficient of the ionized form is zero, a negligible quantity of acids is left in the hydrocarbon phase which is recycled to
the first set of mixer-settlers. The hydrocarbon to aqueous base flow is
4.0/1.0 and volume of the single-stage mixer is 3,857 ft³ and settler is
19,285 ft³.

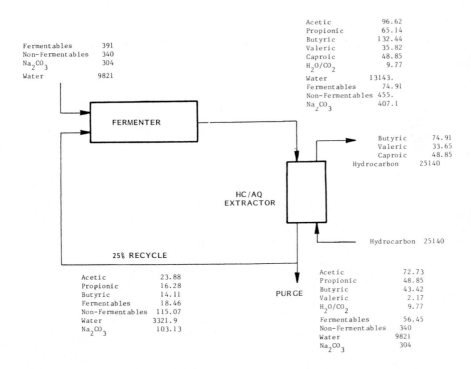

Fig. 6.8. Overall Fermenter Balance (Tons per Day)

Table 6.6

SUMMARY OF ASSUMPTIONS IN DESIGN OF FERMENTER/EXTRACTOR/ELECTROLYSIS

1. 25% of purge stream from extractor is recycled to fermenter.

2. Stoichiometries of conversion for fermentation presented in Table 6.4.

3. Organic acid spectrum determined according to Table 6.5.

4. Feed stream to fermenter contains 3.6% (by weight) fermentable material.

5. 85% of fermentable material converted.

6. Partition coefficients (hydrocarbon/aqueous) are:

acetic:	0.0
propionic:	0.0
butyric:	0.2
valeric:	0.7
caproic:	2.9

7. Initial series of extractors consists of 3 ideal mixer-settler stages.

8. No acids in hydrocarbon entering stream, dilute solutions, and solvents are insoluble in each other.

9. Hydrocarbon to aqueous flow rate is 3.2/1.0.

10. Removal efficiencies in the initial series of extractors are:

acetic:	0%
propionic:	0%
butyric:	56%
valeric:	95%
caproic:	100%

11. Second series of extractors consists of single mixer-settler stage.

12. Removal efficiency in second series of extractors is 100%.

13. Electrolysis is performed at 2.8 V.

14. Current density for electrolysis is 0.15 amp/cm^2.

15. Hydrocarbon to aqueous base flow is 4.0/1.0.

16. Minimum total acid concentration in exit of electrolysis cell is 0.5 moles/ℓ.

Table 6.6 (Continued)

17. Acids are electrolyzed at the following relative rates:

    butyrate: 1.0
    valerate: 2.0
    caproate: 3.0

18. Retention time of 3 min for mixers.

19. Retention time of 18 min for hydrocarbon/aqueous settlers.

20. Retention time of 15 min for hydrocarbon/base settlers.

21. Fermentation rate constant of 2.00 day^{-1}.

22. Conversion of H_2/CO_2 produced in electrolysis to electricity in a fuel cell at 60% efficiency.

23. Conversion of organic acids to olefins by electrolytic oxidation according to the following stoichiometries:

    a. $CH_3CH_2CH_2COONa + H_2O \rightarrow CH_3CHCH_2 + H_2 + CO_2 + NaOH$
    b. $CH_3CH_2CH_2CH_2COONa + H_2O \rightarrow CH_3CH_2CHCH_2 + H_2 + CO_2 + NaOH$
    c. $CH_3CH_2CH_2CH_2CH_2COONa + H_2O \rightarrow CH_3CH_2CH_2CHCH_2 + H_2 + CO_2 + NaOH$

24. Electrolysis is performed at 100% current efficiency and product is 100% olefin.

6.1.7 Electrolysis Cell

The electrolysis cell for the process is designed to convert 157.41 TPD of acids to olefins. This is equivalent to 1.46×10^6 g moles/day of acids converted. The current required is two electrons per mole of acid reacted, or 3.26×10^6 amps. With an optimum current density of 0.15 amp/cm^2 (139.4 amps/ft^2), the required electrode area is 2,174 m^2 (23,403 ft^2). The required voltage for the cell is 2.8 V.

The hydrogen produced in the electrolysis step can be used to produce electricity in a molten carbonate fuel cell. At an efficiency of 60%, the fuel cell will produce 2,429 KW of electricity, enough to supply 26.6% of the requirement for electrolysis. The fuel cell operates at 650°C and will produce steam, which can be used to provide mechanical power for the process. The material flows for the second series of extractors and electrolysis cell are shown in Figure 6.9.

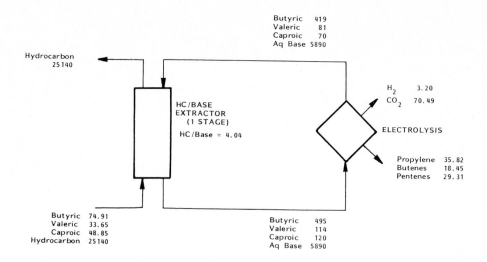

Fig. 6.9. Overall Electrolysis Balance (Tons per Day)

Table 6.7

MIXING REQUIREMENTS FOR VESSELS

VESSEL	SIZE (GALLONS)	TOTAL HP
Solubilization Tanks	412,000	825
Oxidation	No Agitation	0
Fermentation	Mixing Provided by Pumping	0
HC/AQ Mixers	3 x 30,400	182
HC/BASE Mixers	1 x 29,000	58
TOTAL		1,065

6.1.8 Mixing Requirements and Pumps

Mixing is necessary in two of the process steps. These steps are solubilization and liquid-liquid extraction. The mixing requirements for the extractors are assumed to be 2 HP/1,000 gals and for the solubilization tank 2 HP/500 gals. The total power necessary for mixing (Table 6.7) is not more than 1,200 HP (21,500 Kwh/day).

Table 6.8

PUMP AND PUMP HORSEPOWER REQUIREMENTS

	STREAM	N	GPM ea.	ΔP (PSI)	BHP (ea.)[2]	TOTAL BHP
Recycle HX	1A	20	811[1]	20	12	240
Oxidation Feed Pumps	2	20	114	750	62	1,250
Condensor Recycle Pumps	5	20	24	10	0.5	10
Fermenter Feed	11	5	362	20	6	30
Fermenter Recycle	16	5	477	20	7	35
Fermenter Purge	17	5	357	10	3	15
HC/AQ Mixers	15 + 19	3 X 5	2,023	15	22	340
HC/AQ Settlers	19	3 X 5	1,541	15	17	250
HC/Base Mixers	20 + 22	1 X 5	1,923	15	21	105
HC/Base Settlers	19	1 X 5	1,541	15	17	85
Electrolysis Feed	21	1 X 5	382	15	5	25
CW Pumps		2	12,500	20	180	360
Vacuum Pumps		5			10	50
TOTAL						2,800 (3,000)

[1] Recycle Flow = 7.1 x Flow of Stream 2.

[2] BHP = 5.834×10^{-4} (GPM)(ΔPSI)/η.   η = 0.8.

The total pumping requirement for the process is estimated as not more than 3,000 BHP as shown in Table 6.8. Pumps are assumed to operate at 80% efficiency.

6.1.9 Process Energy Requirements - Base Case

The major energy requirements are listed by process step in Table 6.9. The total energy supplied to the process (most of which is due to electrolysis, 48.3%, and O_2 compression, 30.3%) is reduced by recovering energy

Table 6.9

BASE CASE NET ENERGY REQUIREMENT

INPUT:

PROCESS STEP	ENERGY (KWH/DAY)
Peat Dewatering	0.215×10^5
O_2 Compression	1.365×10^5
Electrolysis	2.181×10^5
Agitation	0.214×10^5
Pumping	0.537×10^5
TOTAL INPUT	4.512×10^5

OUTPUT (Excluding Product):

CO_2 Expansion (60% eff.)	0.541×10^5
Fuel Cell (60% eff.)	0.563×10^5
Steam from Fuel Cell (30% eff.)	0.113×10^5
TOTAL OUTPUT	1.217×10^5

Net Energy Required: 3.295×10^5 Kwh/Day
At 32.5% Conversion Efficiency: 3.45×10^9 Btu/day

from H_2 produced in the electrolysis and hot CO_2 exhaust produced in the oxidation. The energy required for process operation is estimated at 3.3×10^5 Kwh/day. If this is supplied from a fuel which is converted to electricity at 32.5% efficiency, 3.45×10^9 Btu/day of fuel are required.

6.2 Product Cost Estimate — Base Case

Cost estimates in this report are based on the production of mixed olefin product (unpurified) suitable for hydration to mixed alcohols. The cost of producing alcohols from olefins, based on the relative market prices of these chemicals, is approximately $0.18/lb olefin converted (from the relative prices of butene and butanol).

The cost of the mixed olefin product produced in this process is calculated from capital and operating costs derived from the base case design. A discounted cash flow routine is used assuming 15% return on investment.

6.2.1 Capital Cost Estimate

The process is designed in parallel trains to reduce the size of some of the equipment. The pretreatment and fermentation are carried out in 20 trains. Eight dewatering presses, two oxygen compressors, and five extraction-electrolysis trains are used. The fermentation-extraction-electrolysis sections of the process may be regarded as five trains with four fermentation vessels in each train. The number, size, and cost of each major equiment item is listed in Table 6.10. The cost of fermenters is 30% of the total of $28,756,000.

Table 6.10

SIZE AND COST OF MAJOR EQUIPMENT

ITEM	UNITS N	P,T	MATERIAL	AGITATED	SIZE	UNIT PRICE ($)	TOTAL ($)
Solubilization Tank Tank	20	atm. 212 °F	304 SS	Yes	20,000 gal.	34,000	680,000
Oxidation Vessels	20	692 psia 482 °F		No	20,000 gal.	154,000	3,100,000
Recycle HX	20	400 psia	C/S	----	400 ft^2	8,250	165,000
Preheat HX	20	800 psia	C/S	----	100 ft^2	4,500	90,000
Condensor	20	692 psia	C/S	----	500 ft^2	11,700	234,000
Pre-Filter HX	20	150 psia	C/S	----	450 ft^2	4,250	85,000
Filters	20	----	----	----	250 ft^2	33,500	670,000
O$_2$ Compressors	2	----	----	----	3,800 BHP	1,100,000	2,200,000
HC/AQ Mixers	5 x 3 = 15	----	EPLCS	Yes	6,000 gal.	16,400	250,000
HC/AQ Settlers	5 x 3 = 15	----	EPLCS	No	36,000 gal.	25,400	380,000
HC/Base Mixers	5	----	EPLCS	Yes	6,000 gal.	16,400	82,000
HC/Base Settlers	5	----	EPLCS	No	29,000 gal.	20,900	105,000
Cooling Towers	2	----	----	----	12,500 GPM	253,000	506,000
Electrolysis Cells	5	----	----	----	4,650 ft^2	465,000	2,325,000
CO$_2$ Expanders	2				1,000 Kw	100,000	200,000
Fermenters	20	atm. 99 °F	C/S	No	125,700 ft^3	440,000	8,800,000
Pumps			SS				1,340,000
Screw Conveyors	20		----				1,000,000
Dewaterers	8	----	SS	----	3 m belt width	600,000	4,800,000
Fuel Cell				----	2,350 KW	200/KW	470,000
DC Transformer/ Rectifier				----	9,100 KW	140/KW	1,274,000
TOTAL							28,756,000

Table 6.11

CAPITAL COST SUMMARY

Major Equipment Costs (Table 6.10)	$28,756,000
Installation Costs (Piping, Electrical, Instrumentation, Insulation, etc.)	11,502,000
TOTAL EQUIPMENT COSTS	40,258,000
Supporting Facilities (10%)	4,026,000
TOTAL FIELD COST	44,284,000
Contractor's Overheat and Profit (10%)	4,428,000
Engineering and Design (5%)	2,214,000
Contingency (10%)	4,428,000
TOTAL PLANT INVESTMENT (TPI)	55,354,000
Interest During Construction (12%)	6,642,000
Start-up (8%)	4,428,000
Working Capital (2%)	1,107,000
TOTAL CAPITAL REQUIREMENTS	67,531,000

Table 6.12

ANNUAL OPERATING COSTS — BASE CASE

Raw Materials

Peat ($3/ton, dry basis)	$1,971,000
Sodium Carbonate ($100/ton)	11,826,000
Nutrients	1,000,000
Other chemicals and buffers	1,000,000
Oxygen ($25/ton)	9,461,000

Utilities

Electric (5¢/Kwh)	5,412,000

Labor

Operating ($7/hr)	840,000
Maintenance (1.5% TPI)	830,000
Supervision (15% Operating and Maintenance)	250,000

Adminstration and Overhead (60% Labor)	1,152,000

Supplies

Operating (30% Operating Labor)	252,000
Maintenance (1.5% TPI)	830,000

Local Taxes and Insutance (2.7% TPI)	1,495,000
Gross Operating Cost	36,319,000

A summary of the capital requirement is given in Table 6.11. Installation costs are assumed to be 40% of major equipment costs. The Total Plant
Investment (TPI) is estimated at $55,354,000 and Total Capital Requirements
including start-up, working capital, and interest during construction is
$67,531,000.

6.2.2 Operating Cost Estimate

Operating costs are estimated on an annual basis. The plant is assumed
to operate 365 days/yr, 24 hrs/day, with a service factor of 0.9. Base case
costs include peat a $3/ton, sodium carbonate at $100/ton, oxygen at
$25/ton, electricity at 5 cents/Kwh, and operating labor at $7/hr. The
annual operating costs are listed in Table 6.12. The gross annual operating
cost has been estimated at $36,319,000.

6.2.3 Base Case Unit Product Cost

The unit fuel cost is calculated using a discounted cash flow method.
The method used assumes an expected rate of return on investment and determ-
ines the unit fuel cost which will give a present worth of zero for that
rate of return.

In order to calculate the present worth (PW) of a process, the annual
cash flows (c) are multiplied by discount factors (d) with the sum of the
products equal to the present worth.

$$PW = \sum_{j=o}^{n} c_j d_j$$

The discount factor, d_j, is defined by

$$d_j = e^{-rj}$$

where r is the expected rate of return. In the base case analysis, the
annual cash flow is calculated as indicated in Table 6.13. The plant is
depreciated straight-line over ten years and taxable income is taxed at 46%.
The present worth calculation is carried out over a ten year period with an
expected rate of return of 15%. The scrap value (10% of TPI) is credited in
the final year of operation.

The annual production of olefins is indicated in Table 6.14A. If these
olefins are hydrated, the potential yield of alcohols is shown in Table
6.14B. The unit product cost is calculated based on total olefins produced

Table 6.13

ANNUAL CASH FLOW

1.	Gross Revenue	XXXX
2.	Operating Costs	- XXX
3.	Gross Profit (1 - 2)	XXXX
4.	Depreciation (TPI/Plant Life)	- XXX
5.	Taxable Income (3 - 4)	XXXX
6.	Federal Tax (Tax rate x 5)	- XXX
7.	After-Tax Income (5 - 6)	XXXX
8.	Annual Cash Flow (7 + 4)	XXXX

Table 6.14

PRODUCTS FROM PEAT PROCESS

A. Olefins

PRODUCT	TPD	lb/yr	Btu/lb	MM Btu/yr
Propylene	35.82	2.35×10^7	19,683	4.63×10^5
Butenes	18.45	1.21×10^7	19,431	2.35×10^5
Pentenes	29.31	1.93×10^7	19,308	3.73×10^5
TOTAL	83.58	5.49×10^7	----	10.71×10^5

B. Alcohols

PRODUCT	TPD	lb/yr	Btu/lb	MM Btu/yr	Gal/yr
Propanols	51.17	3.36×10^7	14,420	4.85×10^5	5.1×10^6
Butanols	24.38	1.60×10^7	15,530	2.48×10^5	2.4×10^6
Pentanols	42.99	2.82×10^7	16,350	4.61×10^5	4.2×10^6
TOTAL	118.54	7.78×10^7	----	11.94×10^5	11.7×10^6

without separation of the different olefin products or their isomers. The base case unit product cost is \$52.20/MM Btu of mixed olefins or \$1.02/lb of mixed olefins. The cost of alkali is 21.1% of total product cost and of raw materials is 54.8% of product cost. Capital cost is 35.1% of product cost.

6.3 Sensitivity Analysis

In this section, the sensitivity of unit product cost to changes in economic and process parameters is presented. Each figure shows sensitivity to changes in the parameter indicated with other parameters the same as those used in the base case. The product cost in each case is expressed as cost per million Btus (MM Btu) of mixed olefins (propylene, butenes, pentenes). This analysis is performed to provide insight into those areas which show the most potential for improving the product cost.

6.3.1 Sensitivity to Capital and Operating Costs

Capital costs for the process may vary due to inaccuracies in the estimates of changes in assumptions used in the base case. As seen in Figure 6.10, there is a linear relationship between capital investment and mixed olefin cost. Doubling the capital investment increases product cost by 45%.

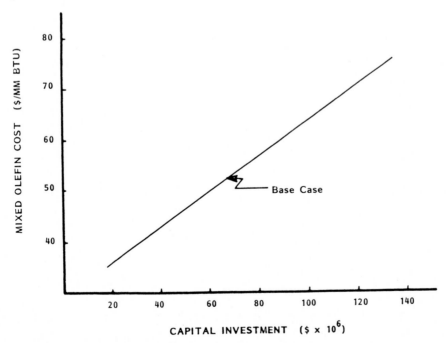

Fig. 6.10. Sensitivity of Unit Product Cost to Capital Investment

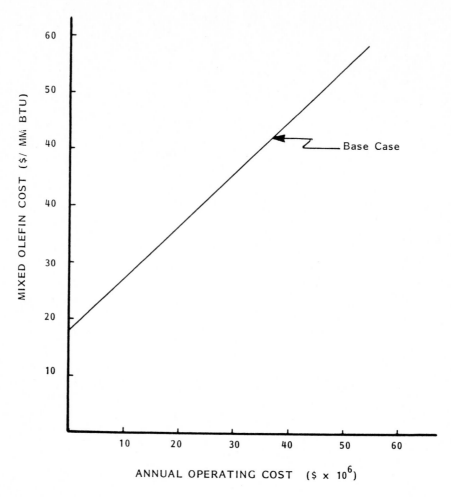

Fig. 6.11. Sensitivity of Unit Product Cost to Annual Operating Cost

The annual operating cost is primarily composed of raw material (alkali, oxygen, and electricity) costs. Figure 6.11 shows a linear relationship between product cost and annual operating cost. Since sodium carbonate is the single largest operating cost, the effect of its cost on product cost is indicated in Figure 6.12.

In the base case design, only 4.6% of the input peat volatile solids is recovered as mixed olefin product. This low weight yield is the result of a limited yield of fermentables from the raw peat feed (21.7%), loss of over 50% of the organic acids formed by fermentation in the purge stream, and loss of 47% of the weight of organic acids in the decarboxylation to form

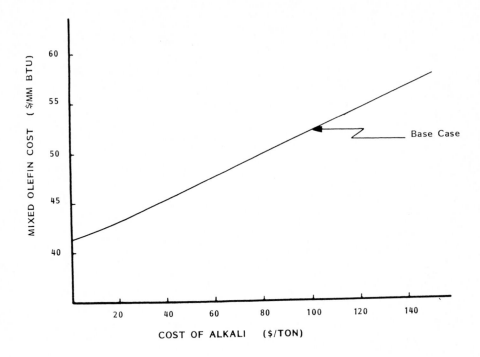

Fig. 6.12. Sensitivity of Unit Product Cost to Cost of Alkali

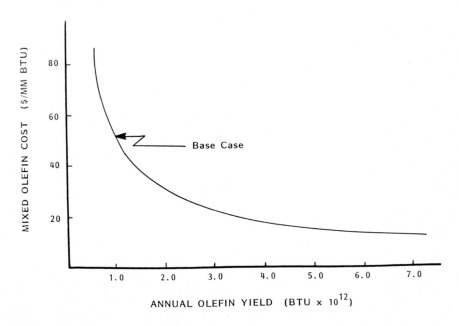

Fig. 6.13. Sensitivity of Unit Product Cost to Product Yield

olefins. Significant opportunity for improving the yield of product exists. This may be accomplished by reducing the loss of product in the purge stream or improving the yield of fermentable material produced in the pretreatment. the plot of annual product yield vs. product cost (Figure 6.13) shows that the base cost is on the steep section of the curve.

6.3.2 Sensitivity to Economic Parameters

The base case calculations assume 15% return on investment. Figure 6.14 shows product cost for returns on investment between 0 and 30%. Doubling the return on investment from 15 to 30% increases product cost by almost 30%.

The base case assumes a plant operating life of 10 years. As seen from Figure 6.15, this life-span falls on the flat part of the curve, indicating little change in product cost for a longer operating period.

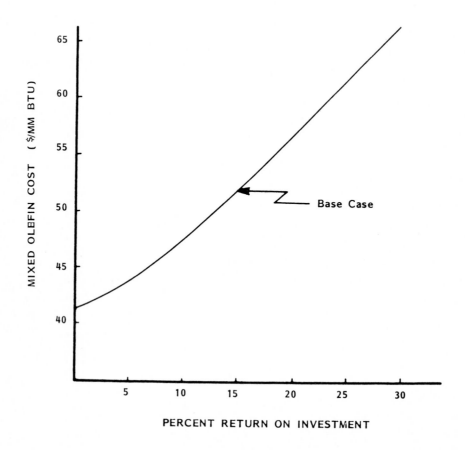

Fig. 6.14. Sensitivity of Unit Product Cost to Return on Investment

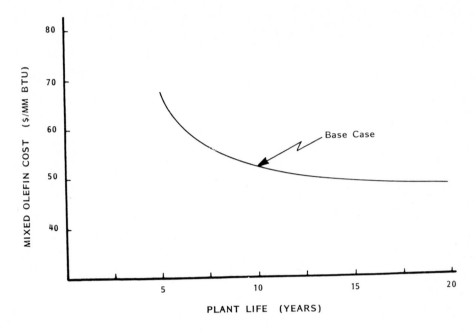

Fig. 6.15. Sensitivity of Unit Product Cost to Plant Life

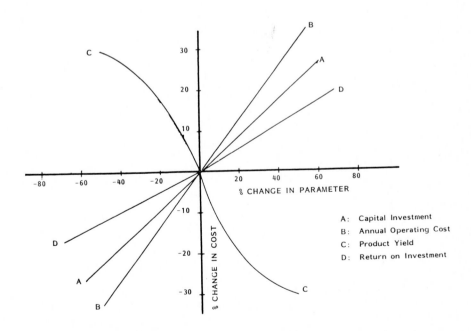

A: Capital Investment
B: Annual Operating Cost
C: Product Yield
D: Return on Investment

Fig. 6.16. Summary of Sensitivity to Changing Parameters

6.3.3 Sensitivity Summary

The effect of changes in the performance and economic parameters discussed above are summarized in Figure 6.16. This figure shows the relative change in product cost due to changes in capital investment, annual operating costs, product yield, and return on investment. The steepness of the slope for each item indicates the relative sensitivity of that parameter to product cost. It can be seen that the product cost can be most influenced by changes from the base case in product yield and annual operating costs.

7. PRELIMINARY PILOT PLANT DESIGN

This section presents a preliminary conceptual design for a pilot plant unit processing one-ton per day peat (dry basis). The same assumptions concerning operating conditions, residence times, and yields as were defined in Section 6 are used in this section. Cost estimates are also given for the major equipment required.

7.1 Peat Dewatering

Peat is assumed to be delivered to the pilot plant in a 3% total solids slurry (the peat is 10% ash). Excess water is removed in two steps. First the peat is screened to 8% solids and then pressed (Vari-Nip type press) to 15% solids. In concentrating one ton/day (TPD) of peat from 3% to 15% solids, 26.6 TPD water is returned to the peat bog and can be used to slurry more peat. Screening and pressing devices similar to those described in Section 6.1.1 can be used. The filtration system can remove 20 gpm/ft², resulting in less that 1 ft² filtration area required. A single press with less than 2% of the capacity of those described for the full-scale process is also required.

7.2 Peat Solubilization

The peat solubilization step is carried out at 212°F with a residence time of three hours in a CSTR reactor. Heat is supplied via an external heat exchanger from the oxidized peat stream. A recycle ratio to the heat exchanger of 7:1 is used. The volume of the solubilization reactor is 27.5 ft³.

7.3 Peat Oxidation

The oxidation step is carried out at 482°F and 692 PSIA. The oxidation step is carried out to combust 40% of the peat volatile solids over a 2.5 hour resident time. Daily oxygen requirement is 1,152 lbs. Preheating of the solubilized peat stream is achieved by heat exchange with the oxidized stream. The volume required for the oxidation vessel is 22.5 ft³.

7.4 Filtration of Oxidized Peat

The undissolved ash and volatile material is removed prior to fermentation by filtration. The filter cake formed is assumed to be five parts water per one part solids. Approximately 15% of the fermentable material is lost in this step. With a filtration rate of 25 gal/hr ft², approximately 2 ft² of filter area are required.

7.5 Heat Exchangers

Heat exchanges are required for the solubilization step, preheating prior to oxidation, and cooling prior to filtration. The heat required for solubilization and oxidation preheating is obtained from the oxidized peat stream. Additional cooling water is required to cool the oxidation gas stream and oxidized peat stream prior to filtration. Table 7.1 lists required heat transfer equipment and Figures 7.1 and 7.2 show the pretreatment material balances.

Table 7.1

HEAT TRANSFER REQUIREMENTS

STEP	q (Btu/day)	TEMPERATURE RANGE (°F)	U (Btu/hr·ft²·°F)	A(ft²)
Solubilization	1.90×10^6	70 – 212	200	4.5
Oxidation	1.21×10^6	212 – 302	250	1.1
Oxidation Gas Stream	2.76×10^6	482 – 300	65	5.0
Pre-filtration	2.03×10^6	253 – 99	300	5.0

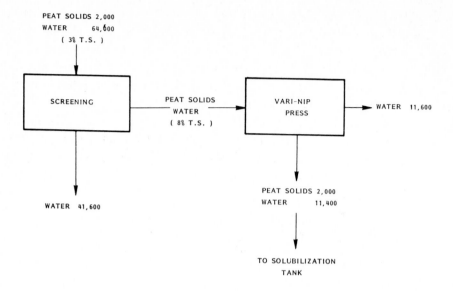

Fig. 7.1. Pretreatment Material Balance (lb/day - Filtration)

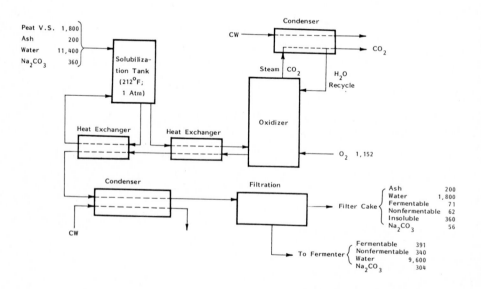

Fig. 7.2. Pretreatment Material Balance
(lb/day - Solubilization/Oxidation/Filtration)

7.6 Fermentation of Oxidized Peat

The pretreated peat is fed to the fermenter which is a packed bed column containing solid supports for microbial attachment. This vessel is designed as a CSTR. The production of organic acids from pretreated peat is assumed to have a first-order rate constant of 2.0 day^{-1} and 85% of the fermentables are converted to product. 25% of the extracted purge stream is recycled to the fermenters and the void volume in the fermenter is 50%. For a retention time of 2.83 days, the total volume of the fermenter is 1,500 ft^3. This could be achieved with two tanks, each 15 ft high and 8 ft in diameter. Figure 7.3 shows the material balance around the fermenters.

7.7 Liquid-Liquid Extraction of Organic Acids

The extraction system for removal of organic acids from the fermenter broth is designed as described in Section 6.1.6. The transfer of butyric, valeric, and caproic acids into hydrocarbon solvent is achieved with a series of three stages of mixer-settlers and a hydrocarbon to aqueous flow rate of 3.2/1.0. For a 3 min. residence time in the mixer, each stage must be 2 ft^3. The settlers, with an 18 min. residence time, must each be 12 ft^3.

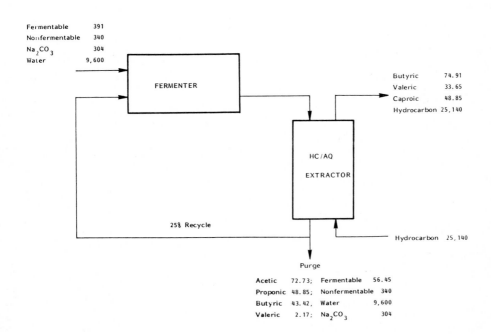

Fig. 7.3. Fermenter Balance

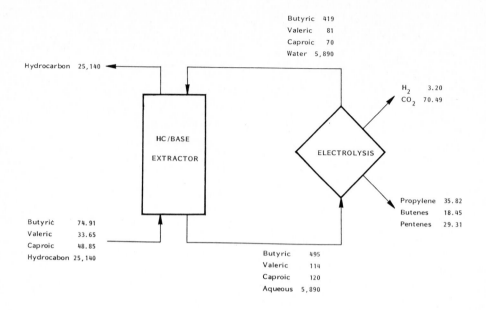

Fig. 7.4. Electrolysis Balance (lb/day)

In the hydrocarbon/aqueous base extractor, the hydrocarbon to aqueous base flow is 4.0/1.0 and a single-stage system is sufficient. The mixer must be 1.9 ft³ and settler 9.6 ft³.

7.8 Electrolysis of Organic Acids to Olefins

The electrolysis cell for the pilot scale process is designed to convert 157.41 lbs of organic acids to olefins. This is equivalent to 727.5 g moles/day of acids converted. The current required is 2 electrons per mole of acid reacted, or 1,630 amps. With a current density of 0.15 amp/cm², the required anode area is 1.085 m² (11.66 ft²). The required voltage for the cell is 2.8V. Hydrogen will be produced at the cathode at the rate of 3.2 lb/day. The material balance around the electrolysis cell is shown in Figure 7.4.

7.9 Pilot Process Yield

If the process is operated at pilot design capacity of one ton peat feed/day, 83.6 lb/day of mixed olefins (propylene, butenes, and pentenes) would be produced. Subsequent hydration would result in a maximum production of 119 lbs of mixed alcohols (propanols, butanols, pentanols). This is equivalent to about 20 gal/day of product, enough to conduct engine costs.

7.10 Cost Estimate of Major Equipment

The cost of the principal equipment items are estimated from costs obtained for full-scale equipment in Section 6. Where applicable, a 0.6 scaling factor is used. The total cost of $130,200 (Table 7.2) reflects only major items and does not include instrumentation, piping, electrical, storage, or installation costs.

Table 7.2

COST OF MAJOR ITEMS FOR PILOT SCALE FACILITY

ITEM	UNITS	SIZE	UNIT PRICE	TOTAL
Dewatering Screen	1	1 ft^2	$ 1,200	$ 1,200
Dewatering Press	1	---	50,000	50,000
Solubilization Tank	1	27.5 ft^3	2,200	2,200
Oxidation Vessel	1	22.5 ft^3	8,800	8,800
Filters	1	2 ft^2	2,000	2,000
Heat Exchangers	4	15.6 ft^2 (total)	---	1,500
O$_2$ Compressor	1	4 BHP	18,000	18,000
Fermenters	2	750 ft^3	20,000	40,000
HC/AQ Mixers	3	2.0 ft^3	500	1,500
HC/AQ Settlers	3	12.0 ft^3	700	2,100
HC/Base Mixer	1	1.9 ft^3	400	400
HC/Base Settler	1	9.6 ft^3	600	600
Electrolysis Cell	1	11.66 ft^2	1,200	1,200
DC Transformer/ Rectifier	1	5 KW	700	700
TOTAL				$130,200

8. ACKNOWLEDGEMENT

This project was carried out under U.S. Department of Energy Contract No. DE-AC02-81ER10914-2 from the Division of Advanced Energy Projects, Office of Basic Energy Sciences, on which D.L. Wise was the Principal Investigator. The author wishes to thank Dr. Riszard Gajewsky, Director, Division of Advanced Energy Projects, for his support, and those who worked on this project, namely, P.F. Levy, S.R. de Riel, E.P. Heneghan, L.K. Cheng, and J.E. Sanderson.

9.   REFERENCES

Aftring, R.P., and Taylor, B.F. (1981). Arch. Microbiol., 130, 101-104.
Allen, M.J. (1958). Organic Electrode Processes, Reinhold, New York.
Bache, R., and Pfennig, N. (1981). Arch. Microbiol., 130, 255-261.
Balch, W.E. and Wolfe, R.S. (1976). Appl. and Environ. Microbiol., 32, 781-791.
Banat, I.M., Lindstrom, E.B., Nedwell, D.B., and Balba, M.T. (1981). Appl. Environ. Microbiol., 42, 985-992.
Brown, and Walker (1891). Annalen, 261, 107.
Buivid, M.G., Wise, D.L., Rader, A.M., McCarty, P.L., and Owen, W.F. (1980). Resource Recovery and Conservation, 5, 117-138.
The Chemical Rubber Company (1969). CRC Handbook of Chemistry and Physics 50th Edition, Cleveland.
Clark, F.M., and Fina, L.R. (1952). Arch. Biochem. Biophys., 36, 26-32.
Colberg, P.J. and Young, L.Y. (1982). Can. J. Microbiol., 28, 886-889.
Conway, B.E., and Dzieciuch, M. (1963a). "New approaches to the study of electrochemical decarboxylation and the Kolbe reaction. Part I: The model reaction with Formate". Can. J. Chem., 41, 21.
Conway, B.E., and Dzieciuch, M. (1963b). "New approaches to the study of electrochemical decarboxylation and the Kolbe reaction. Part II: The model reaction with Trifluoroacetate and comparisons with aqueous solution behavior". Can. J. Chem., 41, 38.
Conway, B.E., and Dzieciuch, M. (1963c). "New approaches to the study of electrochemical decarboxylation and the Kolbe reaction. Part III: Quantitative analysis of the decay and discharge transients and the role of adsorbed intermediates". Can. J. Chem., 41, 55.
Considine, D.M. (1974). Chemical and Process Technology Encyclopedia, pp. 423-427. McGraw-Hill.
Corey, E.J., Bauld, N.L., LaLinde, R.T., Casanova, J., Jr., and Kaiser, E.T. (1957). J. Am. Chem. Soc., 79, 3182.
Decker, K., Jungermann, K., and Thauer, R.K. (1970). Angew. Chem. Internat. Edit., 9, 138-158.
Dickinson, T., and Wynne-Jones, W.F.K. (1962a). "Mechanism of Kolbe's electrosynthesis. Part I. Anode potential phenomena." Trans. Far. Soc., 58, 382.
Dickinson, T., and Wynne-Jones, W.F.K. (1962b). "Mechanism of Kolbe's electrosynthesis. Part II. Changing curve phenomena." Trans. Far. Soc., 58, 388.
Dickinson, T., and Wynne-Jones, W.F.K. (1962c). "Mechanism of Kolbe's electrosynthesis. Part III. Theoretical discussion." Trans. Far. Soc., 58, 400.
Dynatech Report No. 2115 (1981). Peat Biogasification Development Program, DOE Contract No. DE-AC 01-79ET14696, Cambridge, Massachusetts.
Eberson, L. (1973). "Carboxylic Acids" in Organic Electrochemistry, M. Baizer, ed., Marcel Decker.
Evans, W.C. (1977). Nature, 270, 17-22.
Ferry, J.G. and Wolfe, R.S. (1976). Arch. Microbiol., 107, 33-40.
Fichter. (1939). Trans. Amer. Electrochem. Soc., 75, 309.
Flynn, B.L., Jr. (1979). Chem. Eng. Prog., April, 66-69.
Gallo, T. and Sheppard, J.D. (1981). "Wet oxidation of peat" in Symposium Proceedings: Peat as an Energy Alternative II, Arlington, VA, pp. 463-489.
Glasstone, S. and Hickling, A. (1934). "Studies in electrolytic oxidation. VI. The anodic oxidation of acetates: The methanism

of the Kolbe and Hofer-Moest reaction in aqueous solution." J. Chem. Soc., 1878.

Healy, J.B., Jr., and Young, L.Y. (1979). Appl. and Env. Microbiol., 38, 84-89.

Healy, J.B., Jr., Young, L.Y., and Reinhard, M. (1980). Appl. and Env. Microbiol., 39, 436-444.

Hofer, H. and Moest, M. (1902). Ann. Chem., 323, 284.

Keenan, J.D. (1979). Process Biochemistry (May) 9-16.

Keith, C.L., Bridges, R.L., Fina, L.R., Iverson, K.L., and Cloran, J.A. (1978). Arch. Microbiol., 118, 173-176.

Kirk-Othmer (1980). Encyclopedia of Chemical Technology. 3rd Edition, John Wiley, 9, 342-377.

Koehl, W.J., Jr. (1964). "Anodic oxidation of aliphatic acids of carbon anodes." J. Am. Chem. Soc., 86, 4686.

Kolbe, H. (1849). Ann. Chem., 69, 257.

Levy, P.F., Sanderson, J.E., Ashare, E., de Riel, S.R. and Wise, D.L. (1981a). Liquid Fuels Production from Biomass. Dynatech Report No. 2147. DOE Contract No. XB-0-9291-1, Cambridge, MA.

Levy, P.F., Sanderson, J.E., Wise, D.L. (1981b). Development of a Process for Production of Liquid Fuels from Biomass, in Biotech. Bioeng. Symp. No. 11, pp. 239-248, C.D. Scott, Ed.

Levy, P.F., Sanderson, J.E., Kispert, R.G., and Wise, D.L. (1981c). Enzyme Microb. Technol., 3, 207-215.

Levy, P.F., Sanderson, J.E., de Riel, S.R. and Wise, D.L. (1982). Development of a Biochemical Process for Production of Alcohol Fuel from Peat, Dynatech Report No. 2201, Contract No. DE-ACO2-81ER19014-1, Cambridge, MA.

Ljungdahl, L.G. and Wood, H.G. (1969). Ann. Rev. of Microbiol., 23, 515-535.

McCarty, P.L., Young, L., Owen, W., Stuckey, D., and Colberg, P.J. (1979). Heat treatment of biomass for increasing biodegradability, presented at 3rd Annual Biomass Energy Systems Conference, pp. 411-418.

McCarty, P.L., Baugh, K., Bachmann, A., Owen, W. and Everhart, T. (1983). Autohydrolysis for increasing methane yields from lignocellulosic materials, in Fuels and Organic Chemicals from Biomass, D.L. Wise, Ed., CRC Press, in press.

Mensinger, M.C. (1980). Wet Carbonization of Peat: State-of-the-Art Review, IGT Symposium, Arlington, VA. Dec. 1-3.

Oki, T., Ishikawa, H., and Okubo, K. (1978). Mokuzai Gakkaishi, 24, 40-414.

Othmer, D.F. (1978). Combustion, 50, 44-47.

Palmer, J.M., and Evans, C.S. (1983). Phil. Trans. R. Soc. Lond., B300, 293-303.

Perry, R.H. and Chilton, C.H. (1973). Chemical Engineer's Handbook, 5th ed., McGraw-Hill, New York.

Petersen, J. (1900). Elektrolyse der Alkalisalze der organischen sauren. I, II, III. Zeitschr. f. Physik. Chemie, 33, 99, 295, 698.

Petersen, J. (1906). Ziltschr. f. Elektrochem., 12, 22.

Rader, A.M. (1977). Testimony before U.S. House of Representatives Sub-Committee on Environment, Energy, and Natural Resources, Sept. 29.

Randall, T.L., and Knopp, P.V. (1980). J. Water Poll. Cont., 52, 2117-2130.

Ruoff, C.F., Ashare, E., Sanderson, J.E., and Wise, D.L. (1980). Continued development of a peat biogasification process. IGT Symposium, Arlington, VA, Dec. 1-3.

Sanderson, J.E., Garcia-Martinez, D.V., Dillon, J.J., George, G.S. and Wise, D.L. (1979a). Liquid fuel production from biomass, presented

206

at International Solar Energy Society Meeting, Atlanta, GA.

Sanderson, J.E., Garcia-Martinez, D.V., Dillon, J.J., George, G.S. and Wise, D.L. (1979b). Liquid fuel production from biomass, presented at Biomass Energy Systems Conference, Golden, CO, June 5-7.

Sanderson, J.E., Garcia-Martinez, D.V., George, G.S., Dillon, J.J., Molyneaux, M.S., Bernard, G.W., and Wise, D.L. (1979c). Liquid fuel production from biomass, Dynatech Report No. 1931, Contract No. EG-77-C-02-4388-8, Cambridge, MA.

Schall (1896). Z. Electrochem., 3, 83.

Schonheit, P., Kristjansson, J.K., and Thauer, R.K. (1982). Arch. Microbiol., 132, 285-288.

Soper, E.K., and Osborn, C.C. (1922). The occurrence and uses of peat in the United States. U.S. Geological Survey, Bulletin 728, Washington, D.C.

Taylor, B.F., Campbell, W.L. and Chinoy, I. (1970). J. Bacteriol., 102, 430-437.

Teletzke, G.H. (1964). Chem. Eng. Prog., 60, 33-38.

Thauer, R.K., Jungermann, K., and Decker, K. (1977). Bacteriological Reviews, 41, 100-180.

Tibbetts, T.E. (1968). Peat resources of the world: A review, Proc. 3rd Inter. Peat Cong., Quebec, Canada

Utley, J.H.P. (1974). Anodic reactions of carboxylates, in Techniques of Chemistry, Vol. 5., Part 1. Technique of Electroorganic Synthesis. Weinberg, N.L., ed., Wiley, New York. 793.

Waksman, S.A. (1930). Am. J. Sci., 19, 32-54.

Walling, C. (1957). Free Radicals in Solution, New York, Wiley, p. 581.

Weedon, B.C.L. (1952). Quart. Rev., London, 6, 380.

Williams, R.J., and Evans, W.C. (1975). Biochem. J., 148, 1-10.

Winfrey, M.R., and Zeikus, J.G. (1977). Appl. Environ. Microbiol., 33, 275-281.

Wise, D.L., Ashare, E., Ruoff, C.F., and Kopstein, M.J. (1983). Resources and Conservation, 8, 213-231.

Zeikus, J.G., Wellstein, A.L., and Kirk, T.K. (1982). FEMS Microbiol. Let., 5, 193-197.

Zimmerman, F.J., and Diddams, D.G. (1960). Tappi., 43, 710-715.

APPENDIX A

DETERMINATION OF OPTIMUM pH FOR VOLATILE SOLIDS MEASUREMENT

Syringic acid solutions (2.97% by weight) containing 5% sodium carbonate were adjusted to various pH's (1.4 - 11.5) and the apparent volatile solids (VS) content determined by first drying in a 100°C oven to evaporate the water and then ashing in a 600°C furnace.

Theoretical VS measurement at acidic, neutral, and basic pH's differ.

At acid pH

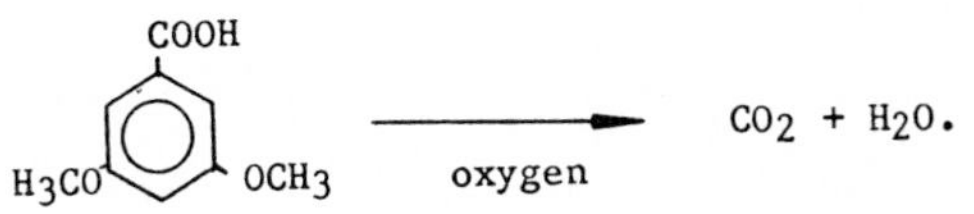

$$CO_2 + H_2O.$$

At neutral pH

$$NaOH + CO_2 + H_2O.$$

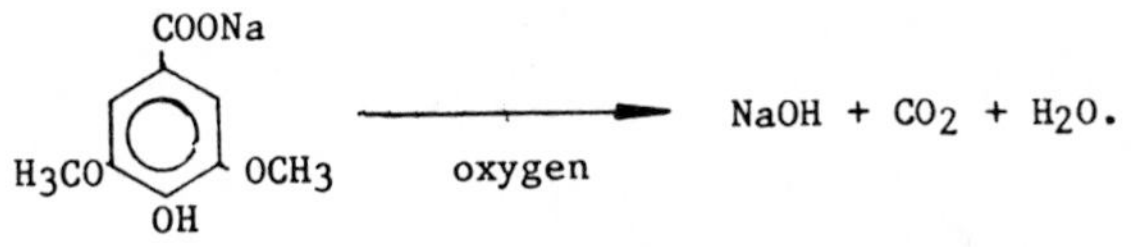

At alkaline pH, the syringic acid may be decarboxylated and

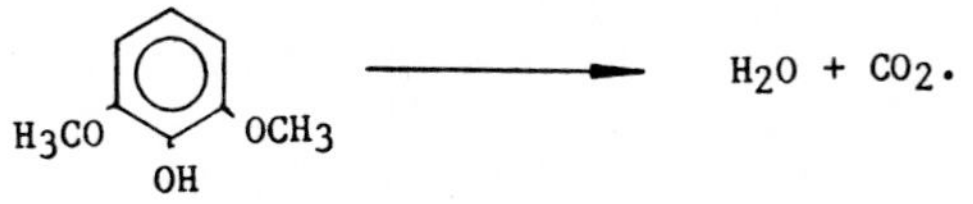

$$H_2O + CO_2.$$

Theoretical VS yields for a 2.97% by weight syringic acid solution, therefore, are:

        a = low pH   : 2.97%
        b = neutral pH: 2.67%
        c = high pH   : 2.31%

Experimentally measured values are as follows:

pH	APPARENT %VS
2.12	3.02
1.43	3.01
4.96	2.86
4.88	2.73
7.05	2.53
6.91	2.52
9.00	2.43
9.00	2.41
11.19	2.33
11.47	2.40

These data are plotted in Figure A.1. In subsequent experiments with peat, all samples were acidified with HCl before measurement of total and volatile solids.

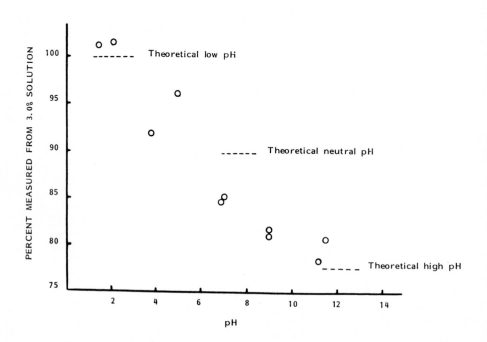

Fig. A.1. Apparent Syringic Acid Recovery as a Function of pH

APPENDIX B

METHANE-SUPPRESSED ANAEROBIC FERMENTATION OF GLUCOSE IN A PACKED BED COLUMN
DIGESTER

A packed bed column digester, as described in Section 4 of this chapter, was inoculated with sewage digester contents and slowly brought into continuous operation over a period of about 6 months. Microbial attachment to the stones used as solid supports was visually apparent and the digester was completely converting a 20 g/l glucose feed solution to methane and carbon dioxide at a 1.8 day retention time. The digester was operated in this manner for one week in which methane production was consistent before suppression of the methanogenic bacteria.

Methane formation was suppressed by addition of 5 x 10^{-4} M 2-bromo-ethane sulfonic acid to the feed solution. The digester was operated continuously on this feed for 2 months. No methane was formed in that period of time. The daily organic acid concentrations in the recirculating

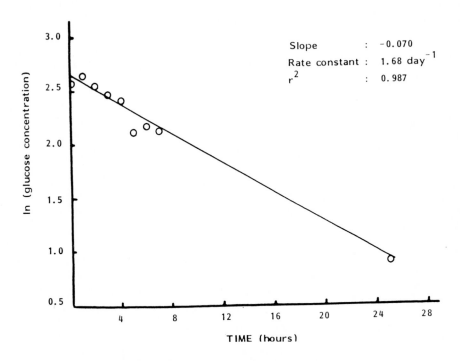

Fig. B.1. Rate Plot of Glucose Degradation in Anaerobic Packed Bed
(Suppressed Methane)

Table B.1

DAILY ORGANIC ACID LEVELS IN METHANE-SUPPRESSED PACKED BED COLUMN DIGESTER

Feed Solution: 20 g/ℓ glucose + 5 x 10^{-4} M BES
Retention time: 1.8 days

DAY	CONCENTRATION IN MEQ/ ℓ					R*	%C*
	ACETIC	PROPIONIC	BUTYRIC	VALERIC	CAPROIC		
4/19	8.7	0.8	7.0	1.3	20.9	117.5	35.3
4/20	11.7	1.1	7.4	1.1	29.1	152.1	45.6
4/21	12.6	1.1	7.4	1.1	29.1	153.0	45.9
4/22	12.1	0.8	10.9	1.5	34.0	181.6	54.5
4/23	16.7	0.0	14.0	0	51.3	256.8	77.0
4/24	16.7	t	9.6	0.8	52.8	254.5	76.4
4/25	15.8	0.8	8.0	1.0	43.3	213.5	64.1
4/26	16.2	0.8	9.7	0.8	43.3	217.7	65.3
4/27	15.8	0.5	10.7	0.7	46.1	230.2	69.1
4/28	15.5	t	10.7	0.6	44.2	221.2	66.4
4/29	14.7	t	11.7	0.6	46.5	231.9	69.6
4/30	18.3	0.5	11.6	0.8	41.6	217.4	65.2
5/1	16.3	0.5	12.8	1.0	42.5	222.4	66.7
5/2	17.9	0.8	16.2	1.0	47.6	253.6	76.0
5/3	17.5	0.8	18.2	1.4	48.7	263.8	79.1
5/4	18.8	0.8	14.2	0.8	38.2	211.0	63.3
5/5	13.7	t	13.3	0	35.1	187.3	56.2
5/6	17.6	t	20.9	0	32.2	198.7	59.6
5/7	11.3	1.3	23.4	2.1	35.6	221.4	66.4
5/8	----------------		N O D A T A; N O T F E D			----------------	
5/9	17.7	1.8	27.9	2.5	37.9	250.3	75.1
5/10	9.3	0.8	16.1	1.8	28.8	171.8	51.5
5/11	8.7	0.0	12.2	1.1	25.8	145.9	43.8
5/12	5.7	0.0	16.1	0.7	27.5	158.1	47.4
5/13	7.7	0.0	29.0	0.5	30.7	204.7	61.4
5/14	8.7	0.0	33.3	0.5	34.3	230.9	69.3

* R represents "acetic acid equivalents per liter" and is calculated from measured concentrations (meq/ ℓ) of acetic (A), propionic (P), butyric (B), valeric (V), and caproic (C) acids:

$$R = A + 1.75P + 2.5B + 3.25V + 4.0C$$

Percent conversion, %C, is calculated from R:

$$\%C = \frac{R \times .06}{S} \times 100$$

Where S is substrate concentration in g/ ℓ.

broth are shown in Table B.1. The steady-state concentration of organic acid was measured on 19 successive days immediately before fresh substrate was added each morning. These acid concentrations are shown in Table B.2. The yield of caproic acid is surprisingly high at over 20% of the weight of glucose fed to the digester. Note that the maximum caproic acid yield, dictated by the stoichiometry of conversion from glucose, is 48%.

Table B.2

SUPPRESSED METHANE FERMENTATION OF GLUCOSE IN PBC REACTOR

Retention Time: 1.8 days
Feed Concentration: 20 g/ℓ
Effluent pH: 5

Average Organic Acid Concentrations:

	MEQ/ℓ	NUMBER OF POINTS
Acetic	14.1 $\pm$ 4.1	19
Propionic	0.5 $\pm$ 0.5	19
Butyric	16.6 $\pm$ 7.1	19
Valeric	0.9 $\pm$ 0.6	19
Caproic	38.4 $\pm$ 7.2	19

Wt yield of 22.3% Caproic Acid
            7.3% Butyric Acid

Maximum Yield Caproic Acid:

$$C_6H_{12}O_6 \rightarrow 0.75 \ CH_3(CH_2)_4COOH + 1.5 \ H_2O + 1.5 \ CO_2$$

$$\frac{116 \times .75}{180} \times 100 = 48.3\%$$

The rate of glucose utilization and organic acid production were estimated by taking successive samples between daily feedings. The glucose concentration and organic acid concentrations were measured in each sample. This data is shown in Table B.3. The ln glucose concentration is plotted vs. time in Figure B.1. to calculate a first-order rate constant of 1.68 day^{-1}.

Hydrogen is also produced in this fermentation. Daily volumetric production (in l at STP) is shown in Table B.4.

Table B.3

GLUCOSE AND ORGANIC ACID CONCENTRATIONS BETWEEN FEEDINGS

TIME (HR)	GLUCOSE CONCENTRATION (g/ l)	ACETIC	PROPIONIC	BUTYRIC	VALERIC	CAPROIC	% CONV
Before Feed	4.36	13.7	t	13.3	0	35.1	56.2
0	13.20	6.1	t	5.9	0	15.6	---
1	14.1	7.6	t	8.1	0	21.0	33.5
2	13.0	8.1	t	8.7	0	24.3	38.0
3	11.9	6.6	t	9.2	0	27.2	41.6
4	11.3	7.1	t	9.8	0	25.5	40.1
5	9.4	7.6	t	9.2	0	26.0	40.4
6	10.2	6.6	t	10.4	0	29.3	45.0
7	9.5	8.6	t	11.5	0	29.7	46.9
25	2.5	17.6	t	20.9	0	32.2	59.6

Table B.4

DAILY HYDROGEN PRODUCTION FROM SUPPRESSED-METHANE FERMENTATION

DAY	ℓ (STP) H_2 PRODUCED
4/19	0.29
4/20	0.66
4/21	0.97
4/22	0.81
4/23	0.89
4/24	0.69
4/25	0.64
4/26	0.80
4/27	0.80
4/28	0.70
4/29	0.71
4/30	0.61
5/1	0.54
5/2	0.55
5/3	0.56
5/4	0.37
5/5	0.50
5/6	0.41
5/7	0.37
5/8	NO DATA
5/9	0.68 (2 days)
5/10	0.68
5/11	0.34
5/12	0.47
5/13	1.00
5/14	0.69
5/15	0.57
5/16	0.37
5/17	0.47
5/18	0.40
5/19	0.41

Energy Recovery from Lignin, Peat and Lower Rank Coals, edited by D.J. Trantolo and D.L. Wise
Elsevier Science Publishers B.V., Amsterdam 1989 — Printed in The Netherlands   215

Chapter 5

ALTERNATE PROCESSES FOR ALTERNATE FUELS

Burl E. Davis
Gulf Research & Development Company
P.O. Drawer 2038
Pittsburgh, Pennsylvania, 15230, USA

During the last fifteen years, we have seen a monumental effort in the United States directed towards developing synthetic fuel processes. Initially the movement developed as a result of the formation of OPEC and curtailment of imported oil. At that time, there were processes under development for conversion of coal and oil shale into products that would replace those from petroleum. These processes utilized high temperatures and hydrogenation to convert solid fossil fuel resources into products to meet markets met by petroleum. Earlier versions of the processes had been operated on a large scale to meet shortfalls of petroleum due to war or political issues. Their process chemistry was relatively well understood and although there were technical problems related to large scale operations, scale up into the world-scale plants appeared to be feasible. However, as the price of crude oil increased, the projected costs of products by the alternative processes also increased significantly. The current estimates of synthetic fuels generally are in the range of $60 to $100 per barrel Crude Oil Equivalent (COE), or higher. The optimism of the mid '70s for synthetic fuels from coal or oil shale has been replaced with projections that such processes will not be competitive until the 21st century.

Fortunately, conservation has proved quite effective in delaying the time when alternative sources of transportation fuels are needed. The question is "What are we going to do with the time gained by conservation?"

As stated earlier, most of the present synthetic fuel processes have attempted to take existing technology and extend it to include new feedstocks such as coal, oil shale, biomass, etc. This is largely due to the fuel technology community in general and their background in coal and petroleum processing. Most engineers feel more comfortable in taking proven technology and modifying it than in striking out in "uncharted regions". This is in keeping with the observation that "Engineers did not invent the wheel; they made it rounder.". The incremental improvements normally sought by engineers will not be sufficient to make synthetic fuel processes economical. Rather, improvements of a factor of two or more are needed to make synthetic fuel processes competitive in the next twenty years. Processes that deviate significantly from existing technology may be required to produce competitive fuels.

Most of the better known synthetic fuel processes have several things in common. They include:

- Extreme pressures and temperatures. This increases both operating time and increases capital costs.
- High cost feedstocks. This can be due to cost of acquisition and handling solid materials.
- Existing markets. The final product is to move into existing markets and its infrastructure.
- Economies of Scale. In order to be competitive, the processes will require world-scale plants with costs of the order of a billion dollars or more to achieve economy of scale.

Each of these factors contributes to the unfavorable economics of synthetic fuel processes. In order to reduce costs significantly, developers of new processes will need to deal with these factors. There are numerous alternatives for processes that can deal with these problems.

1. Process Conditions: Use of high temperatures and pressures create both higher operating costs and increased capital. Utilization of alternative sources of process energy can result in less rigorous process conditions and reduce cost of hardware to contain the process. Chemically catalyzed oxidative processes are available at near ambient temperatures and pressures. Biologic processes, although relatively slow, use similar conditions. Selective conversion processes, rather than the large hammer approach engendered in most cost liquefaction processes could well prove to have significant advantages.

2. Feedstock Costs: In-place energy reserves generally require significant cost to extract and transport. The solid fuels are

mined by the ton and sold by the Btu. The more dilute the energy resource, (i.e. oil shale, lignites, biomass), the higher the cost to the process. Processes that combine extraction with conversion such as underground coal gasification can significantly reduce feedstock costs.

3. Existing Market/Product Compatability: Most of the existing markets have a built-in bias for products from a single feedstock source. The existing infrastructure not only dictates products that are compatible with the conventional product, but the end-use equipment is also designed to deal with process materials with a relatively narrow range of physical properties. Additional processing to meet these requirements creates significant cost increases. The problems encountered with the introduction of gasohol to the U.S. market are examples of incompatability. New processes should focus on dedicated customers willing to modify equipment for the new fuels.

4. Economy of Scale: The large scaling factors inherent in most synthetic fuel processes create large demands for capital that are difficult to obtain for high-risk businesses. This is largely due to the nature of the mining and conversion processes requiring large reactors. Processes with smaller scaling factors will reduce the need for billion dollar scale plants.

The prospects for commercial processes to make conventional fuels by unconventional processes are not good. The petroleum based transportation fuels are firmly entrenched because of the existing infrastructure. However, we do not have to have the processes tomorrow, rather some time in the next ten to twenty years. This gives us time to look at alternative processes and bring them to a level of maturity such that they can be commercialized in that time frame.

The committment to many of these alternative processes also carries with it an acceptance of the downstream problems related to developing a market for the products. We must be willing to follow many different paths, most of them leading to failure, to identify lower cost routes to fuels capable of replacing petroleum products. The workshop held by Dynatech at Cambridge in August of 1984 dealt with many possible routes and feedstocks. These new routes attempt to avoid the problems with conventional synthetic fuel processes in that they utilize milder conditions that can certainly translate into lower costs. Many of them are quite unconventional. Most of them may not be "practical". However, they do show promise and deserve serious consideration.

Energy Recovery from Lignin, Peat and Lower Rank Coals, edited by D.J. Trantolo and D.L. Wise
Elsevier Science Publishers B.V., Amsterdam 1989 — Printed in The Netherlands

Chapter 6

ENHANCED COAL LIQUEFACTION USING LIGNIN

Robert W. Coughlin and Paul Altieri
Department of Chemical Engineering,
University of Connecticut,
Storrs, CT 06268, USA

1. INTRODUCTION

Conversion of the cellulose in biomass to alcohol has received much attention as an alternative energy source for liquid fuels. When the cellulose of vascular plants is converted to sugars, a material known as lignin is left behind as an insoluble residue that could assume very great importance if alcohol from biomass becomes a widely used fuel. The cost of alcohol fuel produced from cellulose cannot compete economically with fuels produced from petroleum under present day economic conditions. To approach economic competitiveness, the cellulose must have a far lower raw material cost. One way to lower the raw material cost of cellulose is to increase the raw material value of lignin. The lignin by-product from paper-making processes is burned as process fuel but lignin has also shown the potential for use in producing liquid fuel, manufacturing activated carbon as well as other specialty products. Thus lignin does have the potential as a raw material of higher value than its value as a process fuel.

Lignin is a complex 3-dimensional polymer composed largely of phenyl-propane sub-units linked by ether bonds. The aromatic rings are also substituted with methoxy and hydroxy groups. Structurally, lignin depends on the plant species from whence it comes, as well as the pretreatment process used in its isolation and purification. Aromatic and alkyl aromatic moieties abundant in the lignin molecule call to mind the following chemical products and intermediates: aromatic chemicals such as phenol, styrene, benzene, toluene, and xylene. This suggests that lignin might be utilized by depolymerizing it to aromatic chemicals.

One of the possible uses for lignin is to convert it to a liquid aromatic fuel. To make such a fuel, the lignin molecules must be dismantled. In the past, lignin has been often used as a solid fuel since it possesses a relatively high energy content. If we take an energy value of $3.50/million Btu, then using a heat of combustion of 11,350 Btu/lb, lignin has an energy value of $0.04/lb. Ultimately, the economic value placed on lignin as a feedstock in liquid fuel production would naturally depend on the properties of the fuels produced: i.e. octane numbers, stability of the liquid fuel etc.

Currently three major chemical products are produced commercially from lignin. They are vanillin, dimethyl sulfide, and methylmercaptan. Assuming a large scale industry was developed for producing alcohol from biomass, vast quantities of lignin would become available for conversion into aromatic fuels and other chemicals. Such utilization of lignin would then increase the economic value of biomass as a raw material.

Our interest in lignin centers on its ability to depolymerize, thus allowing formation of aromatic liquid products. We have observed that adding lignin to a reaction mixture of coal plus hydrogen-donor solvent promotes more extensive depolymerization of the coal at lower temperatures. Using lignin, for example, about 43% of the organic portion of coal was liquefied at 300°C under about 1400 psi of hydrogen, compared to 5% liquefied at similar conditions but without lignin. Thermal depolymerization of the lignin at relatively low temperature is believed to form resonance-stabilized phenoxy radicals which then attack the coal causing scission of aliphatic carbon-carbon bonds in the coal. This hypothesis is also in accord with the observation that using guaiacol as solvent further increases coal conversion (up to about 80%) under similar conditions.

Aside from pyrolysis and indirect processes (via synthesis gas), the liquefaction of coal involves thermal rupture of its molecules into free-radical fragments which are then stabilized by capturing hydrogen atoms usually from a donating solvent or hydrogen gas. Besides transferring hydrogen, the donor-solvent often promotes the thermal bond rupture and thereby permits liquefaction of some coals to occur at temperatures as low as 400°C. Tetralin is the most usual model compound used as a hydrogen donor solvent in laboratory experiments or in theoretical consideration.

In practical processes, however, the donor solvents would ordinarily include a wide spectrum of tetralin-like molecules present in a liquid fraction separated from the reaction product and recycled for use in the liquefaction reaction. The conditions required for such liquefaction reactions usually exceed 450°C and several thousand psi of hydrogen pressure. Some examples include: the liquefaction of high-rank Pennsylvania coals in a recycle solvent at 426°C and 3500 psig by Shah and Cronauer (ref. 1) and

Neavel's (ref. 2) conversion of an Illinois high-volatile bituminous coal in tetralin at 400°C and 4000 psi. Broader discussions of these reaction conditions and processes can be found in papers by Shah and Cronauer (ref. 1), Wen and Tone (ref. 3) and Tsai and Weller (ref. 4).

In 1962 Heredy and Neuworth (ref. 5) reported that bituminous coal could be depolymerized by treating it with BF_3 and phenol at 100°C and atmospheric pressure. This reaction caused more than 60% of the coal to dissolve in the phenol and was interpreted as an exchange of phenol for the aromatic structures of coal upon thermal cleavage of aromatic-aliphatic linkages within the coal molecules. Ouchi et al. (ref. 6) reported similar reactions using p-toluenesulphonic acid as a catalyst. Darlage and Bailey (ref. 7) investigated a variety of other solvents and Friedel-Crafts catalysts for this reaction and found that polymerization is the predominant reaction when non-phenolic solvents are employed, e.g. toluene, xylene or tetralin. Larsen et al. (ref. 8) reported that heating coals in phenol to 425°C within bombs caused a depolymerization accompanied by extensive rearrangement of hydrogen to produce a hydrogen-rich soluble fraction and a hydrogen-poor insoluble residue. Merriam and Hamrin (ref. 9) obtained similar dissolution of coal in both BF_3-phenol and phenol-benzenesulphonic acid but interpreted the results more as a solvation than due to a catalytic effect of either BF_3 or the sulphonic acid.

In this chapter we report low temperature depolymerization of mixtures of coal plus lignin, a widespread "renewable" raw material readily available as a major component of all vascular plants. When coal and lignin are reacted together in a solvent, they depolymerize under mild conditions to produce a filterable liquid product with yields greater than would be predicted based on a simple side-by-side, concurrent liquefaction of both components independently.

2. EXPERIMENTAL

2.1 Materials

Coal used in the experiments was Illinois No. 6 (74-105) bituminous coal obtained from the Pittsburgh Energy Technology Center. It was dried in the same way as the lignin. This coal contains 13.6% ash[10] which was accounted for in computing results on a moisture and ash-free basis (maf).

Lignin: A caustic type obtained from steam-exploded aspen, was purchased from Iotech Corp., Ottawa, Canada, and dried at 85°C for 25 hours under low pressure (10 torr); after drying, the lignin was kept in a desiccator over $CaSO_4$. Some experiments were also done with a sulfite lignin (trade name Indulin) furnished by the Westvaco Corporation.

Catalysts: The different types of catalyst included Lewis acids (BF_3 and $AlCl_3$), $NiO-MoO_3$ (on alumina) hydrocracking catalyst and a $SiO_2-Al_2O_3$ cracking catalyst. Cracking and hydrocracking catalysts were dried and activated at 300°C for 12 hours in a vacuum oven (10 mm Hg).

Reagents used were as follows: Tetralin (practical grade) and acetone (Baker analyzed reagent) from F.T. Baker Co.; $AlCl_3$ from Mallinckrodt, Inc.; BF_3 gas from Matheson; P-dimethoxybenzene and guaicol from Pfalt and Bauer Chemical Co.; and Phenol (99%) from Aldrich Chemical Company. Nickel molybdate on alumina, 3% NiO, 15% MoO_3 pellets approximately 3 mm and silica-alumina (53% and 45% respectively) were purchased from Alfa Products Co.

2.2 Methods

A. High pressure hydrocracking using donor solvent

Tetralin, lignin and coal were reacted in a 1000 ml stainless-steel, magnetically stirred autoclave (manufactured by Autoclave Eng., Ind.). The apparatus was pressurized with H_2 to 1000 psi and then heated to 300°C (572°F). The temperature usually attained 300°C after 30 minutes, during which time the pressure rose from 1000 psi to 1400-1800 psi. Pressure then usually remained constant during the reaction, unless hydrogen was consumed by the process. The temperature was controlled at 300±5°C and the reaction mixture was stirred at 1500 r.p.m. After 3 hours at reaction temperature the mixture was cooled and filtered to separate insoluble residues which were then dried and weighed. The filtrate was further separated into various fractions by vacuum distillation at 10 mm Hg and each fraction was later analyzed by gas chromatography/mass spectrometry (GC/MS). The volatile products distilled in the range 90°C - 135°C under 10 mm Hg and were collected only after the tetralin had distilled off first at about 90°C. In each case about 95±2% of the tetralin was recovered.

Qualitative analysis of hydrocracking products: The distilled fractions were dissolved in acetone and analyzed by gas chromatography/mass spectrometry (GC/MS) using a Hewlett Packard Model 5985 equipped with a flame ionization detector and GC Column DC-200 (Dow Corning silicone oil 200, 12,500 CS methyl) suitable for separation of aldehydes, ketones, esters, fatty acids, hydrocarbons, halogenated and nitrogen compounds.

B. Atmospheric pressure liquefaction of lignin and coal

Measured amounts of pre-dried lignin and coal were reacted in a three-necked flask equipped with a reflux condenser, a thermometer and a rotatable tube for introducing the catalyst. Tetralin and/or phenol were included in the mixtures which were stirred. Before reaction the apparatus was purged with N_2 to remove moisture and oxygen which could degrade the catalyst. When BF_3 was applied as catalyst the gas was conducted into the apparatus by a glass tube during the reaction. Reactions were conducted under reflux for periods ranging from 20 to 44 hours.

Definitions of Yields

Liquefaction data are used to compute liquid yields as follows:

$$Y_F = 100 \, M_F/(M_L + M_C) \tag{1}$$

where:

Y_F = overall yield of filterable liquid produced from a mixture of lignin and coal

M_F = mass of filtrate (filterable liquid)

M_L = mass of lignin reacted in mixture

M_C = mass of coal reacted in mixture, daf basis

$$Y_D = 100 \, M_D/(M_L + M_C) \tag{2}$$

where:

Y_D = overall yield of liquid distilled from reaction products at 10 torr and 150°C

M_D = mass of distillate liquid

$$X_F = \frac{M_F - (Y_F^L/100)M_L}{M_C} \cdot 100 \tag{3}$$

where:

X_F = yield of filterable liquid attributable to the coal in an experiment starting with a mixture of coal and lignin

Y_F^L = overall yield of filterable liquid produced starting with lignin alone

$$X_D = \frac{M_D - (Y_D^L/100)M_L}{M_C} \cdot 100 \tag{4}$$

where:

X_D = yield of liquid attributable to the coal that can be distilled from the reaction products of a mixture of coal and lignin

Y_D^L = overall yield of distillable liquid produced starting from lignin alone

In the following we shall use the special instances of yield given below:

(1) Y_F^{L+C}, Y_D^{L+C}, X_F^{L+C}, X_D^{L+C} – for experiments starting with equal masses of lignin and coal

(2) Y_F^L, Y_D^L – for experiments using only lignin

(3) $Y_F^C = X_F^C$ and $Y_D^C = X_D^C$ – for experiments using only coal.

3. RESULTS

Table 1 summarizes the various experiments in which, for the most part, catalysts were used and conversions were measured mainly by simple filtration.   Several preliminary experiments (1-E through 1-I) were conducted with lignin but with no coal in the reaction mixture using tetralin/lignin ratios ranging from 4 to 8 ml/g. In these experiments the range of conversions of lignins to filterable liquid was 30 – 38% with a mean of 34%. Based on these experimental results for lignin alone, it was then assumed that the conversion of lignin remained at 34% in the other experiments, in which coal was also present in addition to the lignin. With this assumption of constant lignin conversion, the value of Y^L in equation (3) is 34%. The conversions of coal so estimated are summarized in the last column of Table 1.   It appears then that the presence of lignin in the reaction mixture may cause considerable depolymerization and liquefaction of the coal, with conversions to filterable liquids ranging from about 19.1 to 43.2%; under comparable conditions, but with no lignin present, only about 5.2% of the coal was converted (experiment 1-A). Figures 1 and 2 show the values of Y_F and X_F plotted vs. lignin concentration of the reaction mixture.

Experiments 1-K and 1-L conducted with NiO-MoO$_3$ catalyst show a similar conversion for lignin alone (1-L) but no conversion of coal (1-K) when a stabilized lignin depolymerization product (from experiment 1-L) is used instead of lignin. Experiment 1-N with no catalyst gives a similar result; to enhance the coal depolymerization, the lignin added to the reaction mix- ture must itself be originally un-depolymerized. It seems that any favorable effect of lignin in liquefying coal might be attributed to molecular fragments formed by thermally cracking the lignin _in situ_. Experiment 1-M shows the favorable effect of added phenol on depolymerization of lignin by hydrocracking.

Table 1

Reaction Conditions and Overall Conversion for Experiments at High Pressure.[a] Coal - Illinois No. 6 (74/105 m). Reaction Times = 3 hrs. Reaction Temperature = $300^{\circ}C$

Experiment	Lignin Mass (g)	Coal[b] Mass (g)	Solvent Tetralin Volume (ml)	Catalyst Mass (g)	React. Press. (psi)	Overall Conv. Y_F, %	Estim. Coal Conv. X_F, %
1-A	0.0	50	200	$SiO_2\text{-}Al_2O_3$ 5.0	1700	5.21	5.21
1-B	10	40	200	$SiO_2\text{-}Al_2O_3$ 5.0	1550	22.44	19.1
1-C	15	35	200	$SiO_2\text{-}Al_2O_3$ 5.0	1450	28.73	26.11
1-D	25	25	200	$SiO_2\text{-}Al_2O_3$ 5.0	1700	34.33	34.71
1-E	25	0.0	200	$SiO_2\text{-}Al_2O_3$ 5.0	1500	35.0	--
1-F[c]	25	0.0	200	$SiO_2\text{-}Al_2O_3$ 5.0	1600	30.0	--
1-G	50	0.0	200	$SiO_2\text{-}Al_2O_3$ 5.0	1600	31.0	--
1-H	50	0.0	200	$SiO_2\text{-}Al_2O_3$ 5.0	1650	36.0	--
1-I	50	0.0	200	$SiO_2\text{-}Al_2O_3$ 5.0	1650	38.0	--
1-J	35	15	200	$SiO_2\text{-}Al_2O_3$ 5.0	1400	36.49	43.21
1-K[d]	0.0	50	150	$NiO\text{-}MoO_3$ on Al_2O_3 5.0	1650	0.0	0.0
1-L[d]	50	0.0	200	$NiO\text{-}MoO_3$ on Al_2O_3 5.0	1800	37.0	--
1-M[e]	20	0.0	450	---------	1700	63.0	--
1-N[f]	0.0	20	(f)	---------	1700	3.0	3.0
1-O	50	0.0	200	$NiO\text{-}MoO_3$[g] 5.0	1800	38.5	--
1-P	25	25	(h)	$NiO\text{-}MoO_3$	1900	53.0	- 5.0*
1-Q	25	25	(i)	$NiO\text{-}MoO_3$	1500	80.0	- 60*

* Calculated based on 100% conversion for lignin -90%±2 rec

226

(a) In each experiment the initial pressure was 1000 psi applied from a H_2 cylinder. Final pressure ranged from 1400-1800 psi. The products from experiments 1-C, 1-D, 1-J and 1-L were identified by GC/MS, See Table 4.

(b) Mass of coal containing 13.6% ash (26)

(c) 25 g of glass spheres was added to the reaction mixture for experiment 1-F.

(d) 50 ml of the liquid reaction product from experiment 1-L was included in the reaction mixture of experiment 1-K.

(e) 10 g of phenol was added to the reaction mixture of experiment 1-M.

(f) 450 ml of the filtrate from experiment 1-M was used as the solvent for experiment 1-N.

(g) NiO-MoO$_3$ catalyst gives slightly higher liquefaction yield.

(h) solvent was 100 ml tetralin + 100 ml guaiacol

(i) solvent was 200 ml guaiacol

Table 2

Depolymerization of Coal and Lignin at Low Temperature and Atmospheric pressure (without H_2). Coal – Illinois #6 (74/105 μm)

Experi- ment	Lignin Mass (g)	Coal Mass (g)	Phenol Mass (g)	Tetralin Vol (ml)	Catalyst	Temp. °C	Reaction Time hr	Overall Conv. (%)
2-A	25	0	75	0	BF$_3$ gas	110	20	25
2-B	30	0	30	350	BF$_3$ gas	155	44	25
2-C	0	20	0	70	—	155	44	0.0
2-D	0	20	20	230	—	155	44	0.0
2-E	0	20	20	420	—	155	44	0.0
2-F	10	10	150	0	10g AlCl$_3$	120	30	21.5

Notes: 350 ml of filtrate from experiment 2-B was added to reaction mixture 2-C

190 ml of filtrate from repeated experiment 2-B was added to reaction mixture 2-D

Reaction products of experiments 2-A and 2-F were investigated by GC-MS; see Table 4.

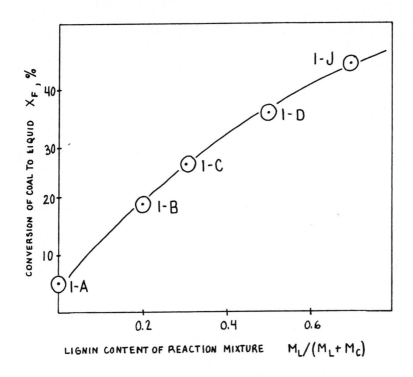

Fig. 1.

Conversion of Coal to Liquid versus Lignin Content of Reaction Mixtures.
Each data point is labeled with the identification of the corresponding
experiment described in Table 1

Table 2 shows some similar results conducted at lower temperatures and
atmospheric pressure (no hydrogen), with $AlCl_3$ and BF_3 as catalysts for some
experiments. Few new conclusions can be drawn from the results, except
perhaps that the filtrate from lignin liquefaction is ineffective when added
to subsequent reaction mixtures. GC/MS investigation of the products of
experiments 2-A and 2-F are discussed below.

Table 3 lists major molecular species detected by GC/MS in the liquid
products from experiments 1-C, 1-D, 1-G, 1-J, 1-L, 2-A and 2-F. Three-ring
aromatic structures are present only in those products from experiments in
which coal was used and are evidence that the coal was liquefied. The
presence of some two-ring structures in the products from experiment 1-L in
which no coal was used may be attributed to the tetralin solvent employed.
GC/MS analysis of products from control experiments under similar conditions

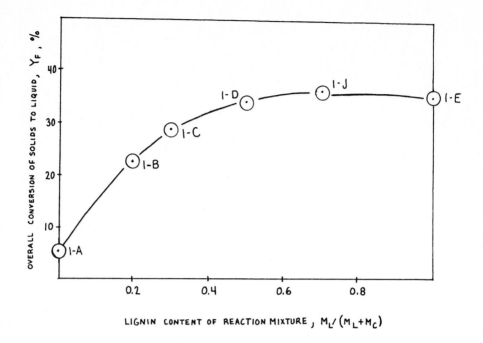

Fig. 2.

Influence of Lignin Content in Reaction Mixture upon Overall Conversion to
Liquid. Each data point is labeled with the identification of the
corresponding experiment described in Table 1

with tetralin alone but with no coal and no lignin showed only a small
amount of tetralol, tetralone and naphthalene. Tetralol and tetralone can
form from the oxidation of tetralin at high temperature and naphthalene is
the product of dehydrogenation.

Table 4 shows the results of a fractional vacuum distillation of the
products of experiment 2-F as well as a listing of the major molecular
components in the fractions as identified by GC/MS. The presence of poly-
nuclear aromatic structures is noteworthy.

Table 5 shows the dependence of overall conversion of lignin (Y_F) on
tetralin-lignin ratio and also on the nature of the pressurizing gas used in
these experiments (5-A through 5-F). It is evident that using H_2 rather
than N_2 as the pressurizing gas did not show any particular advantage at
high tetralin/lignin ratios but became increasingly important at the lower
ratios. It is also clear that the overall conversions increased with
increasing values of this ratio.

Table 3

Products Identified By GC/MS In The Products From Selected Experiments

Exp.	Major Components In Reaction Products
1-C	(chemical structures)
1-D	(chemical structures)
1-G	(chemical structures)
1-J	(chemical structures)
1-L	(chemical structures)
2-A	(chemical structures)
2-F	(chemical structures)

Experimental results shown in Figures 1 and 2 suggest that the co-liquefaction of coal and lignin produces an effect greater than a mere additive result. This can be better seen by comparing the actual results of Figures 1 and 2 with the results of hypothetical ("Gedanken") experiments summarized in Table 6 where the liquefaction of five different mixtures is presumed, ranging from pure coal to pure lignin. In the first set of experiments of Table 6 (No Interaction) the fractional liquid yields from Lignin Y^L and from coal Y^C are presumed to remain constant at 0.5 and 0.1 respectively, independent of the initial composition of the reaction

Table 4

Fractional Distillation Of Reaction Products From Experiment 2-F

Fraction	Boiling Range °C	Density g/ml	Wt. % g	Vol. % ml	Major Components
Collected in trap	30±1	0.95	7.8	9.5	(structures)
1st.	70±2	1.03	51.3	57.4	HO- (structure)
2nd.	78±2	1.08	6.90	6.75	(structures)
3rd.	95-100	1.13	18.6	20.3	(structures)
4th.	110±3	1.16	3.76	3.70	(structures)
Undistilable residue	--	--	11.5	2.35	------

mixture. Then the mean liquid yield Y is computed from the composition and Y^L and Y^C. The other three types of behavior hypothesized in Table 6 are:

<u>L catalyzes C</u>: The presence of lignin is assumed to increase Y^C in proportion to the amount of lignin present; the presence of coal does not influence $Y^L = 0.5$.

<u>C catalyzes L</u>: The presence of coal is assumed to increase Y^L in proportion to the amount of coal present; the presence of lignin does not influence $Y^C = 0.1$.

<u>Mutual, Two-Way Interaction</u>: The presence of coal is assumed to increase Y^L and the presence of lignin is assumed to influence Y^C. The computed data from Table 6 are plotted in Figure 3. It is evident that this type of graph clearly shows the influence of positive enhancement of liquefaction as a positive deviation from the straight line which represents the "No-Interaction" case. Nevertheless, the graph cannot distinguish whether the lignin promotes the liquefaction of coal or vice vera. Comparing Figure 3 to the experimental values of Y in Figure 1 shows that the coliquefaction produces an enhancing effect but the data are insufficient by themselves to decide whether it is lignin that aids coal liquefaction, coal that aids lignin liquefaction, or whether both these effects are present simultaneously. Most of the reasoning of this chapter is directed toward the hypothesis that it is lignin which aids coal liquefaction based on the lower thermal stability of bonds in lignin and the reasonable expectation that in a reaction mixture the lignin would provide more free radicals than the coal. Nevertheless, the inorganic minerals in coal might well catalyze some steps in lignin liquefaction and coal itself would be expected to form some radicals which might attack the lignin and hasten its liquefaction. Based on the GC/MS evidence, the relative thermal stability of lignin and coal, and what is known about the structures of lignin and coal molecules, we believe the effect of lignin on coal liquefaction is significantly greater than the opposite effect.

As discussed in many publications (refs. 11, 12), hydrogen as well as tetralin stabilized intermediate radicals formed in the hydrocracking process. As long as a large amount of one is available, reduction of the other does not have a significant effect on the overall conversion. However, when tetralin and hydrogen are simultaneously reduced, the conversion is lowered as shown by experiments 5-C and 5-F.

Figure 4 shows that overall conversion is also increased by increasing the tetralin-to-lignin ratio. The sensitivity to tetralin content is increased when the gaseous atmosphere is nitrogen rather than hydrogen, as also shown by the data plotted in Figure 4.

Table 5

Dependence of Overall Lignin Conversion on Tetralin-Lignin Ratio and on Nature of Pressurizing Gas

Experiment	Lignin Mass (g)	Tetralin Volume (ml)	Tetralin /Lignin Ratio (mg/g)	Gas	Overall Conversaion (Y) (%)	Reaction Pressure (psi)
5-A	20	450	22.5	H_2	40.0	1700
5-B	20	250	12.5	H_2	37.0	1700
5-C	20	100	5.0	H_2	33.5	1800
5-D	20	450	22.5	N_2	40.0	1700
5-E	20	250	12.5	N_2	32.0	1550
5-F	20	100	5.0	N_2	27.0	1800

*Initial gas pressure applied in these experiments 1000 psi. Final pressure ranged from 1550-1800 psi.

Table 6

FOUR ASSUMED, HYPOTHETICAL INTERACTIONS BETWEEN LIGNIN AND COAL

Lignin Mass, M_L	Coal Mass, M_C	Mass Fract. Lignin, Z_L	No Interaction Y^C	Y^L	$\bar{Y}$	L catalyzes C Y^C	Y^L	$\bar{Y}$	C catalyzes L Y^C	Y^L	$\bar{Y}$	Mutual, Two-Way Interaction Y^C	Y^L	$\bar{Y}$
0	4	0	0.1	0.5	0.1	0.1	0.5	0.1	0.1	0.7	0.1	0.1	0.7	0.1
1	3	0.25	"	"	0.2	0.15	"	0.24	"	0.65	0.24	0.15	0.65	0.275
2	2	0.5	"	"	0.3	0.20	"	0.35	"	0.60	0.35	0.20	0.60	0.40
3	1	0.75	"	"	0.4	0.25	"	0.44	"	0.55	0.44	0.25	0.55	0.475
4	0	1.0	"	"	0.5	0.3	"	0.5	"	0.5	0.5	0.30	0.50	0.5

Notes: Mass is in arbitrary units, Y_C = yield of liquid from coal. Y_L = Yield of liquid from lignin. $\bar{Y}$ = mean yield of liquid from mixture of lignin and coal; $\bar{Y}$ is computed from Y^C and Y^L in each case.

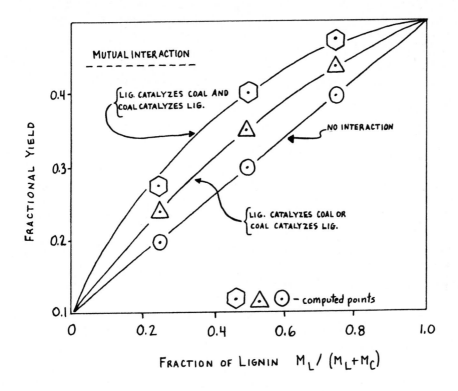

Fig. 3.

Hypothetical Coliquefaction Experiments with Different Kinds of Enhancing
Interaction Between Lignin and Coal

Experimental data collected in Table 7 are from experiments conducted
with no catalyst; in a few instances, sufficient information was obtained
under the same experimental conditions to permit calculation of X_D and X_F
for reaction mixtures of coal and lignin. The initial pressure for each
experiment shown in Table 7 was 1000 psi; final pressures depended on
reaction time, temperature and initial composition of reaction mixture. The
reaction times given refer to elapsed time after the reaction set point
temperature had been reached.

The results for distillate yields are particularly interesting and are
plotted in Figure 5. Note that the overall distillate yield Y_D^{C+L} obtained
in experiments using 50/50 mixtures of lignin plus coal is always significantly larger than the mean of the two yields obtained under identical conditions with coal alone Y_D^C and lignin alone Y_D^L. If the coal and the lignin

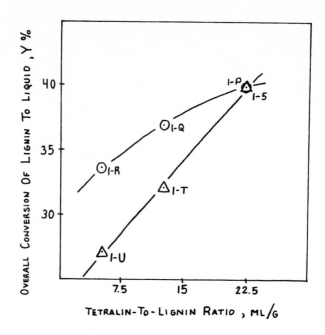

Fig. 4.

Dependence of Overall Conversion on Tetralin-Lignin Ratio Showing Influence
of Pressurizing Gas. Each data point is labeled with the identification of
the corresponding experiment described in Table 1.
(○= H2 atmosphere; △= N2 atmosphere)

reacted independently of each other then we would expect $Y_D^{C+L} = \frac{1}{2}(Y_D^C + Y_D^L)$. That Y_D^{C+L} is greater than would be expected for independent reaction of the lignin and the coal indicates that at least one of the co-reactants promotes the reaction of the other to distillable liquid products. Also noteworthy is the steep increase of X_D^{C+L} with increasing reaction time.

Yields based on filterable liquid products from reaction at 350°C, however, show a rather different pattern as is evident from Figure 6. Here Y_F^{C+L} falls below the mean $\frac{1}{2}(Y_F^C + Y_F^L)$ and this indicates an unfavorable interaction for liquefaction of the mixture, viz. that less filterable liquid is obtained from liquefying the 50/50 mixture of lignin and coal than is produced by liquefying each component separately. Similarly, the computed value of X_F^{C+L} for coal liquefaction yield in the mixture falls significantly below X_F^C, the liquid yield when coal is reacted alone. Thus it appears that for reaction of a 50/50 lignin/coal mixture at 350°C and 2000 psi, the presence of lignin enhances the formation from coal of distillable products

Table 7

Summary of Experimental Results

Reaction times given are times elapsed of mixtures had been heated to
the set point temperature which was usually attained in about 30 minutes.
Other experimental conditions given in footnote a.

Exp. No.	Lignin Mass (g)	Coal Mass (g)	Reaction Temp. (°C)	Reaction Time (min)	Final Press. (psi)	Y_F (%)	Y_D (%)	X_F^{C+L} (%)	X_D^{C+L} (%)
-A	50	--	350	0.0	1500	41.0	12.3	--	--
-B	50	--	350	30	1800	59.0	20.2	--	--
-C	50	--	350	60	2100	67.5	18.0	--	--
-D	50	--	350	90	1900	76.0	21.4	--	--
-E	50	--	350	120	1950	81.0	23.6	--	--
-F	50(b)	--	350	60	1900	76.0	--	--	--
-G	50(c)	--	350	60	2150	42.6	--	--	--
-H	50	--	300	180	1800	30.0	--	--	--
-I	50	--	400	60	2550	90.0	41.0	--	--
-J	--	50	350	30	1850	30.8	--	--	--
-K	--	50	350	45	1850	34.7	4.5	--	--
-L	--	50	350	90	1750	44.0	13.5	--	--
-M	--	50	350	120	1900	45.2	--	--	--
-N	--	50(b)	350	60	1850	38.0	--	--	--
-O	--	50	350	60	1550	29.2	6.5	--	--
-P	--	50(d)	400	60	2100	72.0	17.3	--	--
-Q	--	50	400	60	2350	69.5	29.5	--	--
-Q'	--	50	300	180	1700	5.21	--	--	--
-R	25	25	350	30	1850	40.5	12.6	19.1	3.8
-S	25	25	350	60	2000	48.5	18.2	26.6	18.4
-T	25	25	350	90	2150	52.7	22.7	25.7	24.2
-U	25	25	350	120	1900	56.0	25.2	27.1	27.0
-V	25	25(b)	350	60	2050	54.7	--	--	--
-W	25	25	400	30	2300	68.5	26.0	--	--
-X	25	25	400	45	2400	75.0	--	--	--
-Y	25	25	400	60	2350	80.0	30.5	68.5	18.3
-Z	25	25	400	90	2400	88.0	24.0	--	--
-A	25	25	400	180	2450	90.0	--	--	--
-B	25	25	300	90	2100	26.6	20.5	--	--
-C	25	25	450	90	2800	96.7	28.0	--	--
-D	25	25	300	180	1700	34.33	--	34.7	--

a) All experiments were carried out with no catalyst, 1000 psi initial H_2 pressure, and 200 ml tetrahydronaphthalene. Iotech type of lignin was used in all experiments except I-G in which Indulin was used.

b) 5.0 g. P-dimethoxybenzene was added to the reaction mixture.

c) Lignin used in this experiment was Indulin.

d) 0.5 g. Lignin was added to the mixture.

boiling up to 135°C at 10 torr, but decreases the formation of filterable liquid products under the same conditions. One interpretation of such results might suggest that radicals formed by thermolysis of the lignin eventually react to form unfilterable products but in the course of reaction a significant amount of low molecular weight, low-boiling fragments are also formed from the coal.

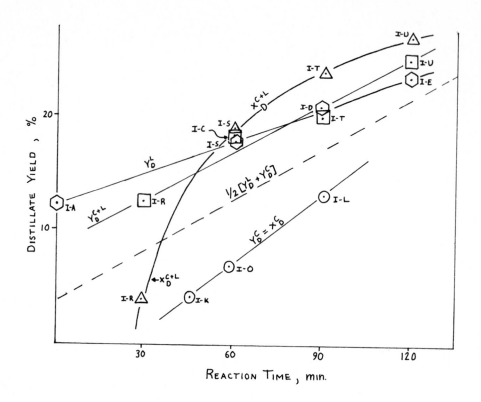

Fig. 5.

Effects of reaction time on the yields of distilate obtained
for liquefaction of only coal $Y_D^C = X_D^C$ (O), only lignin Y_D^L (◯) and
mixtures of equal amounts coal and lignin Y_D^{C+L} (□) and X_D^{C+L} (△);
temperature = 350°C. Reaction mixture always contained 200 ml of tetralin
and 50 g. of solid reactants (lignin or coal or lignin and coal).
Each data point is identified using the same experiment code in Table 7.
Dashed line is computed as shown, based on the assumption of no interaction
of coal and lignin.

A similar decrease in filterable products due to the presence of lignin
is not found after reaction of 50/50 mixtures of lignin and coal for 180
min. at 300°C. The relative yields of filterable liquid products at this
lower temperature bear a relationship to each other that is similar to the
yields of distillable products at 350°C.

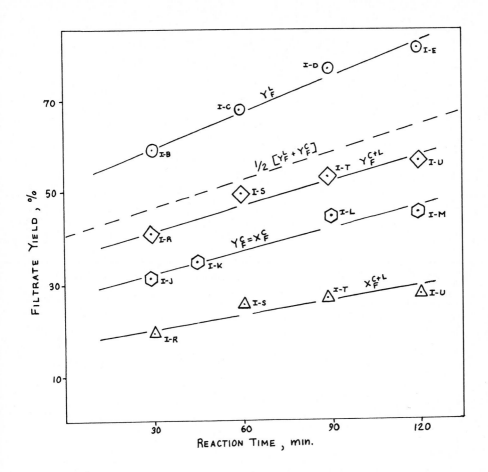

Fig. 6.

Effects of reaction time on the yields of filterable product obtained
in the liquefaction of coal Y_F^C ($\bigcirc$), lignin Y_F^L ($\bigcirc$) and mixture of equal
amounts of coal and lignin Y_F^{C+L} ($\lozenge$) and X_F^{C+L} ($\triangle$). Reaction temperature =
350°C. Broken line was computed from Y_F^C and Y_F^L as in Figure 5. Reaction
mixture always contained 200 ml tetralin and 50 g. solid reactants (coal or
lignin or coal and lignin). Each data point is identified using the same
experiment code in Table 7.

Figure 7 provides a comparison by way of a bar chart for yields
corresponding to the following conditions:

    1500 psi, 350°C, 90 min — filterable and distillable yields

    1500 psi, 300°C, 180 min — filterable yields

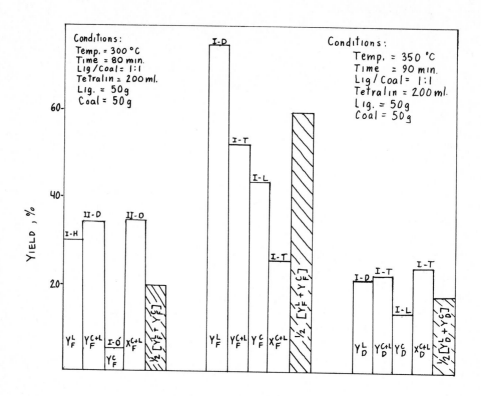

Fig. 7.

Comparison of yields for reaction of lignin Y^L, coal $Y^C = X^C$ and mixtures of equal amount of lignin and coal Y^{C+L} at 300 and 350°C.

Dashed bars are average yields computed from $\frac{1}{2}(Y^L + Y^C)$. Reaction mixture always contained 200 ml tetralin and 50 g. solid reactants. Each bar is identified using the same experiment code in Table 7.

The favorable effects on liquefaction of coal to filterable products caused by lignin at 300°C but not at 350°C, can be hypothesized to result from more extensive polymerization or recombination of radicals at the higher temperatures. Perhaps at 300°C sufficient radicals are formed from the lignin to cause some depolymerization of coal but the radical concentrations may remain sufficiently low that extensive polymerization would not occur.

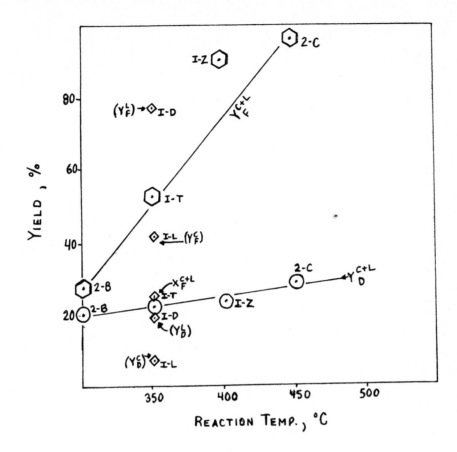

Fig. 8.

Effects of reaction temperature on the yields of filterable products Y_F^{C+L} ($\diamondsuit$) and distillable products Y_D^{C+L} (O) in liquefaction of equal amounts of coal and lignin. Reaction time = 90 min. For comparison other data are as follows: 1-D = Y_F^L and Y_D^L, 1-L = Y_F^C and Y_D^C, 1-T = X_F^{C+L}, all at 350°C. Each data point is identified using the same experiment code in Table 7.

Figure 8 shows the effect of temperature on Y_F^{C+L} and Y_D^{C+L} for a reaction time of 90 min. It is evident that increasing reaction temperature has a stronger effect on increasing filterable product yields than distillable product yields. Other yields available only at 350°C are also shown in Figure 8. The enhancement effect of the lignin is only evident for the distillate and a negative effect may be present for production of filtrate. Data obtained for 60 minutes reaction time indicate that $Y_D^{C+L} > \frac{1}{2}(Y_D^C + Y_D^L)$ at

240

350°C, but $Y_D^{C+L} < \frac{1}{2}(Y_D^{C} + Y_D^{L})$ at 400°C and this suggests that the favorable effects of lignin on distillate yield were confined to lower temperatures. For 60 minutes reaction time $Y_F^{C+L} = \frac{1}{2}(Y_F^{C} + Y_F^{L})$ at both 350°C and 400°C.

As shown in Figure 9, for experiments at 350°C in which 50/50 mixtures of lignin/coal were liquefied, the yield of distillable products from the coal X_D^{L+C} is always lower than the corresponding yield X_F^{L+C} of filterable products. It is noteworthy, however, that X_D^{C+L} rises rapidly as the duration of the experiment increases and for reaction times of about 90 – 120 min, the values of X_D^{C+L} are almost as large as X_F^{C+L}. This suggests that larger reaction times ensure that almost all of the filterable products produced from coal in the presence of lignin are also distillable below 135°C at 10 torr.

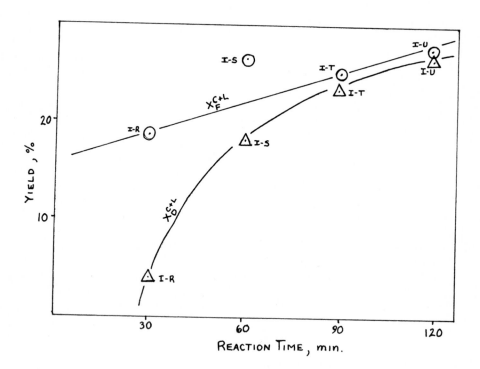

Fig. 9.

Effects of reaction time on the yield of coal liquefaction: X_F^{C+L} (O) filterable products and X_D^{C+L} (Δ) distillable products in a mixture of equal amounts of coal and lignin. Reaction temperature = 350°C. Each data point is identified using the same experiment code in Table 7.

Table 8

SOME YIELDS REPORTED FOR COAL LIQUEFACTION

Ref.	Type of Coal	Solvent	Cataly.	Temp. ^{0}C	Time Min	Press. psi	(a) Yield %
1.	High-rank Pennsylvania	recycle solvent	-	426	90	~3500	>50
4.	W. Virginia	Tetralin	CO/MO/ Al_2O_3	400	50	-	~79
8.	Wyodak	Phenol	-	427	15	~4000	39 Pyridine Soluble
5	Bituminous	Phenol	BF_3	100	24 hrs	1 atm	>80 Phenol Soluble
7.	Pond Creek Seam	Phenol	H_2SO_4	180	24 hrs	1 atm	29 Pyridine & Benzene scl
7.	Pond Creek Seam	Phenol	BF_3	150	24 hrs	1 atm	38
7.	Pond Creek Seam	m-Cresol	H_2SO_4	190	24 hrs	1 atm	36
13.	Bituminous Ill. No. 6	$ZnCl_2$	-	454	-	2700	60-70
6.	Yubari	Phenol	Tuluene Solfunic Acid	180	300	1 atm	90 Benzene, Pyridine Soluble
14.	High Volatile Bituminous, Ill.	Tetralin	-	400	<5	-	90 Pyridine Soluble
15.	West Kentucky	-	-	427	<3.0	-	70-80

a. Yields are computed differently in each experiment. It has been cited
whereever was possible. The degree of coal dissolution in an organis solvent
is usually an indication of the effectiveness of the depolymerization process.

It is very difficult to compare the yields reported here with those of
other workers, some of which are collected in Table 8. Problems arise from
the lack of complete correspondence of coals, catalysts and reaction conditions, as well as different ways of defining liquid yields. It is
interesting to focus on the yield in Table 8 for Illinois No. 6 coal at
454°C in comparison with our results for coal of the same origin at 400°C,

reported as experiments I-W through II-A in Table 7. It is evident that our overall yields of filterable product range from about 69% for 30 minutes reaction to about 90% at 180 minutes. Even if one assumes that all the lignin was liquefied in our experiments, our yields for coal liquefaction using lignin at 400°C would still compare favorably with Struck and Zielke's (ref. 13) results using $ZnCl_2$ as a catalyst at 454°C. The result of Experiment I-Y in Table 7 indicates an estimated conversion of the coal of about 69% to filterable products at 400°C. Such results at lower temperature are encouraging in view of the well-known decrease in liquefaction yields at lower temperature, as illustrated in Table 9 by the results of Kamiya et al. (ref. 16) for Oyubari coal. Here it is seen that a temperature decrease from 460°C to 378°C reduces conversion from about 86% to 32%.

Comparison of the results of experiments I-C, I-F and I-G indicates that Indulin lignin does not liquefy as easily as Iotech lignin and also shows that the addition of p-dimethoxybenzene to the reaction mixture enhances liquefaction. The latter compound is believed to serve as a source of free radicals as discussed below. Comparison of yields for experiments I-K (45 min), I-N (60 min), I-L (90 min) and I-M (120 min) suggests that p-dimethoxybenzene may have little effect during liquefaction of coal alone. Experiment I-0 is ignored because it is inconsistent with experiments I-K through I-M which produce a smooth curve when plotted vs. temperature in Figure 10.

Table 9

EFFECT OF REACTION TEMPERATURE ON LIQUEFACTION OF OYUBARI COAL (a)

Coal Mass g	Tetralin ml	Temp. °C	Time Min	H_2 Pressure MPa	(b) Conversion %
20	60	378	120	0.93	32
20	60.0	407	120	0.98	63.6
20	60.5	435	120	0.98	81.2
20	59.9	460	120	0.98	86.4

(a) See reference 10 for more information.

(b) Conversion was calcualted as a percentage of the maf feed coal by substracting the percentage undissolved to give the conversion

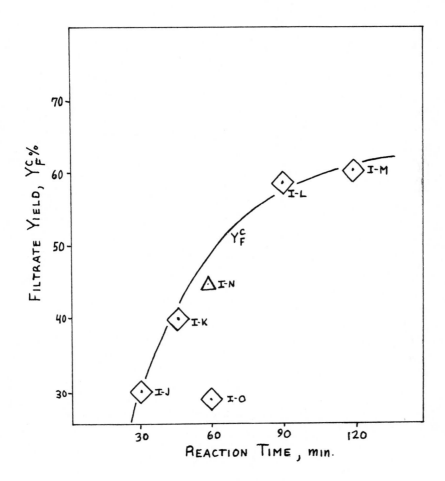

Fig. 10.

Effects of reaction time upon the yields of filterable products, Y_F^C for coal liquefaction at 350°C. Experiment 1-N ($\triangle$) was conducted with 5.0 g. of p-dimethoxybenzene in the reaction mixture. Each data point is identified using the same experiment code in Table 7. Indulin type lignin was used for experiment I-O ($\Diamond$).

244

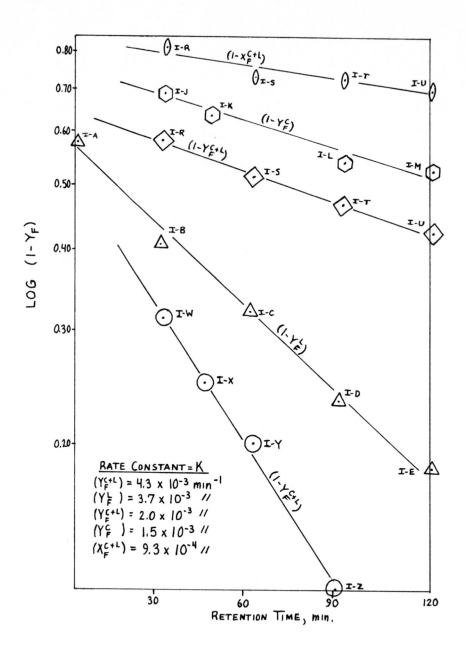

Fig. 11.

First order kinetic behavior with respect to Y_F for liquefaction of coal (Y_F^C), lignin (Y_F^L) and coal and lignin (Y_F^{C+L}) at 350°C. Similar behavior was observed for Y_F^{C+L} at 400°C. Each data point is identified using the same experiment code in Table 7.

For the various yield definitions plotting log (1-Y) versus time gives approximate straight lines as shown in Figures 11 and 12, thereby suggesting a kinetic behavior that is first order in amount of unreacted material. In the case of filtrate yield at 350°C there does not seem to be a strong kinetic advantage due to the added lignin, although the precision with which the slopes or rate constants can be estimated does not rule out such an effect. The better distillable yields caused by the presence of lignin may therefore involve a strong solubilization effect.

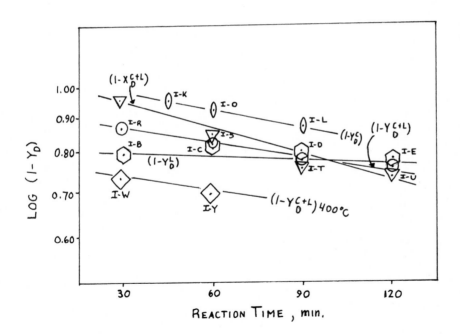

Fig. 12.

First order kinetic behavior with respect to the yield of distillable products obtained in liquefaction of coal (O), lignin (◇) and mixture of equal amount of coal and lignin (◇▽◦) at 359°C. Similar behavior was observed for coal and lignin, Y_D^{C+L} at 400°C (lowest line).

Comparing our pseudo first-order rate constants with those of other work is fraught with the same sort of difficulties as comparing yields. Nevertheless, our observed values agree reasonably well with the value of 10^{-5} sec^{-1} observed (ref. 17) at 400°C for the decomposition of 1,2-diphenylethane in the presence of hydrogen donors.

These values are also in accord with the activation energy (25 Kcal/g mole) and pre-exponential factor (6.5 x 10^{+4} sec^{-1}) fitted (ref. 18) for first-order pyrolysis of a bituminous coal and a lignite.

4. DISCUSSION

In an effort to explain the apparently favorable effects of lignin on the depolymerization of coal, it is helpful to recall that hydrocracking of lignin has been conducted by many investigators (refs. 11, 19 - 21) at temperatures as low as 300°C, i.e. lower than temperatures normally required for coal liquefaction (400 - 500°C).

Hydrogen donor solvents such as tetralin have been found to produce favorable reaction environments (ref. 22) and typical hydrocracking catalysts such as $NiO-MoO_3$, $SiO_2-Al_2O_3$ and Pd/alumina have been found to accelerate the depolymerization reactions of lignin. For example, it was found (ref. 22) that hydrocracking of kraft lignin in tetralin at 350 - 400°C and high pressure (1500 psi) produced significant yields of phenols. Boomer et al. (ref. 11) found that in tetralin at 350°C and 3850 psi, 85% of a whole wood sample (British Columbia fir wood sawdust) was converted into liquids and gases.

Yields of monophenols from lignin hydrocracking vary widely depending on process conditions. Earlier work in Japan (ref. 19) and in the U.S. by Goheen (refs. 23, 24) demonstrated yields of distillable phenols up to about 22% whereas later work (ref. 25) indicates yields as high as 38 - 42%.

Thermal and hydroliquefaction of coal and of lignin is believed to proceed by a free-radical mechanism (ref. 2, 22, 26) in which the primary step involves thermal homolytic bond rupture and formation of radical fragments. Free radicals have also been shown to exist in naturally occurring coal (refs. 14, 27) as well as in lignin extracted from wood (refs. 28 - 30).

In the experiments reported here, thermal degradation of lignin would be expected to produce substituted phenoxyl radicals as well as other types of radicals at temperatures (300°C) too low for substantial thermolysis of the coal molecules (scheme 1):

SCHEME I

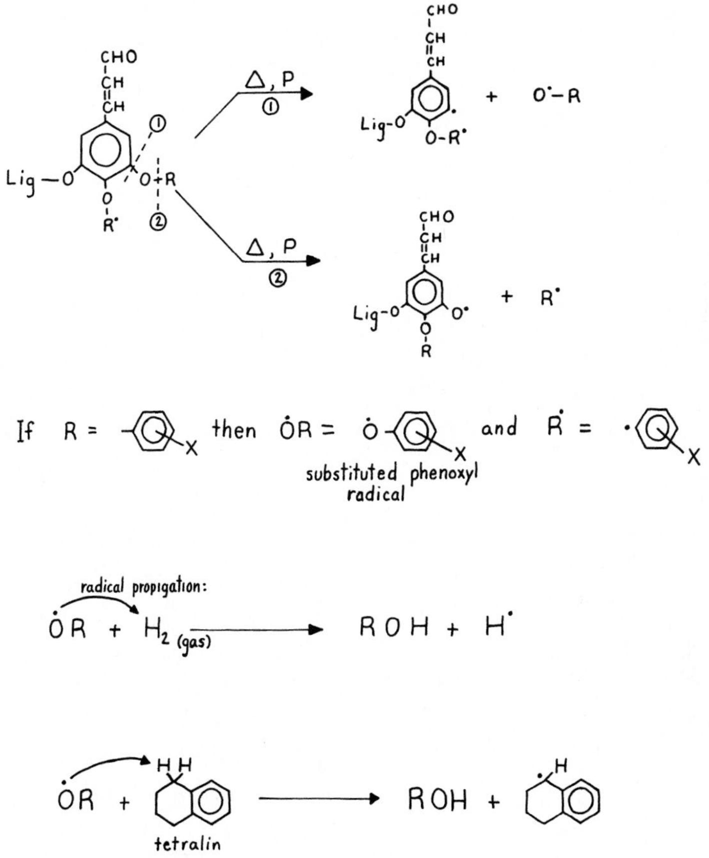

Formation of products such as toluene, o-methylphenol, guaiacol and p-methoxyanisole observed in most experiments can be attributed to primary thermal depolymerization (scheme 2):

SCHEME 2

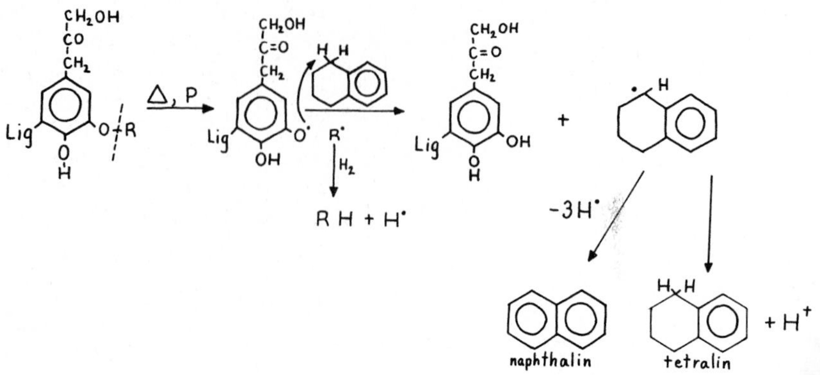

If R represents

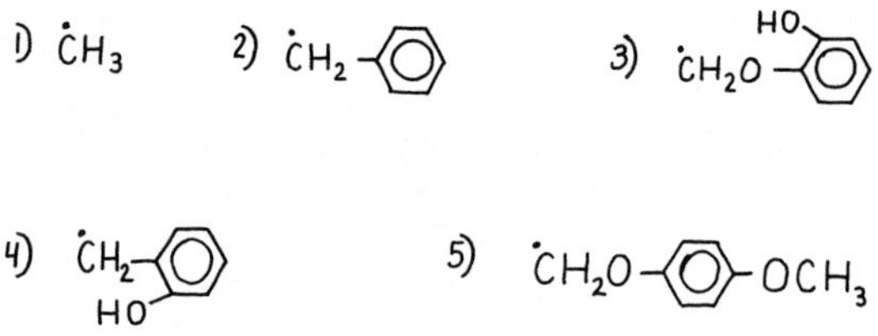

1) $\dot{C}H_3$ 2) $\dot{C}H_2$—⬡   3) $\dot{C}H_2O$—⬡—OH

4) $\dot{C}H_2$—⬡—OH   5) $\dot{C}H_2O$—⬡—OCH_3

then these radicals can abstract hydrogen atoms to form the following stable products:

(1) methane,
(2) toluene,
(3) guaiacol,
(4) o-methylphenol,
(5) p-methoxyanisole.

It is believed that the phenoxyl radicals then can further react with coal to cause hydrocracking of the latter (scheme 3):

SCHEME 3

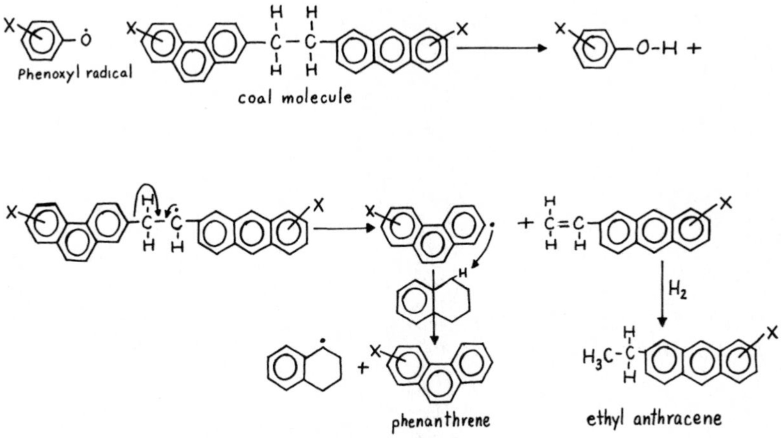

Another alternative reaction path can be as follows:

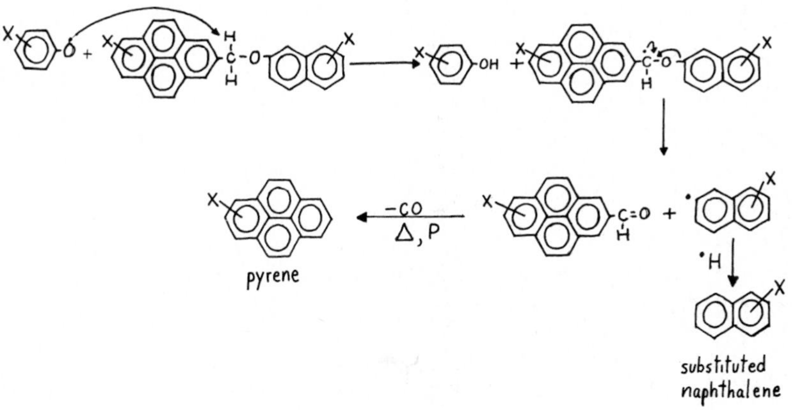

The inherent resonance stabilization of the phenoxy radicals would ensure they existed sufficiently long enough to encounter and react with coal molecules (scheme 4):

SCHEME 4

RESONANCE STABILIZATION OF SUBSTITUTED PHENOXYL RADICALS

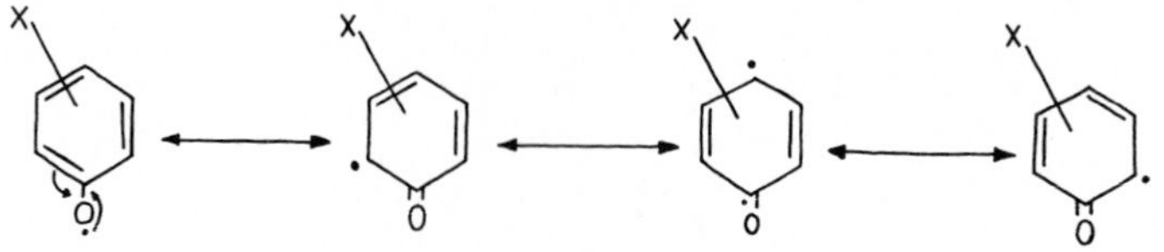

DEPOLYMERIZED LIGNIN IS ALSO RESONANCED STABILIZED

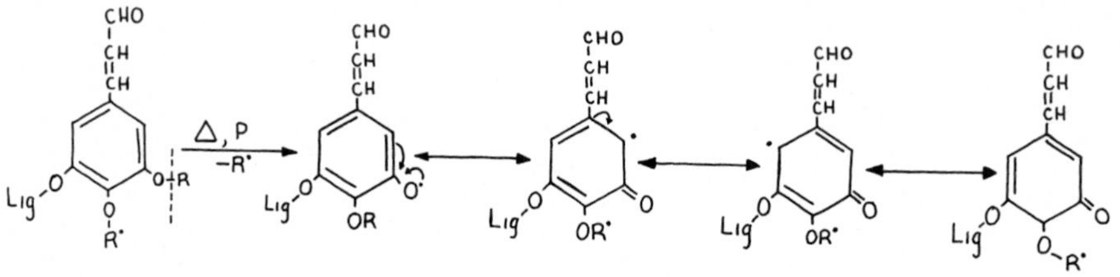

In a typical bituminous coal, aromatic and hydroaromatic clusters, containing two to three rings per cluster, are joined by methylene linkages one to three carbon atoms in length (see Figure 14). On the other hand, lignin contains repeating structural units related to phenylpropane (Figure 15); in the lignin polymer, monoaromatic rings are linked by various C-C and C-O bonds. The ether bonds linking the monoaromatic nuclei in lignin have dissociation energies of about only 60 - 75 kcal/mol whereas the energies for the methylene bridges in coal (Figure 14) fall in the range of 100 - 110 kcal/mol.

As shown in Figure 13, a typical guaiacyl moiety within lignin contains $-O-CH_3$ and $-O-R$ bonds of low energy. Preferred thermal dissociation of these bonds eventually leads to formation of phenoxyl radicals. It is believed that such phenoxyl radicals are effective and active intermediates that underly improved liquefaction of coal in the presence of lignin.

It seems reasonable that the substituted phenoxyl radicals formed by thermolysis of lignin might react with coal to cleave the methylene bridges thereby forming the anthracene, naphthalene and phenanthrene derivatives identified in the products of experiments 1-C, 1-D and 2-F as shown in Table 3.

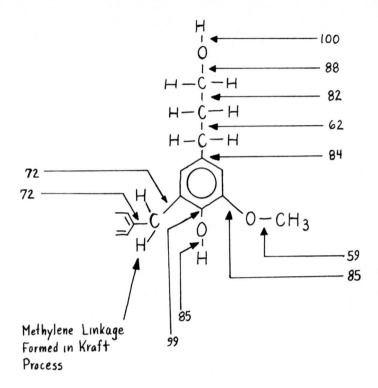

Fig. 13.

Guaiacyl Moiety Occurring in Lignin Showing Bond Energies, kcal/mole

(from ref. 31)

No coal liquefaction was observed when, instead of as-received lignin, the reaction products from independent lignin depolymerization experiments (1-K, 1-N, 2-C, 2-D) were substituted. Presumably, insufficient phenoxy radicals were present or could be formed in these reaction products to attack the coal to a significant extent; this can be attributed to deactivation of these radicals by reaction with hydrogen or by polymerization, either during experiments 1-K, 1-N, 2-C and 2-D or during subsequent separation prior to contact with the coal. When unreacted lignin was mixed directly with the coal (experiments 1-B, 1-C, 1-D) the phenoxy radicals were available for immediate reaction with the coal as soon as they were produced in the reaction mixture. This idea is also supported by the higher overall conversions (Y = 53 and 80% in experiments 1-P and 1-Q, respectively) observed in the presence of guaiacol (O-methoxyphenol). The very high overall conversion (80%) to liquid in experiment 1-Q using pure guaiacol

252

instead of tetralin indicates that at least 60% of the coal was converted to liquid (60% is the value computed based on assuming that 100% of the lignin was converted to liquid). About 90% of the guaiacol was recovered from the mixture after experiments P and Q. Guaiacol is a source of phenoxyl radicals which helps bring about depolymerization; it also provides a more polar reaction medium to help dissolve the reaction products. It is also believed that the polar phenoxy products formed from the lignin may enhance dissolution of coal depolymerization products in the reaction mixture.

Increasing the ratio of lignin to coal in the reaction mixture appears to depolymerize a larger portion of the coal. This is shown in Figure 1 where the fraction of coal liquefied (X) in experiments 1-A, 1-B, 1-C, 1-D, 1-J is plotted versus the corresponding lignin-to-coal ratio.

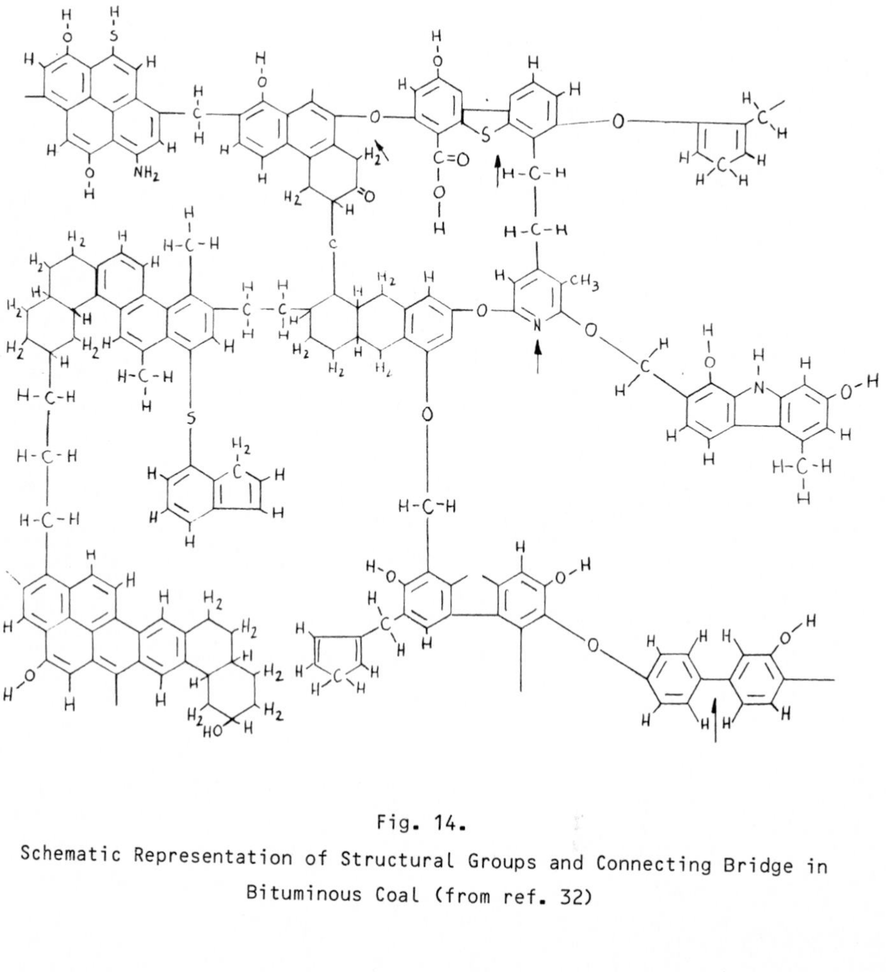

Fig. 14.
Schematic Representation of Structural Groups and Connecting Bridge in Bituminous Coal (from ref. 32)

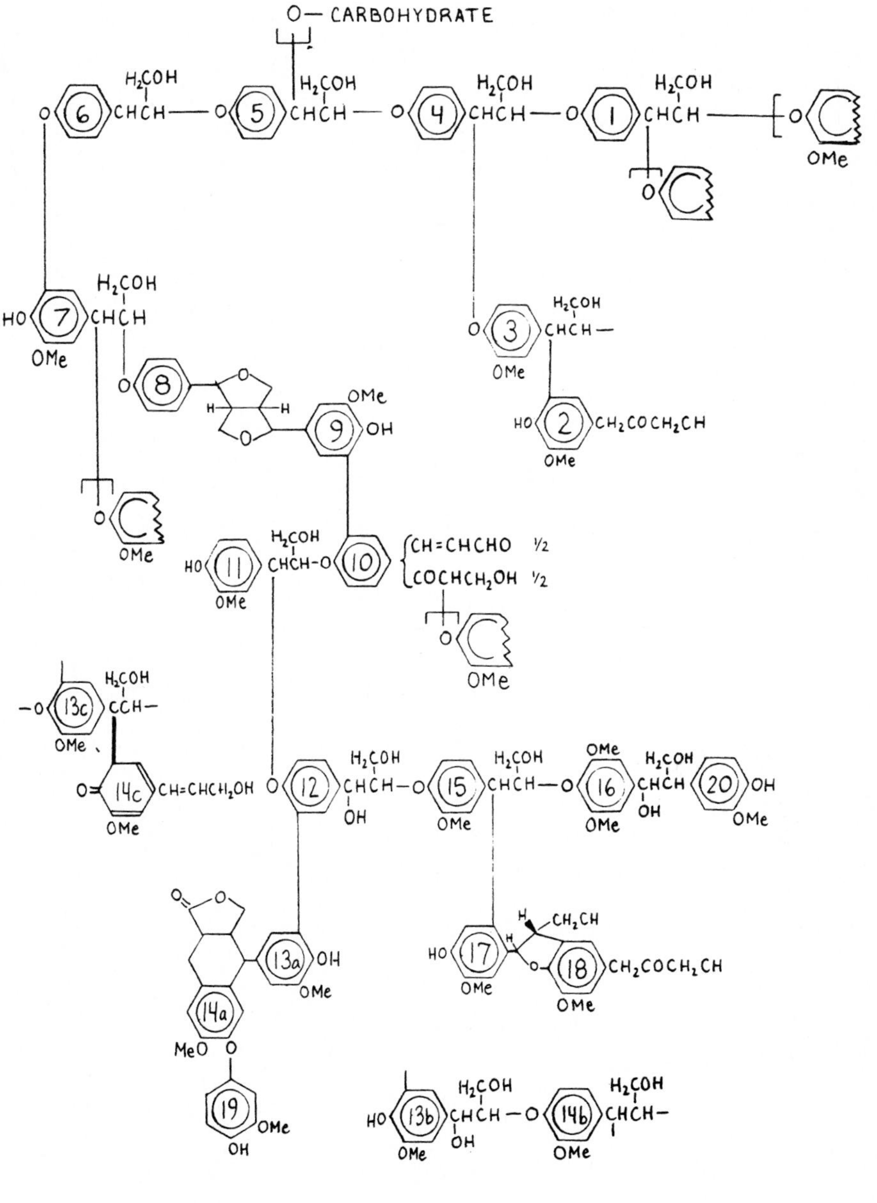

Fig. 15.

Schematic Formula for Lignin Structure

(from ref. 33)

Although Figure 1 does not indicate any optimum lignin/coal ratio for coal conversion X, Figure 2 suggests that the overall conversion Y is not improved further by increasing the lignin/coal ratio beyond about 0.7. It does seem reasonable to expect some optimum ratio for coal conversion because increasing the lignin beyond some point ought to cause such large populations of radicals that their removal by polymerization and self-condensation should begin to compete with the attack of these radicals upon coal molecules.

Figure 2 shows that overall conversion (Y) of coal + lignin increases as the fraction of lignin in the reaction mixture increases, whereas the opposite is true for increased amounts of coal in the reaction mixture, as could readily be seen by plotting yields vs. initial coal concentration.

It is noteworthy that in Figures 1 and 2 the points corresponding to the various experiments define smooth curves which span the entire range from pure lignin to pure coal; this suggests that similar mechanisms and reaction pathways may govern the depolymerization process over the entire lignin-coal composition range.

5.   CONCLUSION

The results compiled for the co-liquefaction of coal and lignin, as presented above, show that the products of reaction have the potential to be shaped into useful fuel products. Currently we are attempting to characterize further some of the filterable and distillable products that have been obtained using the reaction conditions reported earlier in this chapter. Various methods of chemical and physical analysis will be used to determine the usefulness of such derived products.

ACKNOWLEDGEMENT

We are grateful to the National Science Foundation for support of this research under Grant No. CPE 8014523.

REFERENCES

1.   Shah, Y.T. and D.C. Cronauer, Catal. Rev. Sci. Eng., 20 (2), 209 (1979)
2.   Neavel, R.C., Fuel, 55, 237 (1976)
3.   Wen, C.Y. and S. Tone, Amer. Chem. Soc., 56, 56 (1978)
4.   Tsai, M.C. and S.W. Weller, Fuel Proc. Technol., 2, 313 (1979)
5.   Heredy, L.A. and M.B. Neuworth, Fuel, 41, 221 (1962)
6.   Ouchi, K.I., K. Imuta and Y. Yamashita, Fuel, 43, 2a (1964)
7.   Darlage, L.J. and M.E. Bailey, Fuel, 55, 205 (1976)

8. Larsen, J.W., T.L.Sams and B.R. Rodges, Fuel, 60, 335 (1981)
9. Merriam, J.S. and C.E. Hanrin Jr., Fuel, 60, 542 (1981)
10. Coughlin, R.W. and M. Farooque, Amer. Chem. Soc. Ind. Eng. Chem. Process Des. Dev. Vol. 19, No. 2, 211 (1980)
11. Boomer, E.H., G.H. Argue and J. Edwards, Can. J. Res. B., 13, 337 (1935)
12. Nowack, G.P. and M.M. Johnson, U.S. Patent, 3, 787 and 912 (1977)
13. Struck, R.T. and C.W. Zielke, Fuel, 60, 795 (1981)
14. Neavel, R.C., Fuel, 55, 237 (1976)
15. Mobil R&D Corp. "The nature and origin of asphaltenes in processed coal", EPRI-AF-252, Annual Report, February (1976)
16. Kamiya, Y., H. Sato and T. Yao, Fuel, 57, 681 (1978)
17. Stein, S.E., Fuel, 59, 108 (1980)
18. Kobayashi, H., J.B. Howard and A.F. Sarofim, 16th Int. Symp. on Combustion, the Combustion Institute, Pittsburgh, pp. 411-425 (1977).
19. Oshima, M. and K. Kashima, T. Tabata and H. Watanabe, Bull. Chem. Soc. Japan, 39, 2750 (1966).
20. D'yakov, M.K. and N.V. Melenteva, J. Appl. Chem., USSR, 15, 173 (1942)
21. Kleinert, T., Monatsh, 83, 623 (1952)
22. Connors, W.J., L.N. Johanson, K.V. Skanen and P. Winslow, Holzforschung, 34, 29 (1980)
23. Goheen, D.W., Adv. in Chem. Series 59, 205 (1966)
24. Goheen, D.W. in "Lignin occurrence, formation, structure and reactions", ed. by K.W. Sarkanen and C.H. Ludwig, p. 797, Wiley Interscience, N.Y. (1971)
25. Schuman, S.C. and S. Field, Canadian Patent 851, 709 (1972)
26. Wiser, W.H., DOE Symposium Series, 46, 219 (1978)
27. Petrakis, L. and D.W. Grandy, Fuel, 60, 115 (1981) and 61, 21 (1982)
28. Glasser, V.W. and W. Sandermann, Holzforschung, 3, 73 (1970)
29. Kleinert, T.N. and J.R. Marton, Nature, 196, 334 (1962)
30. Steelink, T.R. and G. Tollin, J. Am. Chem. Soc., 85, 4048 (1963)
31. Parkhurst, Jr., H.J., Derk T.A. Huiberts, and Maurice W. Jones, Symposium on Alternative Feedstocks for Petrochemicals, Amer. Chem. Soc., Div. of Petroleum Chem., August 24, 660 (1980)
32. Wiser, W., preprint, Fuel Division ACS meeting, 20 (2), 122 (1975)
33. Freudenberg, K., Holzforschung, 8, 3 (1964)

Energy Recovery from Lignin, Peat and Lower Rank Coals, edited by D.J. Trantolo and D.L. Wise
Elsevier Science Publishers B.V., Amsterdam 1989 — Printed in The Netherlands

Chapter 7

CONTINUOUS ACID HYDROLYSIS OF CELLULOSIC BIOMASS

Hans E. Grethlein
and
Alvin O. Converse
Thayer School of Engineering,
Dartmouth College, Hanover, NH, USA

INTRODUCTION

The appeal of continuous dilute acid hydrolysis of cellulosic biomass is the relatively simple process flow sheet compared to enzymatic and concentrated acid hydrolysis. However, the maximum of glucose yield that can be obtained in one stage of hydrolysis is limited to 50 to 60% of the potential glucose in the substrate. Because of by-product formation, it is recognized that the process must be integrated with useful by-product recovery. In a one-stage process the useful products are glucose and furfural; in a two-stage process the useful products are xylose from the first stage and glucose from the second stage. The remaining solid residue after hydrolysis is mostly lignin.

While this chapter focuses on a continuous hydrolysis process, the same kinetics are encountered in a batch process, such as the Scholler process (ref. 1) and a semi-batch or percolation process such as the Madison Process (ref. 2) or the New Zealand process (ref. 3). In the semi-batch process, fresh, hot (150 to 180°C) dilute sulfuric acid ($\frac{1}{2}$ to 1%) is percolated through a packed bed of woodchips. By having an open packed bed, the liquid resident time is in the range of 10 to 30 minutes while the chips remain in the reactor for about three hours. The glucose yield can be as high as 70% of the potential glucose when the residence time of the liquid phase is kept short, but as a result, the total sugar concentration in the accumulated percolate is of the order of 5% (ref. 2).

By contrast, the liquid and solid phases in a continuous process have the same residence time in the reactor. Thus, the feed to the reactor must be a pumpable slurry which requires particle size reduction to 1 mm or

smaller.   This may be achieved through mechanical size reduction, steam
heating and explosion, or disc refining.   Special solids handling pumps
developed by Church and Wooldridge (ref. 4) and the extruder used by the NYU
process (ref. 5) can handle larger pieces of biomass from 1 to 2 mm, and
concentrated slurries.

The Dartmouth continuous process operates in the range of 230 to   260°C
with $\frac{1}{2}$ to 2% acid for 6 to 20 seconds. The high temperature, besides
reducing the reaction time, also increases the rate of glucose formation
relative to its decomposition. Because cellulosic slurries are thixotropic,
the velocity profile in a pipe is flat which gives the material plug flow
characteristics - an advantage in controlling the reaction. Naturally, a
continuous steady state process allows for convenient heat integration in
the design of the plant.

When dealing with substrates such as waste paper, or agricultural
residues such as corn stover or wheat straw, a sufficiently permeable packed
bed, required for the percolation process, may not be obtained.   In these
cases a plug flow reactor may be the only practical one.

EQUIPMENT FOR HYDROLYSIS

A bench scale continuous plug flow reactor was developed for process
studies for acid hydrolysis of cellulosic biomass (ref. 17). A schematic
diagram of the equipment is given in Figure 1.   Cellulosic biomass is
suspended as small particles (40 mesh or smaller have been used on the bench
scale, but larger particles of $\frac{1}{2}$mm would be suitable for a larger scale
operation) in water and pumped continuously through a flow reactor. The
pressure is increased above the saturation pressure of water for the chosen
reaction temperature. A positive displacement moving cavity pump (Robbins &
Myers, Springfield, Ohio) is used to handle the slurry.   Concentrated
sulfuric acid is injected into the slurry to give a final acid concentration
in the slurry of about 1% based on the water. If the slurry pump is acid
resistant, the acid can be added to the slurry in the slurry feed tank. The
slurry is heated by the injection of live steam into the slurry, although
indirect heating of the slurry by heat exchange is also possible, which
exposes the slurry to a longer temperature-time history. The temperature is
regulated by the steam flow to the reactor. The reactor is an insulated 1 m
tube made of Zirconium with an inside diameter of 1.18 cm which provides
about 6 to 12 seconds reaction time. The hot slurry is quenched by an
adiabatic flash across a 1.0 mm orifice to 1 atmosphere.   The sudden
temperature drop arrests the hydrolysis and the decompression of the solids
acts as a continuous steam explosion.   Hence, the biomass is subjected to

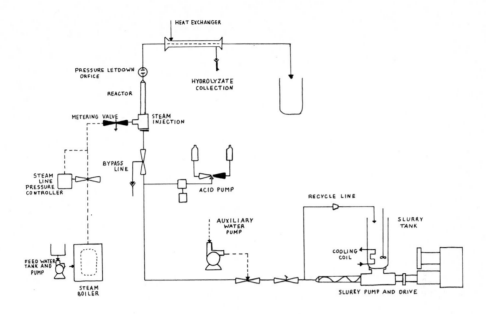

Fig. 1. Schematic Diagram of Laboratory Scale Continuous Plug Flow Reactor
for Dilute Acid Hydrolysis of Cellulosic Biomass

both hydrolysis and steam explosion. The reacted slurry is cooled to room
temperature by a heat exchanger.

This reactor has been operated up to 15 weight percent solids in the
feed slurry to the pump. For commercial processes the solids concentrations
should be as high as possible, 15% if not 20% or 30%. Work is under way to
develop such a process. However, most of our laboratory studies were done
at about 5% solids since the effectiveness of the hydrolysis did not depend
on the solids concentration.

After the slurry is cooled, the acid is neutralized with NaOH or
CA(OH)$_2$. In order to keep the salt level within the range tolerated by the
microorganisms used to ferment the final hydrolyzate, it is desirable to use
lime for neutralization which forms Ca_2SO_4. The precipitated Ca_2SO_4 does
not interfere with the subsequent fermentation.

It should be noted that the equipment in Figure 1 can be used for acid
pretreatment or conventional acid hydrolysis of cellulosic biomass. The
only difference is the temperature at which it is operated. Generally, 180
to 220°C is used for pretreatment, where the hemicellulose is hydrolyzed,
and 230 to 260°C is used for acid hydrolysis of cellulose. This same equipment can be used to study the kinetics of hydrolysis of lignin, peat, or

lignite using acid or base catalyst. The reactor has been modified to also allow air injection into the reactor so simultaneous hydrolysis and oxidation of biomass can be studied.

HYDROLYSIS KINETICS FOR CELLULOSE

Cellulosic biomass is really a mixture of cellulose, a linear polymer of glucan, hemicellulose, a branched polysaccharide of mixed anhydro-sugar moeties, and lignin, a branched polymer of mixed phenyl-propanes. The major reactions of cellulosic materials in dilute acid are shown in Figure 2. The hemicellulose, being amorphous, is hydrolyzed much faster than the cellulose which is partly crystalline. When conditions are chosen to hydrolyze the hemicellulose, little cellulose is hydrolyzed; and when conditions are chosen to hydrolyze cellulose, no hemicellulose is left, and the pentose is mostly converted to furfural and tars. Most of the lignin remains insoluble during acid hydrolysis, with a reduction in its average molecular weight and increase in its solubility in solvent such as ethanol (ref. 6) or dioxane (ref. 7). A minor portion is solubilized to mono- and di-aromatic ring compounds, the so-called acid soluble lignin (ref. 8). When the temperature is 260°C, the fraction of the lignin that is solubilized can be increased so that in certain biomass materials such as corn stover no solid residue is left.

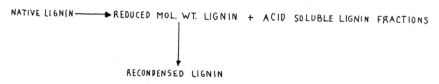

Fig. 2. Reactions of Biomass in Dilute Acid

For the purpose of process analysis, the carbohydrate in wood and agricultural residue can be thought to consist of glucan and xylan, the major sources are hexose and pentose respectively. The other minor sugars can be taken as a glucose or xylose equivalent. The following model based on the work of Saeman (ref. 9) is used for the formation of glucose.

$$
\begin{array}{c}
\text{resistant cellulose} \xrightarrow{\ k_1\ } \\
\text{glucan} \qquad\qquad\qquad\qquad \text{glucose} \xrightarrow{\ k_2\ } \text{decomposed glucose} \\
\text{available cellulose} \xrightarrow{\ k_0\ }
\end{array}
\qquad (1)
$$

Saeman found that the hydrolysis of cellulose in dilute acid can be modelled as a homogeneous system of consecutive pseudo-first order reaction in series by expressing the cellulose as potential glucose. That is:

$$
\frac{dC}{dt} = - k_1 C \qquad (2)
$$

$$
\frac{dG}{dt} = k_1 C - k_2 G \qquad (3)
$$

See Nomenclature at the end of chapter for definition of symbols.

In practice the rate k_0 is large compared to k_1, so that the glucose from available glucan is taken as free glucose at the initiation of the hydrolysis. Each rate constant k_i has the Arrhenius form with a modification for the effect of the acid as shown in equation 4:

$$
k_i = P_i A^{n_i} \exp(-E_i/RT) \qquad (4)
$$

The parameters are the pre-exponential factor P_i, the acid exponent n_i, and the activation energy E_i, which are estimated from isothermal reaction studies. Since the glucose decomposition is a homogeneous reaction, and independent of the cellulose hydrolysis reaction, it can be studied separately and the parameters of the rate constant k_2 can be determined accurately. These parameter values can then be imposed when evaluating the parameters in k_1. While McKibbins (ref. 10) determined these parameters from batch reaction condition, Smith found that in the continuous flow reactor a larger rate constant for k_2 was needed to predict the rate of glucose decomposition (ref. 11) than given by McKibbins. Although the first order glucose decomposition model is an over-simplification, it is adequate for process design studies when the focus is on glucose, but is not adequate if one is interested in the decomposition products of glucose.

If one measures the potential glucose equivalent of the unreacted cellulose, the parameters for the rate constant k_1 can be determined by known methods for first order reactions (ref. 12). This method is, furthermore, independent of the parameter values of k_2. However, it is more accurate to measure the glucose in the hydrolyzate than residual potential glucose. The parameters for k_1 have therefore been determined by a method according to the following procedures.

For a given temperature and acid concentration the integration of equation (3) gives the predicted glucose at time t.

$$\hat{G}(t) = G_1 \left\{\frac{k_1}{k_1 - k_2}\right\} \{\exp(-k_2 t) - \exp(-k_1 t)\} + G_0 \exp(-k_2 t) \qquad (5)$$

Note that G_0 is the glucose fraction at t = 0 from the available glucan and is another parameter to be estimated for a given cellulosic material. The value of $\hat{G}(t)$ depends only on the parameters assigned to k_1 and k_2. The parameters of k_2 are assigned from prior work of Smith (ref. 11) and the parameter P_1, n_1, E_1, and G_0 are found by a non-linear search technique that minimizes the sum of squared differences between the observed and predicted glucose yield; namely:

$$\sum_{i=1}^{n} (G(t_i) - \hat{G}(t_i))^2 \qquad (6)$$

where i is summed over the number of data points.

The kinetic parameters and the evaluation of the rate constants at 200° and 240°C for 1% acid are given in Table 1 for a number of substrates studied in the Dartmouth plug flow reactor. Note the last entry is for glucose and relates the parameters for k_2.

While at first there appears to be a wide range of values for P_i and E_i for the various substrates, in fact the parameters P_i and E_i are highly correlated. Hence, the rate constant, k_1, for a specific acid concentration and temperature (as given in the last two columns of Table 1) is remarkably similar for all these cellulosic substrates. Moreover, since the activation energy $E_1 > E_2$ and the acid exponent $n_1 > n_2$, the selectivity, defined as k_1/k_2, increases as the temperature or the acid concentration are increased. A typical set of glucose yield curves is shown in Figure 3 for Solka Floc BW-200 using the parameters from Table 1.

The kinetic parameters in Table 1 are for the first pass through the plug flow reactor. It is of interest to know whether the kinetics are altered for the residual cellulose on a second pass through the reactor.

Data from Currier (ref. 13) show essentially the same rate constants are estimated for steam exploded poplar when it is run through the flow reactor for a second time. Thus, one can use the kinetic parameters in Table 1 to simulate processes with one or more hydrolysis steps; or where residual cellulose is recycled to the inlet of a reactor.

From the work of Thompson (ref. 14) it was demonstrated that the kinetic parameters do not depend on the solid concentration in the reactor over the range of 5% to 14%. This is because the acid is a catalyst and it is the hydrogen ion concentration that fixes its activity, not the acid to solid ratio. When biomass such as wood has a certain ash content which neutralizes some of the acid, the acid concentration used in the model is the net acid concentration after the ash neutralization. At some point, as the solids concentration increases, the acid to solid ratio will become important because there is less and less water. We expect the kinematic model used here to hold up to 20%, if not 30%, solids.

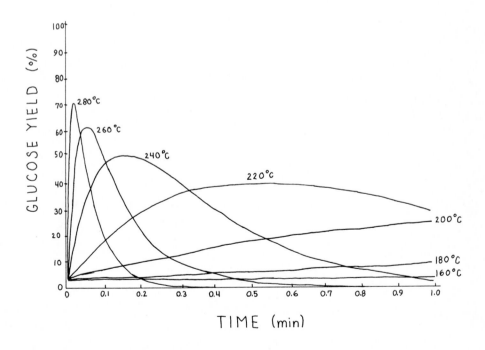

Fig. 3. Glucose Yield Predicted by Equation 5 for the Parameters from Table 1 for Solka Floc BW-200 for 1% Acid

Table 1: Kinetic Parameters for Acid Hydrolysis of Various Substrates in Equations 2 and 3 and Instantaneous Glucose Yield

Substrate	G_0	P_i min^{-1}	n_i	E_i cal/gmol	k_i min^{-1} for 1% acid at 200°C	at 240°C	Ref.
Corn Stover	.146	9.62×10^{14}	1.40	32,800	0.670	10.2	17
Solka-Floc BW-200	.030	5.15×10^{16}	1.14	37,000	0.428	9.22	14
Poplar	.011	6.12×10^{15}	0.987	35,150	0.350	6.47	
White Pine	.043	7.80×10^{13}	0.963	30,170	0.880	10.78	
Mixed Hardwood 90% Birch 10% Maple	.006	1.45×10^{15}	1.16	33,720	0.380	6.24	
Newsprint	.038	1.18×10^{17}	0.696	38,300	0.237	5.68	13
Computer Paper	.039	3.86×10^{15}	0.763	34,850	0.306	5.51	13
Causted Extracted Steam Exploded Poplar	0	2.875×10^{17}	0.816	39,090	0.246	6.36	13
Kraft Pulp Solka Floc BNB-100	.015	1.68×10^{15}	1.20	33,860	0.381	6.33	13
Douglas Fir Prehydrolyzed	0	1.73×10^{19}	1.34	42,900	0.260	9.15	9
Glucose		3.96×10^{8}	0.569	21,00	0.775	4.43	11

HYDROLYSIS KINETICS FOR HEMICELLULOSE

In order to predict the xylose and furfural yields, a model of hemicellulose hydrolysis was developed which is applicable to the plug flow reactor (ref. 15). The hydrolysis of the hemicellulose is taken to follow the analogous sequential pseudo first-order reactions as in cellulose to give xylose and decomposed xylose. The xylose decomposition is modelled after the work of Root (ref. 16) in order to predict the furfural yield. The reaction scheme is given in equation (7):

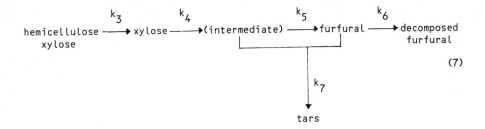

$$\text{hemicellulose} \xrightarrow{k_3} \text{xylose} \xrightarrow{k_4} \text{(intermediate)} \xrightarrow{k_5} \text{furfural} \xrightarrow{k_6} \text{decomposed}$$

(7)

Since all the reactions are homogeneous except for xylan hydrolysis, independent estimates from the plug flow reactor can be obtained for the decomposition of furfural, k_6 and xylose, k_4 by following the loss of pure furfural and pure xylose respectively over the temperature and acid concentration range of interest. Finally, k_3 was obtained by measuring the appearance of xylose from mixed hardwood (90% birch and 10% maple). The xylose appearance was measured simultaneously with the glucose appearance for the reaction conditions used to find k_1. Mixed hardwood (90% birch and 10% maple) is being studied now to complete the hydrolysis model of hemicellulose.

A typical set of observed data for glucose, residual glucose, xylose and furfural is given in Figure 4 for mixed hardwood along with the predicted values from the integration of Equations (2), (3), and a preliminary model for the xylose and furfural. Note that at temperatures above 240°C the furfural yield is 80% of the potential xylan, while the glucose yield is 55% of the potential glucose.

CONCLUSION

Together, the kinetic models for cellulose and hemicellulose can give a useful prediction of the glucose, xylose and furfural yields as a function of acid concentration, temperature and time. We view the development of the hemicellulose model as an important addition to the cellulose model for describing the entire hydrolysis of the carbohydrate in the biomass.

A final word about the hydrolysis of lignin type materials. Lignin in wood is made of a random branched polymer of phenyl propane groups connected through carbon-oxygen and carbon-carbon bonds. As wood decays in nature the carbohydrate is consumed by microorganisms and the original lignin is evident in the succeeding peat and lignite.

When acid or base is used as a catalyst, the carbon-oxygen bonds in the lignin matrix can be hydrolyzed with a corresponding reduction in molecular weight of the residue. Since the flow reactor allows high temperature-short reaction times, the breakdown of lignin type components can be studied with-

266

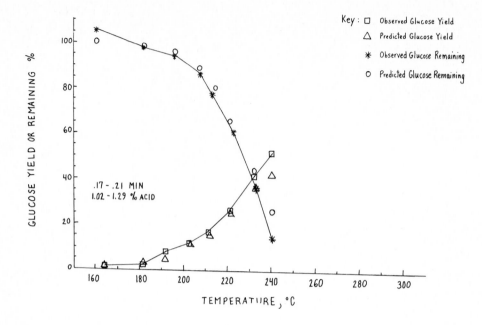

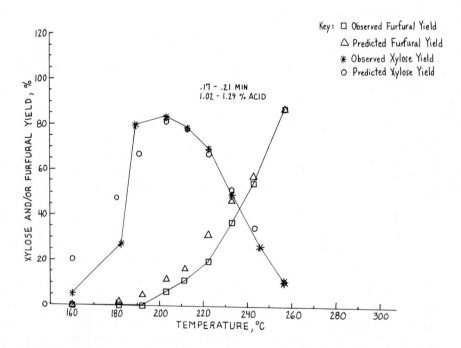

Fig. 4. Observed and Predicted Yields as Percent of Potential Values
in Plug Flow Reactor for Feed Solids of 6.6%

out the recondensation of these products into tars. The use of the flow reactor for lignin type breakdown into aromatics is promising and should be exploited.

NOMENCLATURE

A	weight % H_2SO_4 in the aqueous phase
C	fraction unreacted cellulose (remaining potential glucose)
E_i	activation energy of reaction i cal/gr mol °R
F	fraction potential xylose as furfural, F x 100 is percent furfural yield
G, G(t)	fraction cellulose as glucose at time, t, G x 100 is percent glucose yield
G_1	fraction cellulose as unavailable at t = 0
G_0	fraction cellulose converted to glucose at t = 0
$\hat{G}$	fraction cellulose as glucose at time, t, predicted by model in Equation 5
I	fraction potential xylose as intermediate xylose decomposition product
k_i	rate constant for reaction i, min^{-1}, reactions 0, 1, 2, 3, 4, 5, 6, and 7 defined in Equations (1) and (7)
n_i	dimensionless power on acid concentration in Equation (4) for reaction i
P_i	preexponential factor in Equation (4) for reaction i, min^{-1}
t	time

REFERENCES

1. H. Scholler, Zellstoff Faser, _32_, 65-74, 1935
2. E.E. Harris and E. Belinger. "Madison Wood Sugar Process." Ind. Eng. Chem., _38_, 890, 1946
3. B.K. Guha and A.L. Titchner. "Acid Hydrolysis of Wood." Report No. 56, New Zealand Energy Research and Dev. Committee, January, 1981
4. J.A. Church and D. Wooldridge. "Continuous High-Solids Acid Hydrolysis of Biomass in $1\frac{1}{2}$ in Plug Flow Reactor." Ind. Eng. Chem. Prod. Res. Dev., _20_, 371-375, 1981
5. B. Rugg, P. Armstrong, and R. Stanton. "Preliminary Results and Economics of NYU Process." Developments in Ind. Microbiology, _22_, 131, 1981
6. H. Grethlein et al. "Annual Report on Acid Hydrolysis of Cellulosic Biomass." SERI, March, 1980 to March, 1981
7. J.H. Lora and M. Wayman. "Delignification of Hardwoods by Autohydrolysis and Extraction." TAPPI, 6, (6), 47-50, 1978.
8. D.A. Stanek. "A Study of Low-Molecular Weight Phenols Formed upon the Hydrolysis of Aspen." TAPPI, _41_, (10), 601-609, 1958. K.T. Sears et al. "Southern Pine Prehydrolyzates." J. Polym. Sci., Part C. No. 36, 425-447, 1971

268

9. J.F Saeman. "Kinetics of Wood Saccharification." <u>Ind. Eng. Chem.</u>, 37, 43, 1945

10. S.W. McKibbins, J.F. Harris, J.F. Saeman and W.K. Neil. "Kinetics of the Acid Catalyzed Conversion of Glucose to HMF and Levulinic Acid." <u>Forest Products Jr.</u>, 12, 17, 1962

11. P. Smith, H.E. Grethlein and A.O. Converse. "Glucose Decomposition at High Temperature Mild Acid, and Short Residence Time." <u>Solar Energy</u>, 28, 41, 1982

12. J.W. Moore and R.G. Pearson. "Kinetics and Mechanisms." 3rd Ed. pp. 19, 68-70, John Wiley & Sons, New York, 1981

13. P. Currier. "Investigations of Acid Hydrolysis of Delignified Cellulose by Multiple Pass through a Plug Flow Reactor." B.E. Thesis, Thayer School of Engineering, Hanover, NH, August, 1982

14. D. Thompson and H.E. Grethlein. "Design and Evaluation of a Plug Flow Reactor for Acid Hydrolysis of Cellulose." <u>Ind. Eng. Chem.</u>, Product <u>Res. Devl.</u>, 18, 250, 1979

15. I. Kwabena Kwarteng. "Kinetics of Acid Hydrolysis of Hardwood in a Continuous Plug Flow Reactor." Ph.D. Thesis, Thayer School of Engineering, Hanover, NH, July, 1983

16. D.F. Root, J.F. Saeman, and J.F. Harris. "Kinetics of the Acid Catalyzed Conversion of Xylose to Furfural." <u>Forest Products Jr.</u>, 9, 158, 1959

17. J.J. McParland, H.E. Grethlein and A.O. Converse. "Kinetics of Acid Hydrolysis of Corn Stover." <u>Solar Energy</u>, 28, 55, 1982

Energy Recovery from Lignin, Peat and Lower Rank Coals, edited by D.J. Trantolo and D.L. Wise
Elsevier Science Publishers B.V., Amsterdam 1989 — Printed in The Netherlands

Chapter 8

SIMILARITIES AND DIFFERENCES IN PRETREATING WOODY BIOMASS
BY STEAM EXPLOSION, WET OXIDATION, AUTOHYDROLYSIS, AND
RAPID STEAM HYDROLYSIS/CONTINUOUS EXTRACTION

Tor P. Schultz,
Gary D. McGinnis,
and Christopher J. Biermann
Mississippi Forest Products Laboratory,
Mississippi State University, MS 39762, USA

1.   INTRODUCTION

1.1 Introduction and Review of Literature

Since about 1970 the need for alternative sources of energy and
chemicals currently obtained from petroleum has led to extensive research
efforts to develop economical processes for converting agricultural,
forestry, and municipal wastes into fuels and chemicals. This research has
involved studies of both old and new processes for biomass conversion. Some
of the older processes, such as acid hydrolysis and fermentation to produce
ethanol, are being reinvestigated by use of newer technology to determine
whether these processes can be economically competitive with those based on
petrochemicals for production of certain organic chemicals.   At the same
time, studies are also proceeding on new processes for biomass conversion.
One example is the development of enzymatic processes made possible with the
development of strains of fungi, yeast, or bacteria that produce enzymes
which react at faster rates with polysaccharides, tolerate greater temperature fluctuations, and convert a wider variety of sugars into ethanol or
other products than those enzymes used previously.

Cellulose, lignin, and hemicellulose are the major constituents of
woody biomass and are present in quantities of approximately 50, 25 and 25%,
respectively.   Cellulose and lignin are the two most abundant organic

Reprinted with permission from Energy from Biomass and Wastes VIII, D.L.
Klass, Ed. (Inst. of Gas Technology, Chicago, IL, Sept. 1984) pp. 1171-1198.

compounds on earth. The structure of cellulose is similar in all plants and consists of repeating (1-4) linked D-glucopyranose units. The structures of the hemicelluloses and lignin vary, depending on the plant material.

Proposed processes for making ethyl alcohol from cellulosic materials (Hajny, ref. 11) consist of three major steps: a pretreatment step to partially break down the biomass; a hydrolysis step to convert the poly-saccharides into monosaccharides; and a fermentation step to convert the monosaccharides into ethyl alcohol. The primary purpose of the pretreatment is to disrupt and break down the lignin and hemicellulose polymers and over-come the resistance of crystalline cellulose to hydrolytic cleavage. In other words, the pretreatment should increase the rate and extent at which the cellulose is hydrolyzed.

In addition, the pretreatment should:

(a) utilize inexpensive chemicals and require simple equipment and procedures;
(b) solubilize and fractionate the lignin and hemicellulose; and
(c) be suitable for pretreating a wide variety of woody material and agricultural wastes.

The only way to attract large-scale investments from the private sector is to produce ethyl alcohol from cellulosic material at a cost comparable to that of gasoline. Consequently, each step in a biomass conversion process must be relatively inexpensive and any chemicals or equipment used must be inexpensive.

The ability to fractionate lignin and hemicellulose components from the biomass in the pretreatment process is important for several reasons. The yield of alcohol is dependent on the cellulosic content of the biomass since the hemicelluloses of hardwood and agricultural wastes consist mainly of five-carbon sugars which are not fermentable to ethyl alcohol using commer-cial strains of yeast. Thus, pretreated wood with an 80% cellulose content would give a higher ethanol yield than nontreated wood with a cellulose content of 45%. Another advantage in removing the lignin and hemicellulose is the general enhancement of the hydrolysis step when these components are removed (Lipinsky, ref. 13; Goldstein, ref. 10). This is particularly true if enzymatic hydrolysis is used since steric factors caused by the lignin and hemicellulose polymeric network reduce the hydrolysis rate. Finally, if large-scale biomass conversion units are ever to become economically feasible, some markets for the lignin and hemicellulose must be found since these materials make up approximately 50% of the plant material (Cowling and

Kirk, ref. 7). Therefore, at some point in the process, these components need to be isolated and separated.

The difficulty of hydrolyzing cellulose makes it important to decrease the crystallinity and molecular size of the cellulosic material. Even cellulose that contains no lignin or hemicellulose is difficult to hydrolyze witout pretreatment (Millett et al., ref. 24 ; Emert and Katzen, ref. 8; Lipinsky, ref. 13). For acid hydrolysis, disrupting the crystalline structure of cellulose can improve the rate of cellulose conversion to glucose. For enzymatic hydrolysis, the hydrolysis rate can be enhanced by increasing the surface area on which the cellulases can work.

One of the more important characteristics of a pretreatment is that it should be effective with a wide variety of biomass materials. Many pretreatments are very dependent on the type of feedstock used. In reviewing past research, one finds that a large amount of biomass research has been conducted on aspen. Not only is aspen relatively rare in the United States, but Millett et al. (ref. 25) and Lipinsky (ref. 13) have both noted that aspen is uniquely easy to pretreat. Therefore, promising results obtained using aspen do not necessarily indicate that the pretreatment will be successful with other woods. In addition, hardwoods are generally more susceptible to pretreatment than softwoods. In order to compare different pretreatments, a realistic feedstock should be used. For economic reasons, a biomass plant would prefer to use the lowest cost cellulosic material available in large quantities in each particular region. In the southeastern United States, the feedstock source of choice would probably be mixed hardwood chips obtained by total-tree chipping of the so-called trash species.

1.2 Review of Selected Pretreatments

Autohydrolysis: The reaction between water and woody biomass has been studied for at least fifty years. Since the pioneering work of Richter (Rydholm, ref. 26), most of the early work has been directed toward developing a commercial process as a first stage in pulping wood. The treatment of wood with water at temperatures up to 185°C causes the formation of organic acids by the cleavage of labile ester groups present on the hemicellulose fraction of the wood. The liberated organic acids, mainly acetic acid, catalyze the hydrolysis of the hemicelluloses to soluble oligosaccharides. In a typical reaction, 15 - 20% of hardwood hemicelluloses and 10 - 15% of softwood hemicelluloses can be solubilized by autohydrolysis at 160 - 170°C for 45 minutes (Rydholm, ref. 27). The rate and effectiveness of this reaction can be improved by addition of small amounts of inorganic acid (sulfuric or hydrochloric). Currently, autohydrolysis is used as the first stage of the prehydrolysis-Kraft process, a commercial process for pulping both hardwoods and softwoods.

When the autohydrolysis reaction is run under more severe conditions of higher temperatures or longer times, the solubilized carbohydrates undergo a series of secondary reactions to form furfural (2-furancarboxaldehyde) and 5-hydroxymethyl furfural. These reactive compounds can undergo further self-condensation reactions to form polymers or can react with lignin to form a copolymer.

Although the hydrolysis of the hemicellulose is one of the major reactions of autohydrolysis, recent work has shown that lignin in certain hardwoods also undergoes considerable degradation and modification during autohydrolysis. More than forty years ago, Aronowsky and Gortner (ref. 1) found that 45% of the lignin present in aspen wood was soluble in ethanol-benzene when autohydrolyzed with water at 185°C for four hours. More recent studies by Lora and Wayman (ref. 14) indicate that autohydrolysis causes a considerable change in the overall structure in the lignin molecule, especially in hardwoods. They found that after autohydrolyzing aspen, poplar, or eucalyptus at 175 - 220°C, 60 - 92% of the lignin becomes soluble in aqueous dioxane or dilute base. Wayman and co-workers (refs. 4, 5, 15, 34) characterized the lignin obtained from autohydrolysis using nitrobenzene oxidation, infrared and ultraviolet studies, chemical fractionation and molecular-weight distribution studies. Unfortunately, the autohydrolysis reaction of lignin appears to be effective with only certain hardwoods, e.g. aspen (Wayman and Lora, ref. 35).

Wet Oxidation: When air or oxygen is present during the autohydrolysis, the process is called wet oxidation. Schaleger and Brink (refs. 28, 29), at the University of California, found that the initial reaction during wet oxidation is the formation of acids. These acids are formed by the solubilization of the acidic hemicellulose components, de-esterification of the acetate groups on the hemicelluloses and oxidation. As the acid concentration increases and the pH drops, hydrolytic reactions become favorable. More and more of the hemicelluloses are broken down into lower-molecular-weight fragments which dissolve in the water. This reaction is not limited to the hemicellulose fraction of the wood. At the lower temperatures of 120 - 170°C, the cellulose and lignin are affected but are not solubilized to a large extent. According to Brink and Schaleger (refs. 28, 29), the rate of acid hydrolysis of the cellulose is increased substantially after a wet oxidation pretreatment. Thus the accessibility of the cellulose is increased by wet oxidation. The major difference between wet oxidation and autohydrolysis is that the wet-oxidation reaction with the hemicelluloses occurs at a lower temperature and is more rapid and complete due to the higher concentration of acids and the presence of oxygen. One of the major advantages of the wet-oxidation process is that it appears to be suitable for a wide range of woody biomass, including both hardwoods and softwoods.

Most of the earlier work was done at relatively low oxygen partial pressures. Recent work by McGinnis and co-workers (refs. 20, 21, 22) indicates that higher oxygen pressures and the addition of metal catalysts lead to a more specific reaction, both at low and high temperatures.

<u>Steam Explosion</u>: The steam explosion process was originally developed by Mason in 1925 and has been extensively used in the manufacture of hardboard (Spalt, ref. 33). The commercial process involves first filling a vertical cylinder with wood chips. Once filled with chips, the cylinder is sealed and pressurized with saturated steam at pressures up to 1,000 psig, and the chips develop high internal pressures. When the bottom of the cylinder is opened, the wood chips are defibrated by the sudden decompression. This steam explosion causes considerable chemical changes as well as physical changes.

In 1978 the Iotech Corporation Ltd. of Canada started using steam explosion for production of feed for ruminants. In view of the early results which showed the high digestibility of steam-exploded aspen, Iotech decided to explore its use as a method for pretreating aspen (Foody, ref. 9). Iotech reported that poorer results were obtained when other lignocellulosic materials, such as oak, were exploded.

Since Iotech reported their initial results, a few other investigators have also studied the steam explosion process as a biomass pretreatment for hardwoods (Marchessault et al., refs. 17, 18; Marchessault, ref. 19; Schultz et al., ref. 31). These investigators have found that the following chemical changes occur during steam explosion: (1) The lignin is broken down into products with a molecular weight range of from 150 to 7000. These products retain the basic lignin structure and are moderately condensed. Since the lignin is extensively depolymerized by cleavage of the α- and β-aryl-ether bonds, it is partially soluble in alkaline solutions or certain organic solvents. (2) The hemicelluloses are partially broken down and are predominantly soluble in hot water. In addition, some degradation products are formed which apparently condense with lignin, thereby increasing the lignin content. (3) Steam explosion causes a large increase in the accessibility of the cellulose to enzymatic hydrolysis. Jurasek (ref. 12), Foody (ref. 9) and Schultz et al. (refs. 31, 32) determined that the steam-explosion pretreatment results in a dramatic increase in the susceptibility of hardwoods and agricultural wastes to enzymatic hydrolysis. However, Marchessault and co-workers (refs. 17, 18, 19) report no noticeable changes in the crystallinity of the cellulose, although changes in the size/perfection of the cellulose crystallites were observed. Thus the increase in the rate of enzymatic hydrolysis may have been caused by disruption of the lignin polymer, rather than decrystallization of the cellulose.

Rapid Steam Hydrolysis/Continuous Extraction: The Rapid Steam Hydrolysis/Continuous Extraction procedure, called RASH, is similar to the steam-explosion process. Unlike the steam explosion, autohydrolysis, or wet-oxidation processes, however, soluble or gaseous products are continuously removed from the reaction zone, thus minimizing further degradation.

The RASH process consists of adding saturated steam to a reactor filled with the biomass material. As with the other three pretreatments, the high temperature causes hydrolytic depolymerization of the hemicelluloses and lignin. As part of the steam cools, it condenses and flows down to the reactor bottom. The depolymerized carbohydrate and lignin fragments which are soluble in the hot steam condensate are also carried down to the bottom of the reactor, where a transfer line continuously removes and cools the products. The use of catalysts is not necessary for the RASH pretreatment. However, a variety of gaseous, liquid, or solid catalysts can easily be used if it is shown that the catalysts provide a particular benefit.

The RASH process has been studied utilizing hardwoods (Biermann et al., ref. 3; Schultz and McGinnis, ref. 32) and rice hulls (Lowrimore et al., ref. 16).

This work has shown that: (1) The hemicelluloses are depolymerized and extracted from the biomass with little degradation; (2) up to 95% of the lignin is extracted by aqueous alkali or organic solvents; (3) the remaining solid residue is considerably enriched in cellulose; (4) the enzymatic hydrolysis rate of the cellulose increases dramatically; and (5) little or no increase in the rate of acid hydrolysis is observed. It has also been determined that the addition of selected catalysts ($AlCl_3$, $Fe_2(SO_4)_3$, H_2SO_4, etc.) or reagents (oxygen, butanol, acetic acid) can give greater selectivity in extracting and altering the lignin and polysaccharides (McGinnis, ref. 23).

A summary of the four pretreatments is given in Table 1.

Table 1. Summary of the four pretreatments.

Pretreatment	Advantages	Disadvantages
Autohydrolysis	Used commercially.	Products formed undergo further reactions. Requires relatively long reaction time.
Wet oxidation	Relatively low reaction temperature. At higher temperature is a direct conversion process into carboxylic acids. Works with both softwoods and hardwoods. Increases rate of acid hydrolysis. Exothermic reaction can generate some process heat.	Exothermic reaction is potentially explosive. Relatively costly process. Requires relatively long reaction time.
Steam explosion	Used commercially. Relatively rapid process. Defibrates wood chips.	Water-soluble products are not fractionated during pretreatment.
RASH	Relatively rapid process. Continuously extracts products, thus minimizes degradation. Steam requirement may be lower than steam explosion.	Does not defibrate product.

2. EXPERIMENTAL PROCEDURE

2.1 Raw Material

Unscreened, mixed southern hardwood chips, produced by total-tree chipping, were obtained from a commercial wood yard. The length of the chips varied from about 0.1 to 15 cm. Approximately 82% by weight of the mixed chips were actual wood chips, with the majority being oak and gum species. Pine chips accounted for approximately 1% of the total chips. The remainder consisted of small twigs, leaves, bark, dirt and other material. For the steam-explosion runs, the green hardwood chips were used as is. For the autohydrolysis, wet-oxidation, and RASH runs, the chips were air-dried and ground with a 6 mm screen.

2.2 Pretreatments

Steam Explosion: A Masonite® 1-cubic-foot pilot gun, equipped with a pressure gauge and digital thermometer, was used in the steam explosion process. Prior to a run, a weighed bag of green chips was added to the reactor. The reactor was sealed and steam was gradually added so that the reactor reached the desired temperature in 1 minute. Then additional steam was added, as necessary, to keep the temperature as stable as possible for the reaction time of 1.0 minute. At the end of the reaction, the contents were blown out of the bottom of the reactor and through a cyclone. Material balances were calculated from the material recovered. However, the cyclone was not designed for high pressure runs, where the fibers are formed into many small fragments. Thus the material balance for the high-pressure runs was unrealistically low.

After steam explosion, the fibers were air-dried and ground in a Wiley Mill with a 2 mm screen. The grinding procedure was necessary to mix the fiber thoroughly. The air-dried ground samples were then stored in a refrigerator prior to analysis.

Autohydrolysis: A 300-ml Mini-Parr reactor with thermocouple was employed for the autohydrolysis runs. The water-to-wood ratio was 10:1, with 15 grams (oven-dry basis) of the air-dried 6-mm wood chips used per run. Since heat-up time was approximately 20 minutes, the temperature was maintained for only 5 minutes. The reaction was continuously stirred during the heat-up period and reaction time. At the end of the 5-minute reaction time, the reactor was cooled in water.

The solid residue was isolated and air-dried. Then the material balance was determined, and the solid residue was ground with a 2-mm screen and stored in a refrigerator. A portion of the liquid condensate was also saved for later analysis.

Wet Oxidation: The procedure for the wet-oxidation runs was similar to the autohydrolysis runs. However, 240 psig oxygen (room temperature) was added to the cold reactor prior to the wet-oxidation run. The heat-up time was approximately 10 minutes, with the temperature maintained for 30 minutes.

RASH: The RASH pretreatment was carried out using 30 grams of air-dried 6-mm chips. The chips were placed in a holder, then sealed into the RASH reactor. Steam was added to the reactor, with the run typically reaching the desired temperature in approximately 30 seconds. The reaction time was

1 minute. Steam condensate, products soluble in the condensate, partible products and gaseous products were removed from the bottom of the reactor through a tube equipped with a restriction valve. The restriction valve ensured that the necessary reaction pressure was maintained. The steam condensate and products were cooled and collected. Approximately 300 ml of condensate was collected for a 260°C 1-minute run. At the end of the 1-minute reaction, the steam line into the reactor was closed, and the condensate removal line was fully opened. Typically, cool-down time was about 40 seconds.

The solids were removed, air-dried, and weighed to determine the moisture content and material balance. The air-dried, solid residue was then ground with a 2-mm screen and stored in a refrigerator. The steam condensate volume was measured and a portion was saved for later analysis.

Table 2 summarizes the reaction conditions used for each pretreatment.

Table 2. Summary of the reaction conditions used for each pretreatment.

Pretreatment	Reagent(s)	Approximate heat-up time, min.	Reaction time, min.	Reaction temp. range, °C	Chip size
Autohydrolysis	Hot water	20	5	160-260	6-mm
Wet Oxidation	Hot water and oxygen	10	30	148-172	6-mm
Steam Explosion	Steam	1.0	1	190-235	Total-tree chips
RASH	Steam	0.5	1	170-258	6-mm

2.3 Chemical Analysis of the Pretreated Material

The steam-exploded fiber was extracted with hot water for 1 hour, then filtered and air-dried. This water extraction was done because the other three pretreatments fractionated the water-soluble products; therefore, a comparison of the results would be more meaningful if the steam-exploded fiber were also water extracted. (The RASH process probably does not completely remove all water-soluble products.) The analysis of non-extracted, exploded hardwood fiber is given in a previous paper (Schultz et al., ref. 30).

The solids were analyzed in the following manner: The insoluble lignin was determined by the Klason procedure, with the soluble lignin measured by ultraviolet at 205 nm. The sulfuric acid hydrolyzate was analyzed for sugars by gas chromatography (GC) by the aldononitrile acetate procedure (Chen and McGinnis, ref. 6). A portion of the hydrolyzate was analyzed separately for glucose content using a glucose analyzer, YSI Model 23A, to verify the results obtained by GC. The acid hydrolysis rates of the pretreated fiber were determined by adding 1.5 grams of the fiber to 100 ml of 20 volume percent sulfuric acid. The samples were heated in an ethylene glycol bath set at 105°C. After 6 hours, a sample was removed, neutralized, and analyzed with a glucose analyzer. Since the samples had different amounts of cellulose present in the starting fiber, the results were based on the total available glucose originally present in the fiber. For the rates of enzymatic hydrolysis, approximately 0.13 gram of the air-dried ground fiber was added with 10 ml of enzyme solution to a test tube. The enzyme solution consisted of sodium acetate-acetic acid buffer (pH 5.0), 10 mg/ml of cellulase enzyme (Miercellase CESB, from Trichoderma viride, purchased from Meiji Seika, Ltd., Tokyo, Japan) and 0.32 mg/ml of sodium azide to inhibit microbial organisms. The Meicelase CESB enzyme had a cellulolytic activity, measured by the Somogyi method, of 1130 m/g. The test tubes were placed in shaking water bath set at 40°C. The glucose formed was measured after 24 hours using a YSI glucose analyzer. In addition to the fiber samples, two control samples containing Whatman no. 1 filter paper were also included. The cellulase enzyme was chosen because it could be stored for long periods, and thus the results from different runs could be directly compared.

The liquid portion from the autohydrolysis, wet-oxidation, or RASH runs were analyzed as follows: 3 ml of 72% sulfuric acid were added to 50 ml of liquid. The samples were refluxed 4 hours. The carbohydrates were then analyzed by GC using the aldononitrile acetate procedure of Chen and McGinnis (ref. 6). The insoluble lignin values were determined by filtering the acid hydrolyzate, and the soluble lignin measured by an ultraviolet spectrophotometric procedure.

All samples were run in duplicate, with the average value reported.

3. RESULTS AND DISCUSSION

The percent solids recovered from the wet oxidation, RASH, and autohydrolysis pretreatments are given in Figure 1. The percent solids recovered are not shown for the steam explosion runs because of the material losses that occurred in the cyclone.

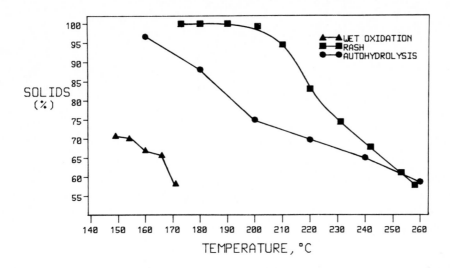

Fig. 1. Solids recovered after pretreatment, based on the oven-dry weight
prior to pretreatment of the mixed hardwood chips

As expected, the wet-oxidation pretreatment solubilized the greatest
amount of material at the lowest temperature. At 170°C, approximately 45%
of the mixed hardwood chips were solubilized. Above 175°C, wet-oxidation is
no longer a pretreatment but is considered a direct wood-to-chemical
conversion process (McGinnis et al., ref. 22).

Autohydrolysis showed an almost linear decrease in recovered solids as
the reaction temperature increased. At 260°C, approximately 40% of the
starting material was solubilized. Unlike autohydrolysis, no weight loss
was observed for wood pretreated by the RASH process below 200°C. At
temperatures above 200°C, the weight loss after the RASH pretreatment
occurred fairly rapidly with about 60% solids recovered at a RASH reaction
temperature of 260°C.

The total glucose recovered from both the solid and liquid portions,
based on the starting oven-dry weight, is shown in Figure 2. (The glucose
recovered was measured by hydrolyzing the recovered solids and liquid
fraction and is, therefore, actually a measure of the recovered cellulose
and glucose oligomers.) The wet-oxidation and autohydrolysis pretreatments
both show a significant decrease in the total glucose recovered at the
higher pretreatment temperatures. As discussed earlier, soluble oligomers
can undergo further degradation reactions in these two pretreatments. The
RASH process, however, shows only a slight decrease in the glucose recovered

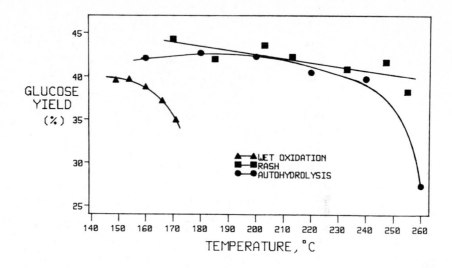

Fig. 2. Percent glucose recovered from both the solid and liquid fractions after pretreating mixed hardwoods. The glucose yield is based on the dry weight of the hardwood chips prior to pretreatment

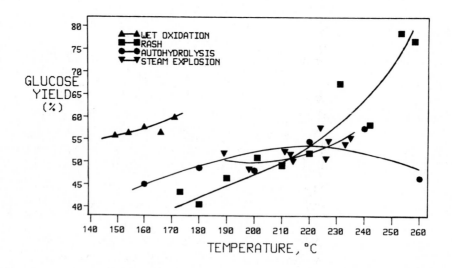

Fig. 3. Glucose content of the solids after pretreating mixed hardwood chips. The steam-exploded samples were extracted with hot water prior to analysis

(Figure 2) which means that when solubilized glucose oligomers are continuously extracted and rapidly cooled, the formation of degradation products is minimized.

As mentioned in the introduction, the pretreatment should increase the cellulose content of the recovered solids by extracting and fractionating the hemicellulose and lignin fractions, thus leaving behind a solid residue which is considerably enriched in cellulose. All solids recovered after pretreatment showed at least a slight increase in the glucose content (Figure 3). (The glucose content, which was measured by hydrolyzing the solids, is an indirect measurement of the cellulose content.) The untreated, water-extracted, mixed hardwood chips had a glucose content of 44%. The largest glucose content, approximately 75%, was for 260°C RASH-pretreated solids. Interestingly, the solids from an autohydrolysis pretreatment at 260°C show a significantly low glucose content of approximately 45%. This low glucose content is believed to be caused by both recondensation of the lignin and degradation of the carbohydrates. Water-extracted steam-exploded fiber has a maximum glucose content of approximately 55%. Wet-oxidation solids have a maximum glucose content of about 60%.

Any pretreatment that heated the hardwood chips above 190°C caused depolymerization and degradation of the hemicelluloses, as can be shown by analysis of the recovered solids for xylose content (Figure 4). The xylose content of the solids after autohydrolysis shows that a greater amount of xylose was degraded by autohydrolysis than by steam explosion at the same reaction temperature. However, this may be due to the longer reaction time of autohydrolysis (5 minutes) than steam explosion (1 minute). The depolymerization of the hemicelluloses, as discussed earlier, is due to the hydrolytic action of liberated acids. Since the RASH process continuously extracts liberated acids, it was not unexpected that the RASH pretreated solids have the highest xylose content at any given reaction temperature (Figure 4). Analysis of the liquid fraction from the autohydrolyzed runs showed little xylose present. Thus the soluble xylose oligomers had undergone further degradation reactions. However, analysis of the RASH condensate shows that up to 60% of the starting xylose can be accounted for in the condensate fraction (Biermann et al, ref. 3; Schultz and McGinnis, ref. 32).

The lignin content of the pretreated solids (Figure 5) shows that the lignin content increased as the autohydrolysis and steam-explosion temperatures increased. The RASH pretreated solids, however, show only a slight increase in the lignin content as the reaction temperature was increased. Analysis of the RASH condensate suggests that the steam/pyrolysis

282

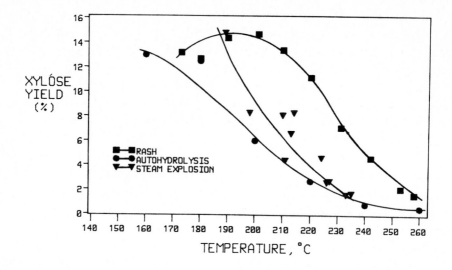

Fig. 4. Xylose content of solids after pretreating mixed hardwood chips. The steam-exploded samples were extracted with hot water prior to analysis

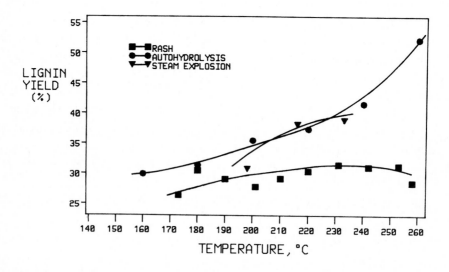

Fig. 5. Lignin content of the solid material remaining after mixed hardwood chips were pretreated. The steam-exploded samples were extracted with hot water prior to analysis

RASH treatment depolymerized, extracted, and removed a considerable portion of the lignin. By rapidly cooling the extracted lignin, it remained fairly uncondensed (Biermann et al., ref. 3).

It would be beneficial if the lignin remaining in the solid material could be fractionated by extraction with aqueous alkali. Figure 6 shows the glucose content after base-extracting the pretreated solids. The solids were boiled in a 2% sodium hydroxide solution, filtered, washed, and air-dried. The glucose content of the RASH and steam-exploded solids are similar, with a maximum glucose content of about 75%. The autohydrolyzed solids, after base extraction, also show a maximum glucose content of about 75%. However, above 230°C the glucose content decreases rapidly for auto-hydrolyzed samples. Surprisingly, analysis of the solids from wet-oxidation, steam-explosion, and RASH pretreatments all show that a significant portion of the glucose (about 30%) is removed by the base-extraction procedure (Schultz et al., refs. 30, 31, 32); Biermann et al., ref. 3; McGinnis, ref. 23). This implies that the cellulose has been at least partially depolymer-ized, but not necessarily decrystallized.

In addition to the lignin being soluble in aqueous alkali after the RASH pretreatment, a significant portion of the lignin was also soluble in acetone (Biermann et al., ref.3; Schultz and McGinnis, ref.32). This solub-ility suggests that the lignin is moderately uncondensed after a RASH pre-treatment, a finding later confirmed by nitrobenzene oxidation experiments on white oak and sweetgum samples pretreated by RASH (Schultz and McGinnis, ref. 32). Lignin of steam-exploded fiber was found to be more condensed than the lignin in the RASH samples (Schultz and McGinnis, ref. 32).

A pretreatment should not only depolymerize and fractionate the major chemical components, but should also increase the rate and extent of cellulose hydrolysis. The amount of glucose formed by enzymatic hydrolysis after 24 hours is shown in Figure 7. Since the pretreated samples contained different amounts of cellulose, the yield of glucose formed by enzymatic hydrolysis is based upon the total available glucose. The percent glucose formed from untreated wood is 2.9% and is shown as a solid horizontal line on the bottom of Figure 7.

The relative enzymatic hydrolysis rates show a dramatic increase for all the pretreated samples. In addition, the enzymatic hydrolysis rates increase as the reaction temperature increases. The enzymatic hydrolysis rates of samples pretreated at the higher temperatures of 240 to 260°C approach that of filter paper. (The hydrolysis rate of filter paper was 23.4%; the value is shown as a solid horizontal line at the top of Figure 7 for comparison purposes.) This increase, as compared with the untreated

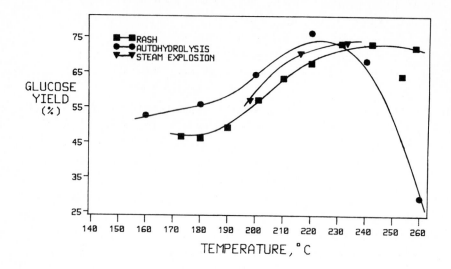

Fig. 6. Glucose content of solids that have been extracted with hot aqueous
alkali

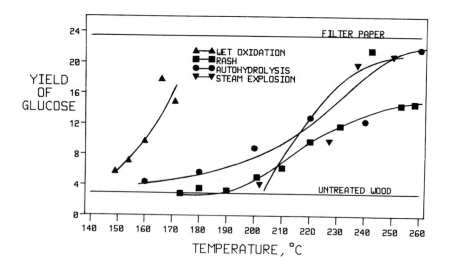

Fig. 7. Enzymatic hydrolysis rates of mixed hardwoods pretreated by the RASH
process. The enzymatic hydrolysis rate of untreated wood is shown as the
solid horizontal line on the bottom, while the hydrolysis rate of filter
paper is shown as the solid horizontal line on the top

sample, was approximately an 8-fold increase. As mentioned earlier, the enzymatic hydrolysis procedure was run for only 24 hours.

The pretreatment of pure hardwood and softwood species by wet-oxidation, steam explosion, or RASH has also been examined at our laboratory (McGinnis, ref. 23; Schultz and McGinnis, ref. 32). Dramatic increases in the enzymatic hydrolysis rates were observed for all wood species which were pretreated.

The addition of different Lewis acid catalysts to wood prior to a RASH pretreatment causes a further increase in the enzymatic hydrolysis rate (McGinnis, ref. 23). Wood pretreated by this catalyzed RASH process hydrolyzed at a relative rate of up to twice that of filter paper.

The pretreatment of selected agricultural wastes by steam explosion has also been studied at our laboratory (Schultz et al., ref. 31). Rice hulls and sugar cane bagasse both show increased enzymatic hydrolysis rates after steam explosion. However, corn stalks were an exception in that untreated corn stalks hydrolyzed at a rapid rate. Thus, the relative rate of enzymatic hydrolysis is similar for both untreated and pretreated corn stalks.

Extended enzymatic hydrolysis experiments were performed on both water oak and sweetgum planar shavings (Biermann, ref. 2). These enzymatic hydrolysis experiments were run continuously for 10 days. A 10-fold greater concentration of cellulase enzyme was used for the 10-day extended runs than was used for the 24-hour enzymatic hydrolysis. The other hydrolysis conditions were the same.

The results of the extended enzymatic hydrolysis are given in Table 5. The total glucose yield is based upon both the starting total available glucose prior to pretreatment and the total available glucose of the pretreated solids. Untreated hardwoods showed very little cellulose hydrolysis (Table 3). The cellulose of pretreated hardwoods, however, was extensively hydrolyzed. Thus, even though a small fraction of the cellulose was degraded by the pretreatment, the total glucose yield based on the starting available glucose content was much greater for pretreated hardwoods.

The acid hydrolysis rates (Figure 8) of the pretreated samples did not show the same dramatic increase as enzymatic hydrolysis rates. The only pretreatment that showed an increase in acid hydrolysis rates was wet oxidation. (The rate of acid hydrolysis of untreated wood was 17.6% and is shown in Figure 8 as a solid horizontal line.) The wet-oxidation pretreatment has previously been shown to increase the rate of acid hydrolysis of pine and hardwoods (McGinnis et al., ref. 20). When steam-exploded, RASH, or wet-oxidized fiber was refluxed in 2% aqueous alkali, one-third of the cellulose was solubilized. It is believed that all four pretreatments degraded the

Table 3. Extended enzymatic hydrolysis of pretreated water oak and sweetgum. The percent glucose hydrolyzed is based upon both the total available glucose prior to pretreatment and the total available glucose of the pretreated solids. The hydrolysis was run for 10 days. A 10-fold greater amount of cellulase enzyme was used for the extended run than was used for the 24-hour enzymatic hydrolysis experiments [Biermann (2)].

| | | | | Glucose hydrolyzed, % | |
Wood species	Total available glucose prior to pretreatment,%	Pretreatment	Glucose[a] recovered, %	Based on glucose content of the pretreated solids	Based on glucose content prior to pretreatment
Water oak	46.0	---	100	5.9	5.9
Sweetgum	42.7	---	100	4.7	4.7
Water oak	46.0	Wet oxidation[b]	94.3	62.9	59.3
Sweetgum	42.7	Wet oxidation[b]	94.3	88.8	83.7
Water oak	46.0	Autohydrolysis[c]	96.2	81.5	78.4
Sweetgum	42.7	Autohydrolysis[c]	92.9	91.0	84.5
Water oak	46.0	RASH[d]	92.9	65.4	60.8
Sweetgum	42.7	RASH[d]	97.1	(90-100.0)	(90-97.1)

[a]Percent glucose recovered in the solid material, based on the starting glucose content of the untreated wood.
[b]The wet-oxidation reaction temperature was 170°C with a reaction time of 15 minutes.
[c]The autohydrolysis reaction temperature was 200°C with a reaction time of 15 minutes.
[d]The RASH reaction temperature was 220°C with a reaction time of 1 minute.

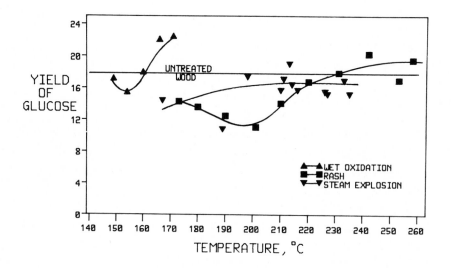

Fig. 8. Rate of acid hydrolysis of mixed hardwoods pretreated at selected temperatures. The acid hydrolysis rate of untreated wood is shown as a solid horizontal line

amorphous sections of the cellulose in the same manner as the hemicelluloses were degraded. This would result in depolymerized cellulose with a slightly enhanced crystallinity. Thus, the pretreated cellulose would have a sufficiently low molecular weight to be partially solubilized by hot aqueous alkali, while the enhanced crystallinity would reduce the accessibility of the cellulose to acid hydrolysis. The increase in the rate of enzymatic hydrolysis was probably caused by disruption of the lignin and hemicelluloses polymeric network. The extent to which the lignin can be solubilized by an extraction procedure also cnfirms that the lignin was depolymerized by each of the four pretreatments.

The acid hydrolysis rates of agricultural wastes pretreated by steam explosion gave some unexpected results. Neither sugar cane bagasse nor corn stover showed any increase in acid hydrolysis rates, while the acid hydrolysis rates of steam-exploded rice hulls was about 70% greater than for untreated rice hulls (Schultz et al., ref. 31). The amorphous silicon (IV) oxide in the rice hulls appears to have acted as a catalyst and decrystallized the cellulose.

Preliminary studies suggest some potential markets for steam-exploded lignin (Schultz and McGinnis, ref. 32). However, it should be remembered that even one commercial-size biomass plant will produce an extraordinarily large volume of lignin. Thus, any feasibility study which assumes that all the lignin can be marketed at a high price should be viewed with caution.

4. <u>CONCLUSIONS</u>

All four pretreatments studied fractionated the cellulose, hemicellulose, and lignin at least partially. The RASH process appeared superior because it fractionated out the greatest amount of the hemicellulose and lignin fractions, minimized degradation, and left a solid residue with a greatly enriched cellulose content.

Dramatic increases in the enzymatic hydrolysis rates were observed for all four pretreatments. The enzymatic hydrolysis rate could be further enhanced by the addition of a Lewis acid catalyst to the RASH process.

Wet oxidation was the only pretreatment which increased the acid hydrolysis rate of mixed hardwoods. Steam-exploded rice hulls hydrolyzed significantly faster with acid than untreated rice hulls, suggesting that amorphous silica may act as a catalyst.

Each pretreatment has unique advantages; however, each also has its unique disadvantages.

This study found that the results reported by researchers who steam-exploded aspen are often significantly different from the results the authors obtained when various hardwoods and other lignocellulosic materials were exploded. These differences may be attributed to the observation that aspen is a particularly easy wood to pretreat (Lipinsky, ref. 13; Millett et al., ref. 24).

ACKNOWLEDGEMENTS

Financial support for this research was provided by the U.S. Department of Agriculture – Science and Education Administration, the U.S. Forest Products Laboratory, and the U.S. Solar Energy Research Institute. The authors thank Bobbie J. Colley and Elizabeth Carter of the Department of Energy – Technical Information Center for their help.

REFERENCES

1. Aronowsky, S.I. and Gortner, R.A., "Cooking Process, Part I", Ind. Eng. Chem., 22 (3), 264-74 (1930)
2. Biermann, C.J., "The Development of a New Pretreatment Method (Rapid Steam Hydrolysis) and the Comparison of Rapid Steaming, Steam Explosion, Autohydrolysis, and Wet Oxidation as Pretreatment Processes for Biomass Conversion of Southern Hardwoods", Ph.D. Thesis, Mississippi State University, Mississippi State, MS, 1983
3. Biermann, C.J., Schultz, T.P. and McGinnis, G.D., "Rapid Steam Hydrolysis/Extraction of Mixed Hardwoods as a Biomass Pretreatment", J. Wood Chem. and Tech. 4 (1), 111-128 (1984)
4. Chau, M.G.S. and Wayman, M., "Characterization of Autohydrolysis Aspen (P. tremuloides) Lignins, Part 1", Can. J. Chem. 57, 1141-49 (1979)
5. Chau, M.G.S. and Wayman, M., "Characterization of Autohydrolysis Aspen (P. tremuloides) Lignins, Part 3", Can. J. Chem. 57, 2603-11 (1979)
6. Chen, C.C. and McGinnis, G.D., "The Use of 1-Methylimidazole as a Solvent and Catalyst for the Preparation of Aldononitrile Acetates of Aldoses", Carbohy. Res. 90, 127-30 (1981)
7. Cowling, E.B. and Kirk, T.K., "Properties of Cellulose and Lignocellulosic Materials as Substrates for Enzymatic Conversion Processes", Biotechnol. and Bioeng. Symp. No. 6 (1976), J. Wiley and Sons, New York, N.Y.
8. Emert, G.H. and Katzen, R., "Chemicals from Biomass by Improved Enzyme Technology", in Biomass as a Non-Fossil Fuel Source, American Chemical Society, Washington, D.C., 1979
9. Foody, P., Optimization of Stream Explosion Pretreatment, Iotech Corp., Ottawa, Ontario, Canada, Final Report to DOE, Contract AC02-79ET23050, 1980
10. Goldstein, I.S., "Chemicals from Cellulose", in Organic Chemicals from Biomass, CRC Press, Boca Raton, FL, 1981
11. Hajny, G.J., Biological Utilization of Wood for Production of Chemicals and Foodstuffs, U.S. Dept. of Agriculture, Washington, D.C., Research Paper FPS-385 (1981)
12. Jurasek, L., "Enzymatic Hydrolysis of Pretreated Aspen Wood", Dev. Ind. Microbiol. 20, 177-83 (1978)

13. Lipinsky, E.S., "Perspectives on Preparation of Cellulose for Hydrolysis", in Hydrolysis of Cellulose: Mechanisms of Enzymatic and Acid Catalysis, American Chem. Soc. - Adv. in Chemistry Series No. 181, Washington, D.C., 1979

14. Lora, J.H. and Wayman, M., "Delignification of Hardwoods by Autohydrolysis and Extraction", Tappi 61 (6), 47-50 (1978)

15. Lora, J.H. and Wayman, M., "Autohydrolysis of Aspen Milled Wood Lignin", Can. J. Chem. 58, 669-76 (1980)

16. Lowrimore, J.T., Schultz, T.P. and McGinnis, G.D., Unpublished Data, Mississippi Forest Products Lab., P.O. Drawer FP, Mississippi State, MS, 39762 (1980)

17. Marchessault, R.H., Coulombe, S., Hanai, T. and Morikawa, H., "Monomers and Oligomers from Wood", Pulp Paper Mag. of Can. Trans. 6 (2), TR 52-56 (1980)

18. Marchessault, R.H., Coulombe, S., Morikawa, H., and Robert, D., "Characterization of Aspen Exploded Wood Lignin", Can. J. Chem. 60 (18), 2372-82 (1982)

19. Marchessault, R.H., "Carbohydrate Polymers: From Cellulose to DNA", Milton Harris 75th Birthday Symposium, April 27-28, 1981, American Chemical Society, Washington, D.C., 1982

20. McGinnis, G.D., Wilson, W.W. and Mullen, C.E., "Biomass Pretreatment with Water and High-Pressure Oxygen. The Wet-Oxidation Process", Ind. Eng. Chem., Product R & D, 22, 352-57 (1983)

21. McGinnis, G.D., Prince, S.E., Biermann, C.J. and Lowrimore, J.T., "Wet Oxidation of Model Carbohydrate Compounds", Carbohy. Res. 128, 51-60 (1984)

22. McGinnis, G.D., Wilson, W.W., Prince, S.E. and Chen, C.C., "Conversion of Biomass into Chemicals Using High Temperature Wet Oxidation", Ind. Eng. Chem., Product R & D, 22, 633-636, (1983)

23. McGinnis, G.D., Unpublished Data, Mississippi Forest Products Lab., P.O. Drawer FP, Mississippi State, MS, 39762 (1984)

24. Millett, M.A., Baker, A.J. and Satter, L.D., "Pretreatments to Enhance Chemical, Enzymatic and Microbiological Attack of Cellulosic Materials", Biotechnol. and Bioeng. Symp. No. 5 (1975), J. Wiley and Sons, New York, NY

25. Millett, M.A. Baker, A.J. and Satter, L.D., "Physical and Chemical Pretreatments for Enhancing Cellulose Saccharification", Biotechnol. and Bioeng. Symp. No. 6 (1976), J. Wiley and Sons, New York, NY

26. Rydholm, S.A., "Chemical Pulping" in Pulping Processes, Interscience, New York, NY, 1965

27. Rydholm, S.A., "Chemical Pulping - Multistage Processes" in Pulping Processes, Interscience, New York, NY, 1965

28. Schaleger, L.L. and Brink, D.L., "Chemical Production by Oxidative Hydrolysis of Lignocellulose", presented at Tappi Conference, Forest Biology/Wood Chemistry, Madison, WI, 1977

29. Schaleger, L.L. and Brink, D.L., "Chemical Production by Oxidative Hydrolysis of Lignocellulose", Tappi, 61 (4), 65-68 (1978)

30. Schultz, T.P., Biermann, C.J. and McGinnis, G.D., "Steam Explosion of Mixed Hardwood Chips as a Biomass Pretreatment", Ind. Eng. Chem., Product R&D, 22, 344-48 (1983)

31. Schultz, T.P., Biermann, C.J. and McGinnis, G.D., "Steam Explosion of Mixed Hardwood Chips, Rice Hulls, Corn Stalks and Sugar Cane Bagasse", J. Agric. Food Chem. 32, 1166-1172 (1984)

32. Schultz, T.P. and McGinnis, G.D., Evaluation of a Steam-Explosion Pretreatment for Alcohol Production from Biomass, Final Report to USDA-SEA, Contract 59-2281-1-2,098-0, Mississippi Forest Products Lab, P.O. Drawer FP, Mississippi State, MS, 39762 (1984)

33. Spalt H.A., "Chemical Changes in Wood Associated with Wood Fiberboard Analysis" in <u>Wood Technology: Chemical Aspects</u>, ACS Symp. Ser. No. 43, American Chemical Society, Washington, D.C., 1977
34. Wayman, M. and Chau, M.G.S., "Characterization of Autohydrolysis Aspen (P. tremuloides) Lignin, Part 2", <u>Can. J. Chem.</u>, <u>57</u>, 2599-2602 (1979)
35. Wayman, M. and Lora, J.H., "Delignification of Wood by Autohydrolysis and Extraction", <u>Tappi, 62</u> (9), 113-14 (1979)

Energy Recovery from Lignin, Peat and Lower Rank Coals, edited by D.J. Trantolo and D.L. Wise
Elsevier Science Publishers B.V., Amsterdam 1989 — Printed in The Netherlands

Chapter 9

CATALYSIS OF LOW GRADE COAL CONVERSION

Kenneth M. Nicholas
Department of Chemistry,
Boston College, Chestnut Hill, MA, USA*

In considering the practical conversion of coal to organic chemicals it
is highly appropriate, in fact central, to investigate the role of catalysts
and promoters as they offer the promise of facilitating these otherwise
sluggish reactions. Indeed, it is fair to say that the more efficient the
catalyst, the more "benign" (and probably the more economical) the process.
Following some introductory remarks regarding catalysts and promoters in
general, we shall examine selected studies of catalyzed coal gasification
and liquefaction, emphasizing the role of the catalyst. We shall conclude
with a discussion of some known homogeneously catalysed reactions of coal
model compounds which suggest future directions for the development of
practical conversion catalysts.

It is important at the outset to define clearly what is meant by the
term 'catalyst' and to distinguish it from related, but distinct, terms such
as promoter or initiator. Bender, who has discussed this issue rather
thoroughly (ref. 1), provides the following operationally useful definition:

"Catalysis of a chemical reaction is an acceleration brought about
by a substance that is not (itself) consumed in the overall
reaction; this substance — the catalyst — usually functions by
interacting with the starting material to yield an alternate set
of species that can react by a pathway involving a lower free
energy of activation to give the products and to regenerate the
catalyst."

The concept is conveniently represented on a reaction profile as in
Figure 1. Key features of a catalyst thus include rate acceleration by means

(*Address correspondence to the author at the Dept. of Chemistry, University
of Oklahoma, Norman, OK 73019)

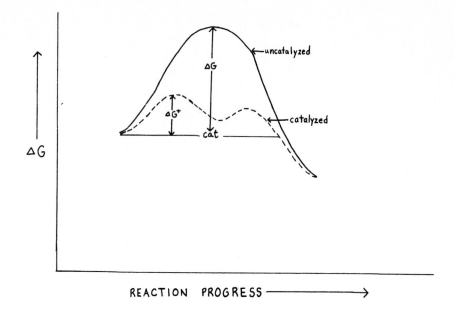

Fig. 1.

of a decrease in ΔG_{act}* and regeneration with product turnover.   It is
important to point out that the catalyst does not affect the position of
equilibrium (defined by K_{eq}) but only the   rate   at   which equilibrium is
attained.

     An example which illustrates the catalyst's role   is   the   industrially
important water gas shift reaction (WGSR):

$$CO + H_2O \xrightleftharpoons{\qquad} CO_2 + H_2 \qquad\qquad (1)$$

Although heterogeneous catalysts (e.g. Fe_xO_y) are presently employed,   there
have   been   several   recent   reports   of   homogeneously catalyzed WGSR by
transition metal complexes including $Fe(CO)_5$, $M(CO)_6$ ($M=Cr,Mo,W$), $Ru_3(CO)_{12}$,
and $HRhL_n$ (ref. 2).   These reactions not only occur under milder conditions
than their heterogeneous counterparts but, more important to the present
discussion, also provide insight into the molecular mechanism of catalyst

---

*Note: This may occur by stabilization of the transition state (shown in
Fig. 1) and/or destabilization of the reactants.

action.   For several systems the catalyst serves the function of activating
by coordination the CO towards nucleophilic attack by H_2O (or OH^-) (Scheme
1).

SCHEME 1:

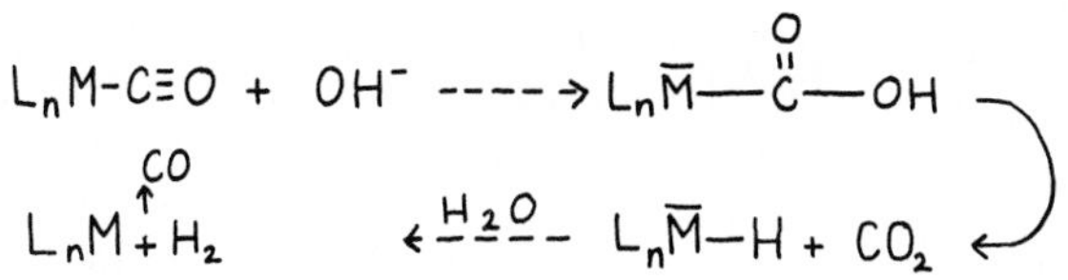

$$L_nM-C\equiv O + OH^- \ ----\rightarrow \ L_n\overline{M}-\overset{O}{\overset{\|}{C}}-OH$$

$$L_nM \overset{CO}{\underset{\uparrow}{+}} H_2 \qquad \xleftarrow{\ H_2O\ } \ L_n\overline{M}-H + CO_2$$

Activation of both H_2O and CO appear to be involved in the Rh catalyzed
process (ref. 3).

A promoter or initiator, on the other hand, while accelerating a given
reaction, actually serves as a reagent in that it is consumed in the
process.   In the familiar "base-catalyzed" (sic) ester hydrolysis (Scheme 2)
hydroxide serves as a promoter rather than a true catalyst since one
equivalent is required.

SCHEME 2:

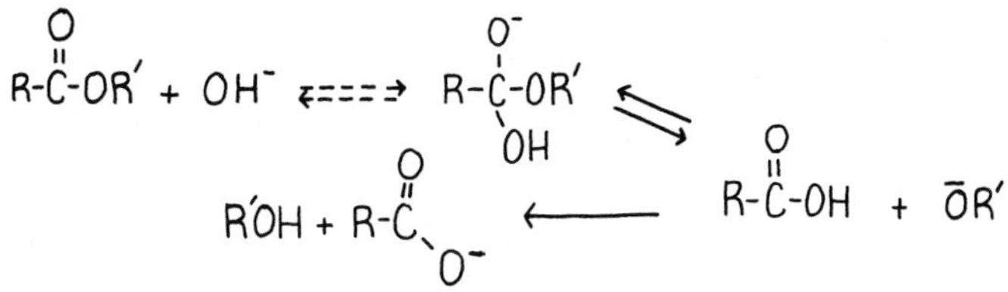

$$R-\overset{O}{\overset{\|}{C}}-OR' + OH^- \ \rightleftharpoons \ R-\underset{OH}{\overset{O^-}{\underset{|}{\overset{|}{C}}}}-OR'$$

$$R'OH + R-\overset{O}{\underset{O^-}{\overset{\|}{C}}} \ \longleftarrow \ R-\overset{O}{\overset{\|}{C}}-OH + {}^-OR'$$

In chain reactions, less than stoichiometric quantities of initiator may be
required, but, nonetheless, it is consumed in the initiation sequence and is
not involved in the product-producing (propagation) steps.

Since economic considerations severely limit the list of prospective
promoters for a commercially viable coal conversion process, we shall
largely restrict our attention to use of true catalysts in coal gasification
and liquefaction.

Gasification

There have been relatively few studies on the catalysis of gasification
of low grade coal (i.e. sub-bituminous, lignite).   Indeed, the rather
forcing conditions of temperature and pressure required for gasification

(even using catalysts), could rightfully disqualify these from consideration as "benign" treatments. Nevertheless, a knowledge of the function of the catalysts in these reactions could be useful for the future development of improved catalysts functioning under milder conditions and therefore are considered here.

A summary of the basic conversion processes is provided in Scheme 3 and includes catalyzed steam-, hydrogen-, and carbon dioxide-gasifications.

SCHEME 3:

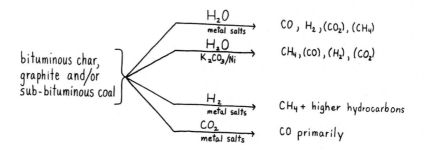

The enhanced reactivity of coal towards steam caused by the addition of alkali salts especially carbonates and oxides, has been known for over 100 years (ref. 4), but there is still no generally accepted explanation for this phenomenon. A recent thorough study by McKee and co-workers (ref. 5) on the catalysis of steam and CO_2 gasification of bituminous char and graphite has provided some insight into possible mechanisms of catalyst action and destruction. The relative efficacies of alkali metal carbonates versus untreated bituminous char are shown in Figure 2. Li_2CO_3 was clearly the most active steam gasification catalyst, with Na_2CO_3 and K_2CO_3 showing lower activity.

Other important observations include the following:

(1) catalyst deactivation occurred during successive steam gasification experiments with the char samples but not with
   graphite;
   and
(2) catalyst activity was retained during successive cycles of CO_2
   gasification both with char and graphite.

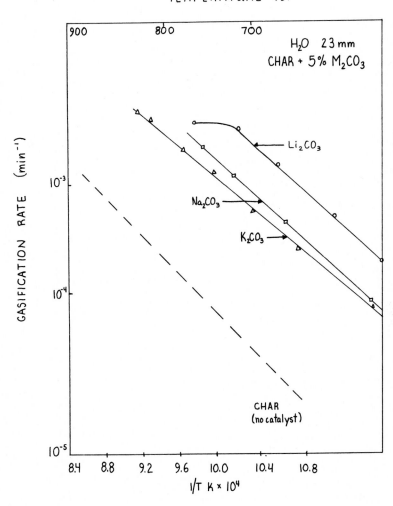

Fig. 2. Catalytic effects of alkali metal carbonates on the gasification of
coal char in water vapor. After charring at 700°C,
5% salt was added (ref. 5).

Based on kinetic and thermodynamic data and chemical precedent, Schemes 4 and 5 have been proposed for the alkali catalyzed gasifications of both char and graphite.

SCHEME 4: $C-H_2O$ reaction

$$M_2CO_3 + 2C = 2M + 3CO$$
$$2M + 2H_2O = 2MOH + H_2$$
$$2MOH + CO = M_2CO_3 + H_2$$
$$\overline{C + H_2O = CO + H_2}$$

SCHEME 5: $C-CO_2$ reaction

$$M_2CO_3 + 2C = 2M + 3CO$$
$$2M + CO = M_2O + CO$$
$$M_2O + CO_2 = M_2CO_3$$
$$\overline{C + CO_2 = 2CO}$$

In the case of Li_2CO_3 the first step in either reaction may involve production of the metal oxide (Li_2O) rather than the free metal. Catalyst deactivation appears to involve reaction with minerals in the char (e.g. illite, kaolin, quartz), a process which is suppressed by carbon dioxide.

A related study conducted at the Bureau of Mines (ref. 6) documents 10 - 90% enhancements of steam gasification rates on high volatile bituminous coal by several additives. Among the most active catalysts were KCl, NaCl and K_2CO_3. It would appear from these results that the metal ion (as well as the anion, above) also may play a role in promoting gasification. No evidence or speculation relevant to the mechanism of catalyst action was presented, however.

An interesting variation on the carbonate-catalyzed gasifications involves the use of a bi-functional catalyst system. Wilson and co-workers (ref. 7) conducted steam gasification of Wyoming sub-bituminous in the presence of K_2CO_3 (15 - 60 wt %) and Ni-3120 (a standard methanation catalyst, 100 wt %). The resulting product gas was methane-rich. Control experiments indicated that the two catalysts performed separate and distinct functions: K_2CO_3 promoted coal gasification whereas the Ni catalyst promoted methanation of the coal gases. Unfortunately, while virtually all the alkali catalyst was recoverable in the ash as K_2CO_3 and $K_2CO_3 \cdot 1.5H_2O$, the Ni catalyst was poisoned by sulfur from the coal making such a process economically unattractive. Similarly, in another study (ref. 6) Raney Ni, which promotes methane production, was also found to be rapidly deactivated when used as the sole gasification catalyst.

Methane-rich gas can be produced without the requirement of transition metal additives if lignite is hydrogenated under severe conditions (e.g. >900°, 1000 psi) as in IGT's HYGAS process (ref. 8). Alkali additives, especially $KHCO_3$ and K_2CO_3, have been found to promote the hydromethanation of bituminous char although rather forcing conditions were still employed (ref. 9). The mechanism of catalyst action was not discussed.

Clearly much room for improvement exists not only with respect to the efficiency of catalyzed coal gasification (especially with low grade coals) but also in expanding our understanding of how specific catalysts actually do their job.

Coal Liquefaction

The most extensive liquefaction research on US lignites has been conducted at the U. North Dakota and the Grand Forks Energy Center (refs. 10, 11). Included in these studies has been important work demonstrating the beneficial effects of both carbon monoxide (a promoter) and hydrogen sulfide (true catalyst). Under optimum conditions (T = 460°C) with recycle of the vacuum bottoms, distillate yields can be as high (on a dry, ash-free basis) as those obtained from bituminous coals. A solvent, consisting of anthracene oil or the liquid product mixture itself, is employed and can be hydrotreated and recycled.

Although carbon monoxide (usually as 1:1 synthesis gas) is not a true catalyst in these studies, the economic feasibility of its large scale use and the availability of information relating to its possible chemical function encourage us to examine its role here. The dependency of distillate yields on coal rank for CO/H_2 versus H_2 liquefaction is seen in the following data:

Texas lignite: 44% (for CO/H_2) vs. 37% (H_2 only);
Wyoming sub-bituminous: 26% vs. 36%;
Ohio bituminous: 33% vs. 51%.

Several mechanisms have been proposed for the molecular basis of the CO effect (ref. 11) including ones involving the intermediacy of alkali metal formates (ref. 12), formic acid, CO radical anion and isoformate, direct reaction with lignite, and sulfur-mediated water gas shift reaction.

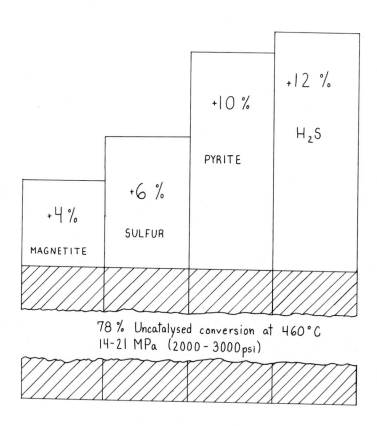

Fig. 3.

Averaged effects of added promoter on conversion of Beulah, North Dakota lignite. Reactions were carried out in the 54 kg day^{-1} (120 lb day^{-1}) GFETC CPU (tubular reactor) on as-received and slurry-dried lignite using either H_2 or syngas. Benefits were more pronounced with syngas and water than with hydrogen (ref. 11)

That the primary role of CO in enhancing liquefaction is to generate formate (via reaction with basic metal oxides) which, in turn, serves as a reducing agent for carbonyl groups present in lignite (ref. 12), has been discredited (ref. 11) by the instability of formate under typical operating conditions, the limited number of carbonyl functions present in lignite, and the observation that excess formate salts are ineffective in promoting conversion in the absence of CO. Formic acid (formed from WGSR) is also thermally unstable under liquefaction conditions but it has significant activity for reductive cleavage of C-C, C-S, and C-N bonds in several model compounds (ref. 13). Low steady-state concentrations of formic acid under reaction conditions could account for significant coal degradation.

The observation that CO reacts preferentially to H_2 with lignite at temperatures to 480°C (ref. 11) suggests that some CO or activated intermediate reacts directly with coal to give volatile products. High energy CO-derived intermediates which have been suggested to potentially perform this function include isoformate $(HO-C(=O)^-$, (ref. 11) and CO^- (ref. 11). Reactions such as (2) and (3) below could be representative of the direct type that could occur between CO and lignite:

$$CO + Lig-OH \ -----> \ Lig-H + CO_2 \qquad\qquad (2)$$

$$CO/H_2O + Lig-O-Lig \ -----> \ 2 \ Lig-H + CO \qquad\qquad (3)$$

These latter suggestions clearly are quite speculative and much experimental work remains before the mechanism of CO promotion of coal liquefaction will be reasonably well understood.

The beneficial effect of sulfur-containing additives on lignite conversion (especially using synthesis gas), pioneered by the North Dakota groups, is illustrated in Fig. 3 (ref. 11). Other important features (with H_2S as catalyst) include: (1) up to a three-fold decrease in reductant consumption; (2) comparable conversion rates at 60°C lower operating temperatures; (3) a decrease in the production of hydrocarbon gases; and (4) decreased coke formation. In general, the order of catalyst efficiency is H_2S> S> pyrite.

Based on the study of reactions between H_2S (and S) and numerous model compounds, several chemical mechanisms have been proposed to account for H_2S (and S) effects on lignite (and lignin) conversion. These include: (1) hydrocracking; (2) insertion reactions in aromatic rings; (3) hydrogen abstraction; and (4) catalysis of the WGSR. Examples of reactions (1) to (3) are given below.

hydrocracking: $Ph-Ph + H_2S \xrightarrow{\text{(ref. 13)}} 2Ph-H + S$

(4)

attack on aromatics:

$$PhCH_2Ph + H_2S \xrightarrow{\text{(ref. 13)}} PhSH + PhCH_3$$

(5)

H-abstraction by S:

$$PhNMe_2 + S_x \xrightarrow{\text{(ref. 14)}} PhNH_2 + ?$$

(6)

It has been demonstrated that H_2S, like the aforementioned homogeneous transition metal complexes, can serve as a homogeneous catalyst for the WGSR at 425°C. Carbonyl sulfide (OCS) apparently is in an intermediate according to the following scheme (ref. 15).

SCHEME 6:

$$CO + H_2S \dashrightarrow OCS + H$$
$$OCS + H_2O \dashrightarrow CO + H_2S$$
$$\overline{CO + H_2O \dashrightarrow CO_2 + H_2}$$

This latter role of H_2S could provide the hydrogen necessary to regenerate H_2S in equations 4 and 5 and could account for decreased reductant consumption relative to the uncatalyzed process as well as improved solvent recycle via hydrotreating.

Final mention should be made of the H-coal process, developed to the pilot stage (600 ton/day) primarily for bituminous coals using hydrogen as reductant (ref. 16). In this process recycle solvent and coal are hydrotreated in a fluidized bed of Co-Mo catalyst. Whether the catalyst serves primarily to hydrogenate the solvent which, in turn, donates hydrogen to the coal or whether direct coal hydrogenation also occurs has not been established. The need for catalysts more suited for low rank coals to overcome higher hydrogen consumption and catalyst deactivation has been pointed out (ref. 16).

New Prospects

Clearly many new opportunities exist for the development of more efficient and more selective catalysts for coal conversion, especially for low grade coals where relatively little effort has been expended. Earlier advances in catalyst (mostly heterogeneous) development in general were made on a largely empirical ("trial and error") basis. More recently an increasing number of homogeneously-catalyzed reactions of simple organics has been discovered which are at the same time more amenable to mechanistic study and frequently more selective and sulfur-tolerant than their heterogeneous counterparts. Furthermore, with continuing advances in the determination of coal structure it now seems possible to pursue the semi-rational design of homogeneous catalysts for coal conversion. In this final section some candidates are suggested based on known model compound reactions.

In order to search for relevant model system reactions, we must have a picture of the structure of lignite, our prototypical low grade coal. While much has been written about the structure of coal, much remains unknown. There is, however, substantial agreement that the primary structure of lignite contains both aromatic and hydroaromatic rings substituted with alkyl, carboxyl, hydroxyl, and to a lesser extent, ether and carbonyl groups. For the purposes of our present discussion it is convenient to refer to the partial structure of lignite (Fig. 4) developed according to Wender (ref. 17) and Sondreal (ref. 11).

Assuming that the macromolecular structure of coal needs to be dissected in order to produce tractable organics, we seek reactions which should lead to cleavage of one or more of the following linkages: $C(aryl)-O$, $C(aryl)-CO_2R$, $C(aryl)-C(alkyl)$, $C(benzylic)-C(alkyl)$, $C(aryl)-C(aryl)$, $C(benzylic)-CO_2R$, $C(=O)-OR$, etc. Indeed, numerous instances of transition metal-catalyzed scissions of such bonds in simple organic compounds are known and form the basis for this final section.

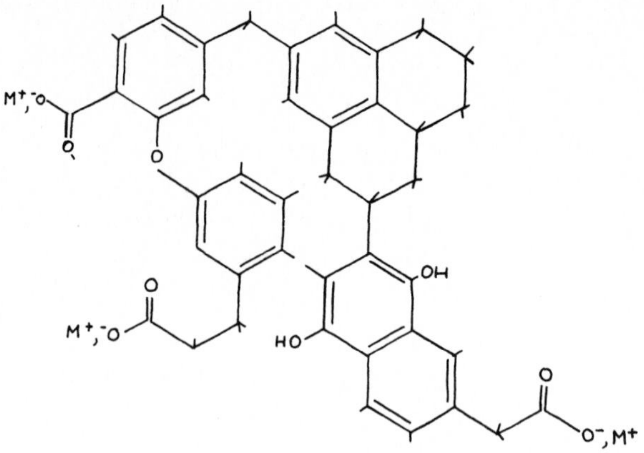

302

Fig. 4.

Partial lignite structure. $C_{42}O_{10}H_{35}$, MW = 699; C, 72; O, 23; H, 5 wt %;
fa, 0.67; %O as $-COO^-$ = 60; %O as $-OH$ = 30; %O as $-O-$ = 10
(ref. 11)

Listed below are selected examples of transition metal-catalyzed reactions of model compounds leading to cleavage of bonds relevant to the problem of lignite conversion.

$$R\underset{NH_2}{CH} - CO_2R' + H_2O \xrightarrow{Cu^{2+}} R\underset{NH_2}{CH} - CO_2H + R'OH \qquad (7)$$

$$Ar\,CH_2CO_2H \xrightarrow[\substack{HOAc \\ (ref.\ 19)}]{O_2/Mn(III)} Ar\,CH_2OAc \qquad (8)$$

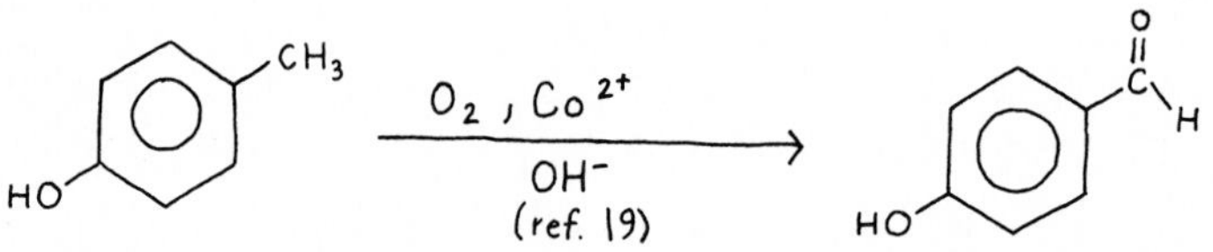

(9)

$$\text{(10)}$$

$$\text{(11)}$$

$$\text{(12)}$$

These reactions can be seen to fall into the categories of hydrolytic, oxidative or reductive cleavages. The specific activating effect of transition metals in the hydrolysis of α-amino acid esters is dramatically illustrated by the rate constant for hydrolysis with various catalysts:

$$H^+ = 1.5 \times 10^{-11}$$
$$OH^- = 5.8 \times 10^{-9}$$
$$Cu^{2+} = 2.7 \times 10^{-3}$$

A requirement for these large accelerations is the potential for chelation of the substrate to the transition metal center (ref. 1). While the α-amino ester function is not present in lignite, other possibilities for chelation do exist, e.g. with ortho substituted phenolic esters:

$$\text{(13)}$$

One might anticipate, therefore, significantly facilitated hydrolysis of ester linkages in lignite in the presence of salts of common transition metal ions such as Cu^{2+}, Ni^{2+}, and Fe^{3+}.

A variety of options exist for oxidative degradation of coal catalyzed by transition metals. In order for such reactions to be practical, however, they should be mild and selective with minimal total oxidation to carbon dioxide, a relatively useless raw material. Fortunately, this feature characterizes many homogeneous transition metal catalyzed oxidations (ref. 24). Particularly activated sites in the lignite structure include the benzylic carbons and phenolic rings as illustrated in equations 8, 9, 10 and 12. Most of these reactions occur for simple organics under mild conditions (T<200°C, 1-10 atm) with molecular oxygen (often as air) as oxidant. More applicable to small scale, specialty chemical production may be reactions utilizing hydrogen peroxide, hypochlorite or periodate as oxidants, e.g. equation 11.

While many transition metal species, both homogeneous and hetero-geneous, are active catalysts for addition of molecular hydrogen to unsaturated functional groups (e.g. C=C, C=O, aromatic rings), such reactions would not result in degradation of the lignite network*. Hydro-genolysis of C-O and C-C bonds, on the other hand, is necessary for this purpose and the former has precedent particularly in the case of the activated benzylic C-O bonds (equation 11). These reactions are again typified by mild conditions and high selectivity.

It would thus appear that exciting possibilities exist for the util-ization of homogeneous transition metal catalysts in the conversion of low grade coals to useful, tractable organic compounds. Experiments are now underway to test the viability of these hypotheses.

---

*These reactions, especially aromatic hyrogenation, could prove valuable for hydrotreating solvents in processes using hydrogen donor solvents. Limit-ations of some heterogeneous catalysts for this purpose were noted earlier. Recently, homogeneous Ru and Co catalysts for arene hydrogenation operating under very mild ambient conditions (<100°C, 1-5 atm) have been found (refs. 25, 26).

REFERENCES

1. M. Bender in "Mechanisms of Homogeneous Catalysis from Protons to Proteins" Wiley, N.Y. (1971), p.8.
2. For leading references see "Catalytic Activation of Carbon Monoxide", P.C. Ford, ed., ACS Symposium Series No. 152 (1981) chap. 5-8.
3. T. Yoshida, T. Okano and S. Otsuka, ref. 2, chap. 5, pp. 79-94.
4. DuMotay, T. British Patent 2458, 1867.
5. D.W. McKee, C.L. Spiro, P.G. Kosky, E.J. Lamby, Chemtech, 624 (1983).
6. W.P. Haynes, S.J. Gasior and A.J. Forney in "Coal Gasification", L.G. Massey, ed., ACS Symposium Series No. 131 (1973) chap. 11, pp. 179-202.
7. W.G. Wilson, L.J. Sealock, Jr., F.C. Hoodmaker, R.W. Hoffman, D.L. Stinson and J.L. Cox in "Coal Gasification", ref. 6, chap. 12, pp. 203-216.
8. Kirk-Othmer "Encyclopedia of Chemical Technology", 3rd ed., vol. 14, Wiley, N.Y.
9. N. Gardner, E. Samuels and F. Wilks in "Coal Gasification" (see ref. 6), chap. 13, pp. 217-236.
10. Review: J.R. Rindt, W.G. Wilson and V.I. Sternberg, "Recent Advances in Catalysis of Lignite Liquefaction", 1983 Lignite Symposium, Grand Forks, N.D., May 1983.
11. E.A. Sondreal, W.G. Wilson and V.I. Sternberg, Fuel, 1982, 61, 925.
12. H.R. Appell and I. Wender, Am. Chem. Soc. Div. Fuel Chem. Preprints, (1968) 12, 220.
13. R.L. van Buren, Ph.D. dissertation, Univ. of North Dakota, 1981.
14. R.J. Baltisbeger, V.I. Sternberg, K.J. Klabunde and N.F. Woolsey, Quarterly Report, Oct.-Dec. 1981, DOE/FC/02101-21, Contract DE-AB18-78FC-02101
15. K. Raman, V.R. Srinivas, R.J. Baltisberger, N.F. Woolsey and V.I. Sternberg, Angew. Chemie, in press.
16. Proceedings of the Low-Rank Coal Technology Development Workshop, Session 6, Liquefaction Section B. Catalysis, San Antonio, Texas, June, 1981, pp. 6-11.
17. I. Wender, Am. Chem. Soc. Div. Fuel Chem. Preprints (1975) 20, 16.
18. M.M. Jones "Ligand Reactivity and Catalysis", Academic Press, N.Y. (1968).
19. R.A. Sheldon and J.K. Kochi, Adv. Catal., 25, 272, 1976.
20. K. Nishizawa, K. Hamada and T. Aratani, Eur. Patent 12, 939 (1979) to Sumitomo.
21. M.M. Rogic, T.R. Demmin and W.B. Hammond, J. Am. Chem. Soc., 98, (1976) 1974.
22. J.A. Caputo and R. Fuchs, Tetrahedron Lett. (1967) 4729.
23. S.G. Morris, J. Am. Chem. Soc., 71, 2056 (1949).
24. R.A. Sheldon and J.K. Kochi, "Metal-Catalyzed Oxidations of Organic Compounds", Academic Press, N.Y., 1981.
25. J.W. Johnson and E.L. Muetterties, J. Am. Chem. Soc., 99, 7395 (1977).
26. L.S. Stuhl, M. Rakowski DuBois, F.J. Hirsekorn, J.R. Bleeke, A.E. Stevens and E.L. Muetterties, ibid, 100, 2405 (1978).

Energy Recovery from Lignin, Peat and Lower Rank Coals, edited by D.J. Trantolo and D.L. Wise
Elsevier Science Publishers B.V., Amsterdam 1989 — Printed in The Netherlands

Chapter 10

DIRECT LIQUEFACTION OF PEAT AND LIGNITE TO BTX-TYPE FUEL

Debra J. Trantolo, Ph.D.,
Raphael J. Cody,
Elizabeth P. Heneghan,
and Donna M. Houmere,
Dynatech R/D Company, Cambridge, MA, USA

SUMMARY

This chapter relates the work on the experimental research and development program for the direct liquefaction of lignite to a BTX-type liquid fuel. The BTX-type liquid fuel process has as the major feature an alkaline hydrolysis pretreatment of the lignite. The goal of this pretreatment is the breakdown of the complex macromolecular lignite structure to simple aliphatic and aromatic compounds. Owing to the geological history of the lignite, these organic compounds are highly oxygenated and, under alkaline conditions, water-soluble to a large extent. It was proposed to investigate the susceptibility of these organics to further chemical treatment and biological treatment for production of BTX-type and alcohol fuels, respectively.

The work directed toward the production of BTX-type fuel is detailed in this chapter. The experimental program involved laboratory scale work in two process steps:

(1)   alkaline hydrolysis for pretreatment; and
(2)   catalytic decarboxylation for conversion of pretreatment
       products to BTX-type fuel.

Both of these areas were approached independently to establish the technical feasibility of the operation and to determine process operating conditions and yields.

The pretreatment experiments called for reaction of lignite, water, and sodium carbonate under batch conditions. The optimal conditions for breakdown of the lignite were found to be 250°C (ϕ hours, 8% volatile solids loading, and 20% sodium carbonate (weight by weight volatile solids)). Under these conditions, 63% of the input lignite volatile solids were solubilized. Longer reaction times and/or higher temperatures resulted in repolymerization, i.e. further coalification, of the hydrolysis products. A decrease in alkali or an increase in solids loading resulted in lower yields. This work is presented in Section 3.

Decarboxylation experiments were carried out using benzoic acid and pretreated lignite. Two procedures for chemical decarboxylation were investigated:

(1) the copper/quinoline method; and
(2) the persulfate/silver ion method.

The former method showed some conversion of benzoic acid to benzene, but quantitation was difficult with the complexity of the product mixture. The latter method showed an identifiable 41% yield of benzene. Because the persulfate/silver ion decarboxylation is carried out in an aqueous medium, it was the preferred method for lignite decarboxylations. Under conditions analogous to the model studies using persulfate/silver, lignite hydrolysate was subjected to decarboxylation. The product mixture showed 6% conversion of the hydrolysate to volatile organic liquid, i.e. 4% overall conversion of the input lignite volatile solids. The decarboxylation work is detailed in Section 4.

1. INTRODUCTION

1.1 The Focus on Peat and Lignite

The diminishing world-wide petroleum reserves and rising crude oil prices have created the need to find other raw materials for the production of fuels and feedstock chemicals. Private industry and the state and federal governments are undertaking a substantial research and development program to utilize other readily available resources to bridge the gap between supply and demand. The sizeable deposits of lignite in the United States may provide raw materials for the production of significant quantities of liquid fuels and organic chemicals as well as fuel gas.

Lignite utilization has been directed largely to producing substitute natural gas - an example being the Great Plains Gasification Plant in Beulah, North Dakota. The great potential for lignite is to be realized only if a full range of fuels, not just substitute gas, is produced from lignite.

The use of selected lower rank coals, especially lignite, should be considered for processing. In this regard, it is of interest to consider the amount and location of lignite in the United States. A breakdown of all the estimated coal reserves in the United States for the four major ranks of coals is given as follows (Aueritt 1967):

Coal Rank	Remaining Recoverable Reserves (billion tons)	Percent of Total
Bituminous	671	43.0
Sub-bituminous	428	27.4
Lignite	447	28.7
Anthracite	13	0.9
	1,559	100.0

If only sub-bituminous coal and lignite are considered as especially suitable to the proposed process, then it is seen that this amounts to 56.1% of remaining recoverable coal reserves.

The location of lignite reserves in the United States is of interest. It is seen that lignite is located largely in Montana, Wyoming and North Dakota. It is interesting to consider the regional impact of the processes utilizing lignite with respect to current coal production. The present leading coal producing states are, in decreasing order of production, West Virginia, Kentucky, Pennsylvania, Illinois, Ohio and Virginia (1968 production figures (National Coal Association 1974)). These states are those with substantial reserves of bituminous coal. Thus, the development of a process for conversion of lignite would have a substantial national impact, because such conversion will take place in areas where coal is not presently mined. Thus, the use of an environmentally acceptable technique would be acceptable.

1.2 Lignite Utilization

The combustion of lignite will release the greatest amount of energy (approximately 10,000 Btu/lb). Lignite has both a high moisture content and a high ash content. Conventional combustion for electrical power − when applied to lignite − has been found economically unattractive.

The problems associated with combustion of low-rank coals can be avoided by taking advantage of alternative utilization methods. Lignite can be used as fuel in several forms. Solid fuel utilization methods traditionally rely on methods for carbonizing, briquetting, or drying. Additional methods for solids beneficiation, such as desulfurization, require some physical or chemical pretreatment. Gasification of low-rank coals is another utilization method, but one which will require some eventual process adaptation for special applications to high moisture, high ash coals. Liquefaction is also a probable technology for utilization of low-rank coals. Here, also, the technology of the major bituminous coal liquefaction processes must be adapted and improved if the methods are to meet with reasonable success using peat and lignite.

As an alternative, it was proposed to wet process lignite and convert this material to liquid and gaseous fuels. Specifically, the objective of this work has been recovery of the energy from lignite in the form of both liquid and gaseous fuels. For each case, the conversion process is carried out in two stages. In the first stage, the coalified material, i.e. the lignite, undergoes hydrolysis to reduce the complex aromatic molecular structure to simple aromatics. In the second stage of the process, the simple aromatics are either anaerobically fermented to fuel gas or otherwise converted to liquid fuel, such as pentanol or benzene. Thus, the conversion process applied to the coalified materials is a combination of an initial hydrolytic treatment ("pretreatment") and a chemical or biological treatment. Chemical treatment of lignite hydrolysates was the focus of the program that will be detailed here.

2. PROGRAM PLAN

2.1 Overview of the Program Plan

Project emphasis was on coordinating alkaline hydrolysis and chemical decarboxylation with the goal of producing a BTX-type liquid fuel from lignite. The intention of this program was to examine the operating parameters of the proposed stages: (1) alkaline hydrolysis of lignite, (2) oxidation of hydrolysis products, and (3) decarboxylation of the final hydrolysate to BTX-type liquid fuel. Those results led to preliminary engineering designs for full and pilot-scale facilities.

2.2 The Program Outline

The lignite liquefaction program to produce a BTX-type liquid fuel included work on pretreatment of lignite, recovery of pretreated lignite material, and subsequent conversion of the lignite products to BTX-type liquid fuel. Included in this program was a preliminary engineering economic analysis for full and pilot-scale facilities to project process costs. The work described herein was to conduct the experimental process development work for the proposed lignite liquefaction process. A complete description of the program tasks specific to this project is as follows:

Task 1: Pretreatment of Lignite to Form Water-Soluble Aromatics

Alkaline pretreatment of peat under oxidative conditions (up to 200°C) promotes degradation of the peat and lignite to low molecular weight, water-soluble aromatics suitable for anaerobic fermentation. Somewhat more severe pretreatment conditions of lignite (up to 300°C) was employed to attain optimum solubilization and conversion to simple aromatics suitable for a liquid fuel. Parameters investigated included pH, temperature, solids loading, oxygen concentration, and time. This liquefaction study was built on the peat work which was oriented toward the goal of pretreatment with fermentation to chemicals. The principal objective of this task was to maximize conditions for yield of soluble material from lignite.

Task 2: Decarboxylation of Water Soluble Aromatics

The alkaline hydrolysis of lignite was to yield water soluble aromatics for recovery as a liquid fuel. This concept was built on the results of the earlier work where this treatment technique was used to break the complex lignin-like, aromatic structure of peat for fermentation. An example of the type of product which was anticipated is benzoic acid. This single-ring aromatic is a benzene ring with one carboxylic acid group capable of salt formation and, thus, water-soluble functionality. Upon further treatment at slightly higher temperatures, the benzoic acid will undergo decarboxylation to yield benzene, a possible liquid fuel. In a similar manner, other single ring carboxylic acids would, upon decarboxylation, yield BTX-type liquid fuels. This was the basis of the liquid fuel recovery technique investigated as part of this program task. The objective of this task was to establish the conditions for the conversion of the water soluble lignite aromatics to BTX-type liquid fuel.

Task 3: Chemical Analysis for Documentation and Verification of Results

Integral to the overall program was the chemical analysis. Methods used in the evaluation of lignite included gel permeation chromatography (GPC), high pressure liquid chromatography (HPLC), gas chromatography (GC) and standard wet chemistry and gravimetric extraction procedures. In addition, viscosities were run using model BTX-type liquid fuels to verify suitability as a liquid fuel and to establish a basis for comparing this material to petroleum-based liquid fuels.

Task 4. Evaluation of Product Recovery Systems

A system was designed for removal and recovery of the BTX-type fuel.

Task 5. Preliminary Process Economics for a Pilot and Full-Scale Facility

This task provides estimates of the capital and operating costs for a pilot and a full-sized facility based on data obtained from the experimental program. Additional research requirements were also delineated. A computer-aided optimization and sensitivity analysis was also performed to determine desired plant operating conditions. Preliminary conceptual designs for adaption of this process to an integrated energy recovery system were prepared and cost estimates made for the BTX-type fuel facility.

3. ALKALINE SOLUBILIZATION AND OXIDATION OF LIGNITE

3.1. Background
3.1.1 The Origins of Lignite

A combination of chemical, biological and physical processes contributes to the natural decomposition of plant material. The various plant constituents decompose at different rates in accordance with their respective susceptibilities to these natural mechanisms of attack. Thus, over time, the deposition of plant materials represents an accumulation of differentiated responses to different geological phenomena.

The plant constituents are accumulated as peat, a material which is composed primarily of macerals, less so of minerals, with water and gases in pore-like interstices. The macerals represent the organic constituents of the plant material, e.g. cellulose and lignin. As all of these materials are variably subjected to decay, they are incorporated into sedimentary strata, and altered by natural physical and chemical processes. As the material is acted upon, it undergoes a molecular metaorphosis manifested in the generation of "new" materials which are physically distinct and unique in their fundamental properties. This process is known as "coalification".

3.1.2 <u>The Structure of Lignite</u>

As peat is subjected to geological alteration, its organic constituents become chemically more complex. The severity of the conditions affects the physical properties of the resultant material. This coalified material is then classified according to rank, with the lignites at the low-rank, followed by sub-bituminous, bituminous, and anthracitic coals.

Because lignite is in the early stages of coalification, it retains some of the characteristics of wood. Pieces of plant debris and remains of cellular structure are visible in electron micrographs. The molecular structure then represents the earliest stages of geological metamorphism of plant material.

The major structural constituents of lignite are derived from three sources: (a) cellulose; (b) lignin; and (c) other plant components, e.g. protein, dispersed in the plant tissues. The coalification process is a deoxygenation/dehydration process. In the early stages, coalification does not create a large number of tertiary bonds. Because these are the bonds which contribute to a strong three-dimensional structure, lignites have loose structures.

The molecular structure of the organic portion of lignite can be represented as shown in Figure 3.1. The major features are summarized as follows. First, in comparison to higher rank coals, lignites have low aromaticity. Lignite is approximately 60% aromatic, peat 50%, and bituminous greater than 70%. Second, aromatic clusters are primarily one and two rings, i.e. aromatics have not yet fused into complex multi-ring forms. In contrast, bituminous coals are comprised of fused ring systems of 3 or more aromatics. Finally, oxygen-containing functional groups are prevalent. These groups are represented by carboxylate, phenolic and ethereal components, groups which chemically account for the complex fused-ring structure of high rank coals.

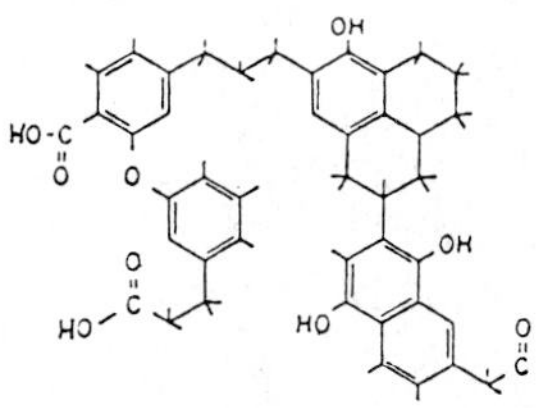

Fig. 3.1. Structural Features of Lignite
(Adapted with modification from Sondreal et al., 1982)

The remaining lignite components are moisture and ash. Again, the relative proportions of these are indicative of the rank of lignite. Lignite typically has a high moisture content when compared with other (higher rank) coals. Water is incorporated into the lignite matrix in an intermolecular hydrogen bonding between the water and the oxygen functional groups in the lignite. The ash-containing portion of lignite can be further divided into the inorganics and the minerals. The inorganics, e.g. calcium and sodium, are to some extent relatively mobile ions, presumably associated with the carboxylic acid functional groups; thus, some lignites can have levels of these inorganics. The mineral matter is primarily represented by clays, pyrite, and quartz. Frequently, the mineral particles are quite small and often appear finely dispersed throughout the carbonaceous material.

3.1.3 The Reactivity of Lignite

Examination of the organic portion of the lignite matrix as represented in Figure 3.1. reveals several important clues to its potential reactivity. A large portion of the aromatic material exists as single rings. These aromatics are joined in a network of carbon-carbon and carbon-oxygen linkages. Aliphatic units are incorporated as aromatic side-chains, alone or in conjunction with other aromatic groups.

This matrix structure is representative of a particular stage of the coalification process, a continuous process of deoxygenation which has resulted in the structural elimination of water. It is this process which contributes to the increasing molecular complexity up through the coal ranks. Thus, it might be expected that reversal of the coalification by hydrolytic reaction could result in considerable degradation of the coal structure into simpler units. This is true especially in the lower ranks of coals, e.g. peat and lignite.

Alkaline hydrolysis can be exploited in conjunction with oxidative processes to break certain well-defined bonds in the lignite. The carbon-oxygen linkages are particularly susceptible to the alkaline hydrolysis. Likening the lignite structure to a polyester/polyether-type material, phenols and carboxylic acids will be the probable products of a base hydrolysis (Figure 3.2a). The carbon-carbon linkages, as found, for example, in the alkylbenzene portions of the lignite matrix, can be effectively cleaved upon oxidation. Generally, the products of an oxidation will be carboxylic acids (Figure 3.2b). Thus, a careful "aqueous alkali oxidation" can effect a reversal of the coalification process particularly in low rank coals.

The effectiveness of an aqueous alkali oxidation will be primarily dependent upon the structural "maturity" of the organic matrix and not upon

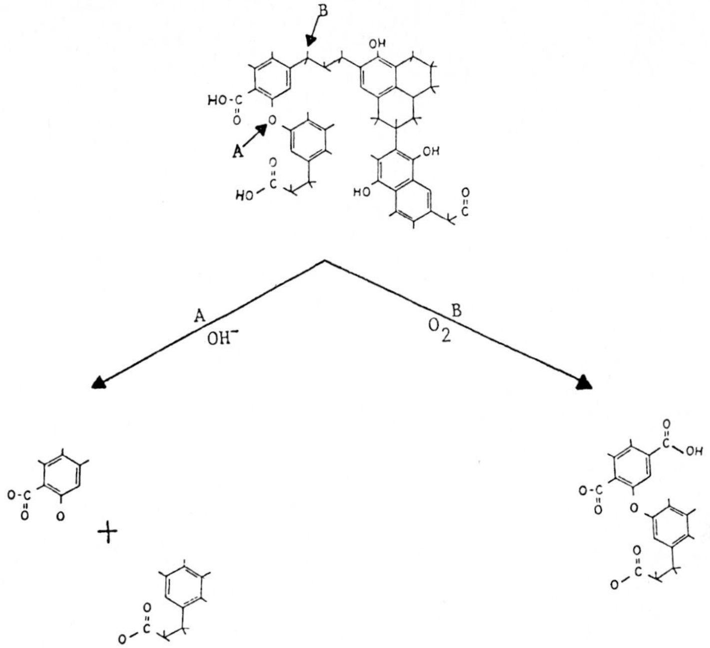

A Alkaline hydrolysis
B Oxidation

Fig. 3.2. Aqueous Alkali Oxidation of Lignite

the ash or moisture characteristics of a ranked coal. A treatment process
which incorporates the elements of an alkaline hydrolysis will obviously
require water and alkali. Lignites obtained from different sources vary
primarily in ash and moisture content; the organic matrices are similar.
Thus, the degree of lignite reactivity, as judged by breakdown potential,
can theoretically be regulated by adjustment of added water and alkali in
consideration of inherent moisture and ash for a particular sample.

3.2 Program Plan for Alkaline Solubilization and Oxidation of Lignite

The objective of the solubilization and oxidation of lignite is the
production of single-ring aromatic compounds which can be treated to yield
BTX-type liquids. Solubilization of lignite has been accomplished under
alkaline conditions. Based on Dynatech's previous work with peat, it was
suggested that oxidation would increase the yield of low-molecular weight
products.

With these objectives in mind, the following program plan was proposed:

- Characterization of raw lignite;

- Investigation of varying solubilization conditions (e.g. solids loading, temperature, concentration of alkali, type of alkali);

- Investigation of varying oxidation conditions (e.g. flow rate of input circulation);

- Characterization of products using standard wet chemical techniques and analytical instrumentation (e.g. extraction, gravimetric analysis, HPLC, GC, GPC, IR, spectroscopy, petrography);

- Evaluations and optimization of experimental procedures;

- Isolation of BTX-type products; and

- Preparation of an appropriate process design.

3.3 Alkaline Solubilization and Oxidation of Lignite - Experimental Procedures

3.3.1 Determination of Lignite Characteristics

Before investigating solubilization and oxidation procedures it was necessary to determine various characteristics of the lignite. Two samples of lignite were received from Meridian Land and Mineral Company (Billings, Montana). These samples were labelled "Beulah" and "Gascoyne", indicating the origin of the lignite. The Beulah lignite was used for this study.

First, the Beulah lignite was sent to Resource Engineering Incorporated (REI) of Waltham, Massachusetts, to be crushed. After being crushed to $\frac{1}{4}$-inch, a sample of the material was retained by REI for ultimate and proximate analysis (report in Appendix). The remainder of the crushed lignite was wet-packed and returned to Dynatech.

The final type of analysis performed on the untreated lignite was the solids analysis done on the stock lignite slurry and the pretreatment slurries. The stock lignite slurry was prepared by wet ball milling the $\frac{1}{4}$-inch wet-packed crushed lignite (as received from REI) and sieving to -100 mesh. Pretreatment slurry was prepared by diluting the stock slurry to the required percent volatile solids, then adding the appropriate amount of alkali.

Two types of solids analysis were performed: (1) total solids, and (2) ash content. Total solids were determined by the following method. Samples were measured into dry round-bottom flasks and the volume reduced _in vacuo_ using a Buchi Rotavapor-R rotary evaporator. The flasks were then placed in a desiccator over anhydrous $CaSO_4$ and dried _in vacuo_ to constant weight.

After drying, the ash content was determined. Samples were measured into dry evaporating dishes and placed in a 100°C Blue M Stabil-Therm Gravity Oven for six hours. The samples were then cooled and weighed. The dried samples were heated at 600°C in a Lindberg M51442 muffle furnace for six hours. After cooling in a desiccator, the residual material was weighed to determine ash content. Volatile solids were determined from the difference between the total solids and the ash.

3.3.2 Preparation of the Lignite Slurry

A stock lignite slurry was prepared by wet ball milling the $\frac{1}{4}$-inch wet-packed crushed lignite sample (as received from REI) and sifting to -100 mesh. Pretreatment slurry was prepared by diluting the stock slurry to the required percent volatile solids, then adding the appropriate amount of alkali on the basis of grams carbonate per gram of lignite volatile solids.

3.3.3 Solubilization Experiments

Solubilization parameters were extensively investigated to determine optional reaction conditions. The factors investigated were:

- Temperature
- Time
- Amount of alkali
- Type of alkali
- Loading of volatile solids.

A. Time/Temperature: A series of treatments was conducted at different temperatures. The charge to the reactor was kept constant at 7.5% volatile solids and 0.2 gram Na_2CO_3/gram volatile solids for this study. The three temperatures studied were 200°C, 250°C and 300°C. Samples were taken at designated intervals during the treatment.

B. Amount/Type of Alkali: The second series of experiments was designed to investigate the effects of different types of alkali and of different amounts of alkali. Two compounds were chosen as alkali: NaOH and Na_2CO_3. A single treatment using NaOH was done at 300°C. The NaOH used was equal in sodium-equivalents to the Na_2CO_3 used in the time/temperature series

described previously (i.e. 3.8 sodium meq/gram volatile solids). In
addition, the effect of varying the amount of Na_2CO_3 was also investigated.
The amount of Na_2CO_3 added was varied from 0.05g Na_2CO_3 to 0.3g Na_2CO_3 per
gram volatile solids. Finally, a control treatment, with no additive, was
also performed.

C.   Solids Loading: The final solubilization factor investigated was the
amount of volatile solids used for the treatment. This amount varied from
4.0% to 11.9% volatile solids (w/v) in the slurry charged to the reactor.

3.3.4 Oxidation Experiments

After evaluating factors which directly affect solubilization, the
process of oxidation and its effects on the overall solubilization were
investigated. A series of experiments was designed to examine the effect of
different air flow rates on solubilization of volatile solids. In addition,
a set of experiments was conducted to investigate the oxidation of solubilized hydrolysis products compared to non-solubilized hydrolysis products.

The first series of oxidations investigated the effect of the input air
flow rate.   Three flow rates were chosen; temperature, concentration of
Na_2CO_3, and lignite loading were held constant for the three oxidations.
The flow rates chosen were 400 mL/min, 740 mL/min and 1000 mL/min. The
oxidations were done at 200°C, with 0.5g Na_2CO_3 per gram volatile solids and
a 4% VS loading. In all three oxidations, samples were taken at designated
intervals for solids analysis.

The second series of oxidations investigated the effects of oxidation
on solubilized vs. non-solubilized hydrolysis products. An initial hydrolysis was conducted to provide solubilized and non-solubilized product. This
treatment was done at 250°C, with 0.25g Na_2CO_3/g VS and an 8% VS loading.
The mixture was brought up to 250°C and then cooled. Solubilized material
was separated from non-solubilized material by centrifugation at 8800 x g
for 15 minutes. The entire batch of treated material was centrifuged. The
supernatant material was designated the "solubilized" and the pellet
material was designated the "non-solubilized".

The next step in this set of experiments was oxidation of these two
materials.   Both oxidations were conducted under the same conditions at
200°C, with 0.5g Na_2CO_3/g VS and a 4% VS loading.

3.3.5 Analytical Methods

Treatment samples were characterized via solids analysis, percent
extractables, and petrography. Solubilized material was analyzed via high

performance liquid chromatography (HPLC) and gas chromatography (GC), and extracted samples were analyzed for molecular weight distribution via gel permeation chromatography (GPC).

A. <u>Solids Analysis</u>: Reaction samples are measured into dry round bottom flasks. The samples are initially reduced <u>in vacuo</u> using a Buchi Rotavapor-R rotary evaporator. The flasks are subsequently placed in a desiccator over anhydrous $CaSO_4$ and dried <u>in vacuo</u> to constant weight for determination of total solids.

Another aliquot of the reaction mixture (pellets, supernatants, or slurries) is measured into a dry evaporating dish. For the determination of oven-dried material, the dish is placed in a 100°C Blue M Stabil-therm Gravity Oven for six hours, cooled, and weighed for oven-dried total solids. Subsequently, the dried material is heated at 600°C in a Lindberg M51442 muffle furnace for an additional six hours. After cooling, the residual material is weighed for the determination of ash content.

B. <u>Percent Extractables</u>: A pellet, isolated from an acidified portion of reaction slurry, is placed in a beaker and dried to constant weight <u>in vacuo</u>. It is stirred with 20 ml tetrahydrofuran for one hour, then stirred with warming for five minutes. The mixture is vacuum filtered into a tared beaker and the filtrate is allowed to dry to constant weight to determine percent extractable solids.

C. <u>Petrography</u>: Petrographic analyses were performed by Resource Engineering Incorporated, Waltham, MA (report in Appendix). Pre- and post-treatment samples were submitted. The pretreatment sample was obtained from a portion of the lignite as provided. The post-treatment sample was obtained after washing and drying the pellet isolated by centrifugation. (The reaction slurry was first acidified to pH 2 with concentrated HCl.)

D. <u>High Performance Liquid Chromatography</u>: The chromatography system used is a Waters unit equipped with a M-45 solvent delivery system, a U6K injector, a free standing model 440 absorbance detector set at 254 nm, and a 10-mv strip chart recorded. The column used was a Regis Octadecyl Workhorse (30cm x 4.6mm).

A monocratic eluent system was chosen on the basis of its ability to separate four characteristic products (benzoic acid, syringic acid, syringaldehyde, and vanillin). This system was 45% MeOH/1% H_3PO_4. Spectro-grade glass-distilled methanol is obtained from Burdick and Jackson Laboratories,

Inc. Phosphoric acid is Mallinckrodt reagent grade. Distilled water is filtered, deionized, and passed through a carbon bed in a system from Sybron/Barnstead. After the solvent system is prepared, it is filtered through a Millipore HA filter and degassed with a Branson sonicator before use.

Solutions of 5-10 µg/ml of the standards (benzoic acid, syringic acid, syringaldehyde, and vanillin) are prepared in 45% MeOH. The retention volumes of these standards are determined by injecting 100 µl of the solutions into the aforementioned chromatographic system. Filtered reaction solutions are diluted with MeOH and characterized with respect to this standardization.

E.   Gas Chromatography: The volatile organic liquid determinations in this phase of the study are made using a Perkin-Elmer gas chromatograph equipped with a Gow-Mac Series 750 flame ionization detector and a Houston Instrument OmniScribe recorder. The separation is caried out on a Chromasorb 101 column ($\frac{1}{4}$ in. x 5 ft) under conditions of linear programming over the temperature range 100 - 200°C (15°C/min) with nitrogen as the carrier gas. The chromatograph is calibrated for analysis of methanol, acetone, acetic acid, propionic acid, butyric acid, and valeric acid. Reaction samples are filtered using a Millipore HA filter before injection.

F.   Gel Permeation Chromatography: The extracts obtained from the extraction are analyzed by gel permeation chromatography in order to determine the size of the molecules which are extractable. Tetrahydrofuran is used as the carrier solvent. The column is restyragel packed, 30 cm long. A Waters chromatography system equipped with a Differential Refractometer R401 detector is used.

3.4   Results and Discussion of Solubilization and Oxidation Experiments
    The extensive investigation of solubilization parameters and the effects of oxidation reveal several interesting conclusions. These conclusions are discussed in the following sections.

3.4.1 Temperature/Time Studies
    Following the established experimental protocol, a testing regime was set up in which lignite was heated with aqueous alkali at various temperatures. The goal was to determine the influence of temperature and time on the degree of lignite breakdown under conditions of alkaline hydrolysis.

This phase of the experimental program can be summarized as a series of three basic temperature/time experiments: (1) 200°C/4 hrs, (2) 250°C/4 hrs, and (3) 300°C/5.75 hrs. In each experiment, a lignite slurry was treated at the specified temperature with samples being withdrawn when the reactor reached temperature, and 0.5 hour, 1.0 hour, 2.0 hours, and 4.0 hours (or 5.75 hours) thereafter. Each of these "hot-drawn" samples was characterized with respect to the percent of the input volatile solids solubilized and the percent isolable, THF-extractable material. In the first quarter of work, this analytical format was established as the one which would best accommodate the needs of the program by providing an informative scientific and engineering data base. In addition, the samples were qualitatively evaluated with respect to physical appearance.

3.4.2 Physical Characterization

A. Appearance: The final product slurries of alkaline solubilizations at 200°C, 250°C, and 300°C compare as follows. The 200°C product is gelatinous (mud-like) and does not settle over time. It is medium brown in color, and does not disperse at all, or generate any surface film when dropped into clean water. By contrast, the 250°C product is a mobile, yellowish brown liquid of high opacity; when dropped into water it rapidly generates a small, stable surface film. The 300°C product is an oily, slightly brownish grey/black mobile liquid of somewhat stronger odor than the others; when dropped into water it generates a very rapidly spreading surface film, which subsequently contracts and wrinkles. Both the 250°C and the 300°C products throw dense layers of solid upon standing; however, in all cases the particulate nature of the products, though not directly visible, is demonstrated by a slow diffusion speed in water.

B. Scanning Electron Microscopy: A suspended particle size analysis was performed to determine the physical effects of Na_2CO_3, temperature, and pressure on the lignite. This was accomplished by analyzing pretreatment samples using scanning electron microscopy (SEM). The specifications of the analysis are as follows: treatment 020031, 8% volatile solids loading, and 0.2 gms Na_2CO_3/gm VS. Samples were taken before carbonate addition, after carbonate addition, and at T = 150°C, 200°C and 250°C. A variety of magnifications ranging from 11 x to 3200 x were used on each sample. First, general particle size distribution was investigated. Then, the analysis concentrated on specific particles to determine the change in physical shape as the treatment progressed. In general, it was shown that although particle size distribution remained constant, particle texture and shape was affected by the treatment. Before the treatment, particles were smoother and more geometric in shape, whereas samples at 200°C and 250°C were chipped, rough, crusty, and much less geometric. An Elemental Distribution Analysis by X-ray (EDAX) was performed on the samples. It is interesting to note that over time there is a decrease of sodium on the particles (corresponding to the "crust") which may have indicated a leaching of entrapped ions. Elements found in later stages of treatment include silicon, phosphorus, sulfur, calcium, and small amounts of sodium.

3.4.3 Solubilization

For each of the experiments, the interim samples were analyzed to determine the percent solubilization of input volatile solids which was effected by the treatment. The percent solubilization was calculated on the basis of the change in volatile solids distribution after centrifugation of a pretreatment slurry (i.e. lignite, alkali, and water) and a post-treatment sample. The results are represented graphically in Figure 3.3 and summarized in the following.

Comparison of the solubilization behaviour within each temperature experiment yields dissimilar trends at the three different temperatures. In the 200°C experiment, successive time samples show an increase in the volatile solids solubilized from 30% at temperature to a maximum of 53% after 4.0 hours. The percent volatile solids solubilized in the 250°C experiment remained relatively constant throughout the course of the experiment having reached a 63% maximum at temperature. Finally, in the 300°C experiment, the observed solubilization decreased over time with a high of 43% volatile solids solubilized at temperature, falling to a low of 17% after 5.75 hours. On the basis of these solubilization figures, the ideal temperature for hydrolytic breakdown under the conditions as specified is 250°C, with extended treatment times having an insignificant influence.

A possible explanation for the observed behavior can be realized upon examining the competing chemistries involved. It has been established (see Dynatech Report No. 2244, Wise et al., 1983) that two processes are evident here: (1) alkaline hydrolysis of the ethereal lignite matrix to yield low molecular weight aromatic material, and (2) repolymerization of free radical reaction intermediates to yield high molecular weight coalified solids. In the light of the experimental results, it would appear that temperature and time can both affect the degree to which either of the competing reactions occurs. During a 200°C treatment, hydrolysis seems to predominate, while during a 300°C treatment, repolymerization predominates. In the 250°C

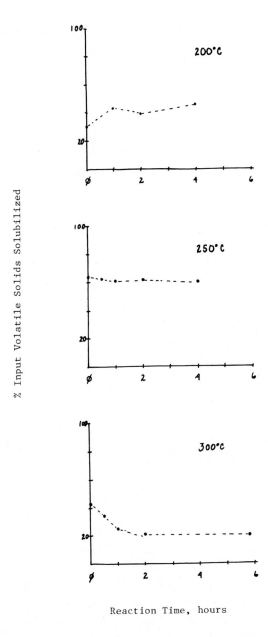

Fig. 3.3. Lignite Solubilization with Sodium Carbonate

experiment, the explanation is less obvious. At extended reaction times, an increase in hydrolysis does not seems evident and repolymerization appears controlled.

3.4.4 Acid Precipitation

The product slurries from each of the temperature/time experiments were initially separated into the base-soluble and base-insoluble fractions. This allowed for the determination of the percent of input volatile solids solubilized as detailed in the preceding section. The procedure can effectively identify the amount of water-soluble species of low molecular weight material which was produced via the hydrolysis.

The solubilized material – particularly the volatile solids – can be further characterized on the basis of its behavior in acidic media. Because of the structural characteristics of the input lignite and the hydrolytic nature of the reaction process, the solubilized material is expected to be high in polar functional groups, such as aromatic hydroxy and carboxylic acid groups.   While this material would remain in solution in a product slurry of pH 10, acidification should precipitate some of the solids to allow for further characterization and identification.

The alkaline solution containing dissolved products was separated from any solids that were suspended in the reaction slurry. This solution was subsequently acidified to pH 2.0 with concentrated hydrochloric acid.   An experimental maneuver such as this is intended for laboratory identification only; by no means is it intended for eventual process work.   Acidification is an effective method of isolation for low molecular weight breakdown material of characteristic oxygen functionality.   The procedure provides sufficient solid for subsequent analytical work.

The data corresponding to the acidifications are presented in Table 3.1. The four columns in the table identify the following: (1) the sample; (2) the percent of input volatile solids solubilized (shown graphically in Figure 3.3.); (3) the percent of the solubilized volatile solids which can be precipitated by acid; and (4) the percent which the latter, i.e. (3), is of the input volatile material. The results indicate that in the majority of the samples, virtually all of the solubilized material can be precipit- ated by acid. This is particularly true for the 200°C and 250°C treatments in which greater than 96% of the solubilized volatile solids were precipitated.   With the samples from the 300°C treatment, a substantial percentage (>82%) of the solubilized solid was isolable after acidification; the cases where the yield was low correspond to samples in which solubilization was also low. It appears, then, that acidification is an

effective experimental procedure if isolation of breakdown material is
desired. This method waa exploited for the large-scale collection of
treatment products.

Table 3.1

ACID PRECIPITATION

SAMPLE		INPUT VOLATILE SOLIDS SOLUBILIZED (%)	ACID-PRECIPIABLE SOLUBILIZED VOLATILE SOLIDS (%)	ACID-PRECIPITABLE INPUT VOLATILE SOLIDS (%)
Temp	Time			
200°C	0 h	30	99	30
	1	45	99	45
	2	42	97	41
	4	53	99	52
250°C	0 h	63	96	60
	1/2	63	97	61
	1	61	97	59
	2	64	97	62
	4	63	96	60
300°C	0 h	43	93	40
	1/2	34	93	40
	1	25	91	23
	2	22	82	18
	5 3/4	17	86	15

3.4.5 Extraction

The low molecular weight material which is produced in the lignite hydrolysis can be further characterized in terms of its solubility in organic solvents, i.e. its "extractability". Generally, tetrahydrofuran (THF) is used as the organic solvent because its reasonably low polarity allows the separation of low molecular weight, non-polar material from a complex product slurry. The product mixture will contain many compounds of dramatically different polarities. Extraction with THF can be used to differentiate products on the basis of polarity.

In this work, product solids were subjected to two different extractions for comparison as shown in Figure 3.4. For the first case, a pellet (1) was isolated from a centrifuged slurry sample which had been acidified (pH 2.0) in toto. For the second, the pellet (2) was obtained from an acidified supernatant only. Pellet (1) is representative of all possible product solids because the isolation protocol includes insoluble product residues. Alternatively, pellet (2) contains only that material solubilized in the hydrolysis. It would seem likely that most of the more non-polar, THF-extractable material would be available in pellet (1). Pellet (2) is primarily aromatic material of characteristic oxygen functionality which tends to be slightly more polar. It is important to note that the material of pellet (2) would be entirely contained within pellet (1).

The results of the extractions are presented graphically in Figure 3.5. The percent extractables relative to pellet total solids are shown for both pellet (1) and pellet (2). The trends in the extraction behaviors can be interpreted in terms of the competing reaction chemistries, in much the same way as those on the solubilizations were. In those cases where a greater percentage of extracted material is in pellet (2) - rather than pellet (1) - it would appear that more of the extractable products are solubilized, low molecular weight compounds of reasonably low polarity with repolymerized macromolecules comprising a major portion of the original reaction residue. Alternatively, when the greater percentage of extracted material is in pellet (1), it would appear that much of the breakdown material is water-insoluble, low molecular weight product which is sufficiently non-polar to allow for extraction by THF. The relative distributions of the various product types are a direct result of the hydrolysis and the repolymerization.

With the 200°C treatment it would appear that most of the extractable material produced initially is the water-insoluble solid which is of reasonably low molecular weight (molecular weight data follows); the solubilized material is not amenable to extraction. The extractables from the 250°C treatment, on the other hand, can be found primarily in the solubilized

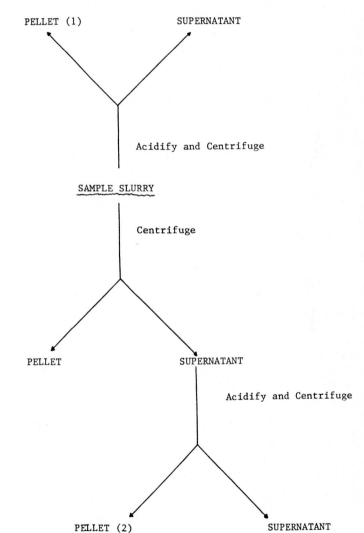

Fig. 3.4. Flow Chart for THF Extractions

product fraction which arises from the hydrolysis; the reaction residue contributes very little to the extractables. This phenomenon was time-dependent with percent extractables increasing over time. Extraction of the products from the 300°C treatment was greatest with the reaction residues rather than the solubilized material. This behavior also showed some time-dependence. Extraction of the residue increased over time while that of the solubilized material decreased.

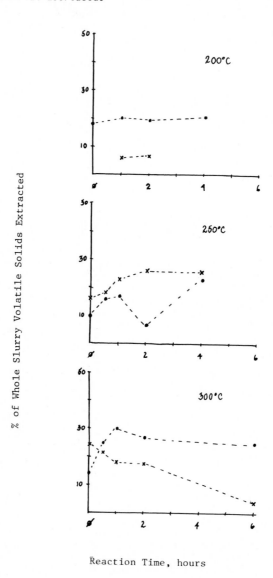

Reaction Time, hours

Fig. 3.5. THF-Extractable Pellet Solids

In comparing these results with those of the solubilization, it seems that there may be some compatible trends in the reaction chemistry which can be inferred from the data. The results suggest that in those cases where hydrolysis is the predominant chemistry, the percent solubilization is highest and the greater percentage of extractable product is in that solubilized material (e.g. the 250°C treatment). Alternatively, when repolymerization successfully competes with hydrolysis, the percent solubilization is lower and the majority of the extractable product is in the reaction residue.

The extracted materials were characterized with respect to the molecular weight distribution. This provides some verification of predicted product characteristics. Gel permeation chromatography (GPC) was used to obtain molecular weight profiles for each of the extracts. Generally, the extraction was effective in separating products of molecular weight less than 1000 with the 50-500 molecular weight range being the predominant one. The GPC profiles that are presented in Figure 3.6 are illustrative of the tracings obtained with each of the extracts. The GPC data shown is for a series of extracts of pellet (1)'s from a 200°C treatment.

3.4.6 Additive Studies

A series of experiments was conducted to investigate the role of the sodium carbonate in the alkaline hydrolysis. Previous work by Dynatech scientists investigating the role of the carbonate in the solubilization of peat had established some of the importance of the sodium cation in successful hydrolyses. In addition, modern knowledge in liquefaction science supports the idea that the size of a sodium cation makes possible diffusion into lignite pores; thus, the cation acts in both the physical and chemical breakdown of lignite. The strong implication that sodium has a major role in the alkaline hydrolysis influenced the direction of this phase of the experimental program.

Three experiments were conducted at 300°C. Using the standard format for reactor loadings, a slurry was prepared which contained lignite, water, and either nothing or one of the following: (1) sodium carbonate or (2) sodium hydroxide. The obvious difference in each of these input slurries is the pre-treatment pH. An initial assumption might then have been that the sodium hydroxide (at an initial pH of 12.0) would be the most effective in promoting the hydrolysis. The pH of the starting and the interim samples is shown in Table 3.2.

The results of each of the different additive treatments were interpreted in terms of the percent solubilization of input volatile solids,

330

Figure 3.6

GPC Tracings of Pellet Extracts

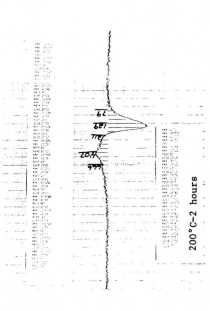

200°C-at temperature

200°C-1/2 hour

200°C-1 hour

200°C-2 hours

200°C-4 hours

Table 3.2

ADDITIVES AND pH

ADDITIVE	TIME, hr	pH
Sodium Carbonate	input	10.10
	0	10.07
	0.5	9.87
	1.0	9.73
	2.0	9.59
	5.75	9.23
Sodium Hydroxide	input	12.02
	0	10.25
	2.0	10.59
	4.0	10.12
None	input	7.85
	0	7.80
	1.0	7.50
	3.0	7.60

as well as in terms of physical appearance. As far as physical appearance, the final product slurries of 300°C solubilizations with Na_2CO_3, NaOH, and no added base, compare as follows.

The product from 300°C solubilization in the presence of Na_2CO_3 is a slightly brownish grey/black mobile liquid which throws a solids layer on standing. The product from NaOH-assisted treatment differs essentially only by being much browner. Both these products are sub-visibly particulate. By contrast, the product from a no-base-present treatment is a pure grey/black, with no trace of brown tint. No solids layer is observed to form on standing, yet the slurry consists of macroscopic, easily visible particles. All three products generate rapidly spreading surface films when dropped into water, including throwing the film upwards from a submerged drop (indicative of hydrophobic constituents). This film expands rapidly to stable final size in the case of the NaOH and no-base products, but shrinks after initial spread with the Na_2CO_3 product, as if an initially hydrophobic layer dispersed into the water over time. The Na_2CO_3- and NaOH-generated products had similar odors, but the no-base product slurry was considerably blander, though not odorless.

The solubilization data are presented graphically in Figure 3.7.

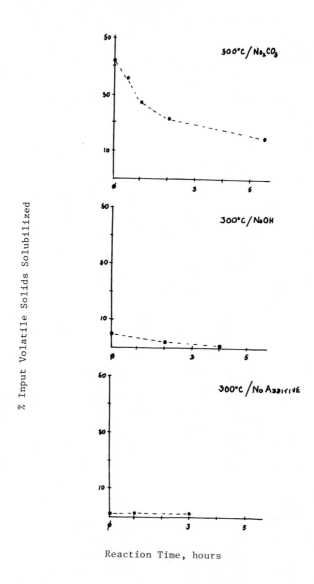

Reaction Time, hours

Fig. 3.7. Lignite Solubilization with Various Additives

In terms of percent solubilization, there are obvious differences in the effectiveness of the treatments. Without any additive, there was no evidence of matrix breakdown. The same was true in the case where sodium hydroxide was added. The most efficient treatment, then, was in the sodium carbonate case. The interpretation of the data is not obvious on the basis of solubilization only. There would appear to be no reason a priori as to why sodium hydroxide was ineffective in promoting hydrolysis.

In addition to the above analyses, the sodium hydroxide and sodium carbonate treatments were evaluated in terms of product extractability. The results are presented in Figure 3.8.

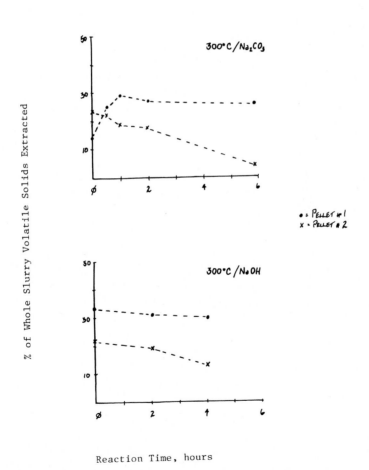

Fig. 3.8. THF-Extractable Pellet Solids: Additive Variations

The data show trends similar to those of the initial time/temperature work. That is, as solubilization decreases, the percent extractables in pellet (1) increases while that in pellet (2) decreases. It is important to note, however, that in terms of extractability it appears that the sodium hydroxide was effective in generating low molecular weight, water-insoluble product.

3.4.7 Oxidation Experiments

Table 3.3 shows the conditions for the oxidations in this series of experiments and the %VS solubilized as a result of oxidation. The %VS in the first supernatant decreases significantly in all cases as the oxidation proceeds. However, the %TS remains the same. Apparently, the oxidation is causing precipitation of the volatile solids solubilized. Table 3.4 shows the %VS in the whole slurry and in the first supernatant. Although both categories show some decreases, the %VS in the whole slurry decreases only very gradually. On the other hand, the %VS solubilized shows a large and sharp decrease as soon as the first sample is drawn. This implies that the solubilized volatile solids are susceptible to oxidation in a way other than combustion. The second set of oxidation experiments was designed to test this theory.

Table 3.3

OXIDATION CONDITIONS* AND %VS SOLUBILIZED

CONDITIONS	SAMPLE	%VS SOLUBILIZED
400 mL/min	input (at 200°C)	2.3
	3/4 hr oxidation	0.5
	2 hrs oxidation	0.5
740 mL/min	input (at 200°C)	3.0
	1/2 hr oxidation	0.6
	1 hr oxidation	0.4
	2 hrs oxidation	0.5
	3 hrs oxidation	0.8
1000 mL/min	input (at 200°C)	2.7
	1/2 hr oxidation	0.5
	1 hr oxidation	0.5
	3 hr oxidation	0.7

* Note: All of these oxidations were done at 200°C.

Table 3.5 compares the volatile solids data from the separate oxidations of solubilized materials and of non-solubilized materials. Again,
the %VS solubilized decreases during oxidation of each type of slurry, while
the %TS remains virtually constant. In addition, the %VS solubilized
decreases sharply once the oxidation of the solubilized material begins, as
observed in the previous series of experiments.

The %VS solubilized does not decrease as drastically in the oxidation
of non-solubilized material as with the solubilized material. This confirms
the theory that solubilized materials are more susceptible to oxidation
phenomena than non-solubilized material. In addition, oxidation does not
increase solubilization of non-solubilized material. This implies that a
two-stage pretreatment process, consisting of solubilization followed by
oxidation, would not significantly increase solubilization above the level
obtained by the solubilization alone.

Table 3.4

TOTAL VOLATILE SOLIDS COMPARED TO % VS SOLUBILIZED

SAMPLE	% TOTAL VS	% VS SOLUBILIZED
400 mL/min		
input (at 200°C)	4.5%	2.3%
3/4 hr oxidation	4.0	0.5
2 hrs oxidation	2.8	0.5
740 mL/min		
input (at 200°C)	5.1	3.0
1/2 hr oxidation	5.0	0.6
1 hr oxidation	4.6	0.4
2 hrs oxidation	4.9	0.5
3 hrs oxidation	4.7	0.8
1000 mL/min		
input (at 200°C)	5.2	2.7
1/2 hr oxidation	5.0	0.5
1 hr oxidation	4.7	0.5
3 hrs oxidation	4.6	0.7

Table 3.5

VOLATILE SOLIDS DATA FROM OXIDATIONS OF SOLUBILIZED AND
NON-SOLUBILIZED MATERIAL

SOLUBILIZED MATERIAL	% TOTAL VS	% VS SOLUBILIZED	%VS sol/ % Tot VS
Input at 200°C 1/2 hr oxidation 1 hr oxidation 2 hrs oxidation	4.1% 5.6 3.8 3.6 3.3	3.1% 2.9 0.1 0.1 0.2	0.76 0.52 0.03 0.03 0.06

NON-SOLUBILIZED MATERIAL	% TOTAL VS	% VS SOLUBILIZED	%VS sol/ % Tot VS
Input at 200°C 1/2 hr oxidation 81 min oxidation 155 min oxidation	6.1% 6.0 5.6 5.5 5.2	1.2% 1.6 0.8 0.8 1.1	0.20 0.27 0.14 0.15 0.21

Oxidation within the prescribed pretreatment regime affects the solubilized hydrolysis products rather than the non-solubilized hydrolysis products. Oxidation should promote degradation of the non-solubilized material. Oxidation, however, does not do this, but rather promotes the precipitation of solubilized hydrolysis product. There are two possible explanations for this: (1) oxidation encourages repolymerization via free radical reaction, or (2) oxidation encourages precipitation via surface charring and agglomeration. In addition, oxidative degradation of non-solubilized material is most likely ineffectual on the "coalified" hydrolysis product (see Table 3.5). Further investigation may be interesting, particularly within the format of the proposed pilot facility which is capable of "fast reaction kinetics" (see Section 5).

4. DECARBOXYLATION OF AROMATIC CARBOXYLIC ACIDS

4.1. Background

The initial products of alkaline oxidation are aromatic aldehydes, alchols, and carboxylic acids with methoxy side-chains. Low molecular weight alkyl carboxylic acids are also formed. These products are depicted in Figure 4.1. Further oxidation leads to degradation of the aromatic nuclei. The methoxy content of the product also decreases with time, ostensibly due to the elimination of side-chains from the aromatic nuclei.

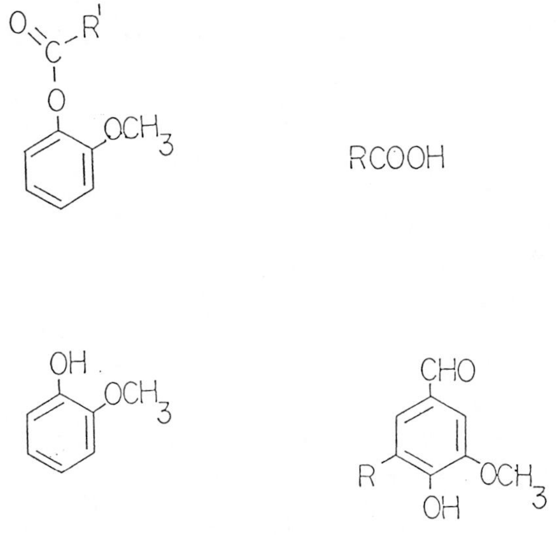

RCOOH

WHERE

$R' = H$ OR CH_2OH

$R. = H$ OR $[CH_3]_3C$ OR OCH_3

Fig. 4.1. Model Compounds
(Alkaline Oxidation Products)

4.1.1 Recovery of Aromatics

Systems for the recovery of the aromatic products are of many different types. Solvent extraction of organic chemicals such as aromatic organic acids from aqueous solution is a technique which has been under consideration for some years. Ethyl acetate, along with diethyl ether, has been suggested as an efficient extraction medium at low acid concentrations. Recently, novel solvent extraction media were developed. This new technology is based on the use of trioctylphosphine oxide in combination with other solvents (Helsen and Dence, 1975). Higher molecular weight acids (butyric or larger) can be efficiently extracted into a liquid hydrocarbon solvent such as kerosene. It would be anticipated that aromatic organic acids could be similarly extractable.

An alternate approach to conventional solvent extraction followed by distillation is solvent extraction followed by re-extraction into an aqueous base. This technique had been successfully employed at Dynatech in work concerned with liquid fuels production from biomass. Organic acids, butyric and higher, are selectively extracted and concentrated in aqueous base. Concentrations of acid salts obtained in the aqueous base have exceeded 1.0 N. Other approaches to recovery of the aromatic organic acids include the addition of an inorganic acid, such as HCl, resulting in the direct precipitation of the aromatic.

4.1.2 Decarboxylation of Aromatic Carboxylic Acids

It has been known for a long time that organic carboxylic acids, particularly aromatic carboxylic acids, can be decarboxylated by heating their salts with excess alkali to produce the parent hydrocarbon. The overall reaction is presumed to be:

$$RCOO^- M^+ + M^+OH^- \xrightarrow{\Delta} R-H + M_2CO_3$$

Kerkovines and Dimrath (1913) found that aromatic carboxylic acids obtained from oxidation of charcoal could be decarboxylated efficiently by heating their barium salts. Shortly thereafter, Fisher (1919, 1921a, 1921b) found that the sodium salts of aromatic carboxylic acids made from coal oxidation could be decarboxylated by heating them with water to approximately 450°C under pressure. Juettner and others (1935) found that Fisher's method could be used to obtain up to a 94 percent yield from a pure sample of aromatic carboxylic acids. Entel (1955) found that the decarboxylation of aromatic acids produced by alkaline oxidation of coal could be carried out at 250°C if the copper salts were used. It is interesting to note that he was able to effect 96-99 percent decarboxylation by this method. Subsequently, workers at Dow Chemical Company (Montgomery and Holly, 1956, 1958) decarboxylated the copper salts of these acids at 265°C using quinoline as a catalyst and claimed improved results. The results of the product analysis resulting from the copper-quinoline decarboxylation of aromatic acids obtained from the alkaline oxidation of Pocahontas No. 3 bituminous coal are presented in Table 4.1. This has related interest to both peat and lignite.

The results of other workers indicate that nearly quantitative decarboxylation is possible by a variety of techniques. A choice between these techniques can be made on economic considerations. It is quite possible that decarboxylation of water soluble aromatic acids will be feasible with peat and lignite hydrolysis products, thus yielding a BTX-type liquid fuel. Alternative methods to recover the aromatics, as described above, should also be practical.

Table 4.1

PRODUCT ANALYSIS OF DECARBOXYLATED AROMATIC ACIDS*

COMPOUND	YIELD[†] g/kg	PORTION OF TOTAL NUCLEI[††] %	COMPOUND	YIELD[†] g/kg	PORTION OF TOTAL NUCLEI[††] %
Mass 280	*	–	Benzophenone	7.6	1.7
Mass 268	*	–	Phenanthrene	15	3.3
Mass 260	*	–	Mass 178 (not		
Mass 258	*	–	phenanthrene)	*	–
Mass 254	*	–	Mass 176		
Mass 244	0.007	0.002	(C_7-benzene)	0.7	0.2
Mass 234	*	–	Methylbiphenyl	2.8	0.6
Mass 232	*	–	Mass 166	0.3	0.07
Terphenyl (m- and p-)	5.6	1.2	Mass 160	0.5	0.1
Mass 228	0.06	0.02	Mass 158	*	–
Mass 226 ($C_{14}H_{16}O_3$)	0.02	0.009	Biphenyl	41	9.0
Mass 220	*	–	C_5-benzene	8.1	1.8
Mass 218	*	–	Methylnaphthalene		
Mass 204			(α and β)	172	38
(Phenylnaphthalene)	5.4	1.2	Butylbenzene	1.8	0.4
Mass 202			Naphthalene	15	3.3
(Not fused aromatic			Mass 122	0.1	0.02
ring system)	2.1	0.5	C_2-benzene	0.2	0.04
Mass 196	0.9	0.2	Toluene	3.4	0.75
Mass 184 ($C_{14}H_{16}$)	0.7	0.2	Benzene	91.4	20.0

* Present † Total yield 378 g/kg   †† Total of all portions 83 per cent.

---

*from Montgomery and Holly, 1958.

4.2 Program Plan for Decarboxylation of Pretreated Lignite

A carefully planned alkaline hydrolysis can effect a reversal of the coalification process, particularly in the low rank coals. It was expected that this reversal of the coalification would result in considerable degradation of the coal structure into much simpler chemical units. A chemical functionality common to the breakdown products is the carboxylic acid group. It was a goal of this program to remove the carboxylic acid group (via "decarboxylation") from lignite hydrolysis products to yield water-insoluble organics suitable for use as fuel.

One of the chemical methods for decarboxylation of aromatic carboxylic acids is the copper-quinoline method. In fact, this technique was demonstrated using coal acids (Montgomery and Holly, 1956). The mechanism by which the acids are decarboxylated is a combination of thermal and chemical actions. The copper salts of the carboxylic acids are combined with quinoline (a high -238°C- boiling solvent) and heated to reflux. The quinoline facilitates solvolysis, as well as temperature control. The method is closely allied to pure thermal methods in which salts of the acids are heated to 250°C. The copper-quinoline technique was thus chosen for investigation on the basis of its proven success.

Based upon overall applicability to the proposed process, as well as technical and economic considerations, the persulfate/silver ion method of decarboxylation (Fristad et al., 1983) was also chosen for investigation. This method uses sodium persulfate in conjunction with a transition metal catalyst to facilitate carboxylic acid decarboxylation. This sodium persulfate is a powerful oxidant which is relatively inexpensive and easily handled. Its high activation barrier makes the use of persulfate alone unreasonable at low temperatures. However, when combined with a transition metal catalyst (e.g. a silver ion) the activation barrier of the persulfate ion is reduced and lower reaction temperatures are feasible. In addition, the persulfate serves as an effective electron shuttle to produce active metal species which can participate in the chemical reactions themselves. The mechanism for decarboxylation by the reagent pair is schematicized in Figure 4.2.

Operating on the premise that the major products of the alkaline hydrolysis would be benzoic acid-like, the program plan was to investigate the conditions for decarboxylation of benzoic acid using the persulfate/silver ion method. Maximum operating conditions were to be identified with respect to solvent composition, reagent concentrations, and reaction temperatures. In the final stage of the decarboxylation work, the hydrolysis products of the lignite pretreatment were to be subjected to the "best case"

$$Ag(I) + S_2O_8{}^{2-} \rightarrow Ag(II) + SO_4{}^{-} \cdot + SO_4{}^{2-} \quad (1)$$

$$Ag(I) + SO_4{}^{-} \cdot \rightarrow Ag(II) + SO_4{}^{2-} \quad (2)$$

$$Ag(II) + RCO_2H \rightarrow Ag(I) + RCO_2 \cdot + H^+ \quad (3)$$

$$RCO_2 \cdot \rightarrow R \cdot + CO_2 \quad (4)$$

$$R \cdot + H\text{-solv} \rightarrow RH + \cdot solv \quad (5)$$

---

* From W.E. Fristad et al., 1983.

Fig. 4.2. Mechanism for Decarboxylation of Organic Acids

decarboxylation for determination of the percent conversion to water-insoluble organics. The products of the reactions were analyzed by GC, in all cases, and HPLC, in those cases where applicable.

4.3 Decarboxylation of Aromatic Carboxylic Acids - Experimental Procedures

4.3.1 Decarboxylation: The Copper/Quinoline Method
A. Decarboxylation of Benzoic Acid

The procedure used to decarboxylate benzoic acid is a modification of that reported by Buckles and Wheeler (1963) in the synthesis of cis-stilbene from α-phenylcinnamic acid. A 1000 ml, three-necked flask is equipped with reflux condenser, magnetic stirrer and thermometer. Benzoic acid (0.205 mol, 25.00 g) and anhydrous cupric sulfate catalyst (.024 mol, 4 g) are added. To this is added freshly-distilled quinoline (2.38 mol, 281 ml) and the system is heated to 210°C using a heating mantle. The temperature of the system is maintained between 210°C - 225°C for 1.25 hr after which the flask is cooled and the reaction mixture is gravity-filtered into a 500 ml round-bottomed flask. At this point the reaction mixture smells strongly of benzene. The flask containing the reaction mixture is fitted with a Vigereaux column and the contents are fractionally distilled.

B. Decarboxylation of Benzoic Acid - Time Study

The procedure described in (A) is repeated with the exception that the amount of each reagent is reduced by 50%. Samples of the reaction mixture are taken before heating, at 100°C, 200°C, and 0.5 hr, 1.0 hr, and 1.25 hr after the temperature reaches 210°C.

4.3.2 Decarboxylation: The Silver/Persulfate Method

A. Decarboxylation of Benzoic Acid

A 250 ml three-necked flask is fitted with reflux condenser, addition funnel, thermometer and magnetic stirrer. Benzoic acid (0.14 mol, 15.88 g) and silver nitrate (2.6 mmol, 0.4 g) are dissolved in a mixture of 100 ml acetonitrile and 33 ml water. This solution is added to the flask and heated to reflux (78°C). Sodium persulfate (260 mmol, 61.9 g) is added to 67 ml water and heated to dissolution. The sodium persulfate is added to the reaction flask over a period of 15 minutes. When addition is complete, the flask is allowed to reflux an additional 5 minutes. After cooling, the reaction mixture is a yellow 2-phase liquid. The layers are separated; the upper layer is washed with saturated sodium bicarbonate solution and the bottom layer is washed with ether. The reaction mixture is extracted with ether and saturated sodium bicarbonate as described above.

B. Decarboxylation of Lignite Hydrolysis Products

A concentrate of lignite hydrolysis products is obtained as follows. The "best case" slurry (250°C/ϕtime) is separated by centrifugation. The solubilized organics are precipitated by acidification) (conc HCl, pH 2.0) of the supernatant. The organics are isolated by centrifugation and then dried. This material is considered the essence of the lignite hydrolysate and in this best case amounts to about 63% of the input volatile solids.

The decarboxylation procedure described in (A) is followed with the concentrated lignite hydrolysate replacing the benzoic acid. The reactant amounts are as follows: lignite concentrate (5.0 g), sodium persulfate (84 mmol, 20.0 g), silver nitrate (.98 mmol, .15 g).

4.3.3 Chromatographic Methods of Analysis

The volatile organic liquid determinations of decarboxylation products are made via gas chromatography (GC). Other organic determinations are made via high performance liquid chromatography (HPLC).

A. Gas Chromatography

The volatile organic liquid samples in this phase of the program are analyzed using a Varian 3700 gas chromatograph equipped with a Varian Model 2010 flame ionization detector and a Spectra-Physics 4270 printer, plotter, integrator. The separation is carried out on an Alltech stainless steel AT-1000 on Graphpac column ($\frac{1}{4}$ in x 6 ft; 80/100 mesh). Samples are run under conditions of linear programming over the temperature range 150°C – 210°C (10°C/min) with nitrogen as the carrier gas (30 ml/min). The injector temperature is set at 220°C, detector temperature at 230°C. Solutions of standards are prepared in ethanol.

B.   High Performance Liquid Chromatography

The organic products are analyzed using a Waters (Milford, MA) HPLC system equipped with a M45 solvent delivery system, a U6K injector, a Model R401 differential refractometer, and a Fisher Recordall Series 5000 recorder. Separation is carried out on a Regis Octadecyl Workhorse column (30 cm x 4.6 mm) using $CH_3OH/1\%$ CH_3COOH (1:1) as the eluent. Solutions of standards are prepared in eluent.

4.4   Results and Discussion of Decarboxylation Studies

The strategy for using the model compounds in the decarboxylation studies was three-fold. First, it was desirable to reproduce the literature work to serve as a verification of results and as grounds for development of pertinent analytical protocols. Second, it was of interest to determine the sensitivities in various parameters of the decarboxylation operation. Third, it was necessary to generate data for a best case analysis in the preliminary process designs. This approach was justified in view of the complexity of the lignite solubilization material. Once the experimental plan was defined with the model compounds, lignite hydrolysate was subjected to decarboxylation.

4.4.1 Decarboxylation of Benzoic Acid

One of the procedures used to decarboxylate benzoic acid was the copper sulfate/quinoline method. Benzoic acid was heated to reflux in quinoline with copper sulfate added as catalyst. At the end of the reaction time, the mixture smelled very strongly of benzene. Distillation of this reaction mixture yielded three fractions: (1) 51° - 56°C (1.95 g); (2) 56° - 80°C (3.73 g); (3) 80°C - 200°C (150 g).

Gas chromatographic analyses showed the major benzene yield in fractions (1) and (2). Quinoline peaks appeared prominently in all three fractions even though the high boiling point of quinoline would seem to preclude it from distilling at the lower temperature ranges. An attempt was made to remove quinoline from the reaction mixture by dissolving it in 10% HCl before the distillation. This resulted in precipitation of a large amount of black, granular solid (possibly decomposed benzoic acid) in a dark orange solution. After filtration and distillation, GC analyses still showed a predominance of quinoline in the samples. This over-shadowed the benzene analyses making benzene quantitation impossible.

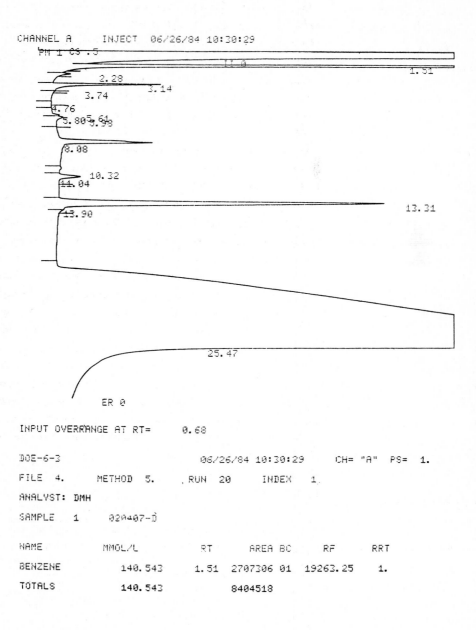

Fig. 4.3. GC Analysis: Base Case Decarboxylation

The second procedure which was used in the decarboxylation of benzoic acid was the persulfate/silver ion method. A mixture of benzoic acid and sodium persulfate was heated to reflux in a solution of 50% acetonitrile/water. At the end of the reaction time, the mixture was a two-phase liquid. However encouraging this two-phase liquid was, a control experiment <u>sine</u> benzoic acid also produced a two-phase liquid. This suggests that the strong oxidizing nature of the reactants may act to promote polymerization and, hence, the H_2O-insolubility of the acetonitrile.

The persulfate/silver ion method was investigated with respect to the effects of reactant concentrations and time on yield of benzene. Both phases of the reaction mixtures were analyzed via GC for the presence of benzene. As would be expected, the majority of the benzene which was generated reported to the upper (i.e. the non-aqueous) layer.

The GC analysis of the "base case" decarboxylation (50% acetonitrile/H_2O; 20 minutes) is shown in Figure 4.3. The yield of benzene from this reaction was 38%. A time study of this decarboxylation showed varying yields with time, a maximum benzene yield of 41% achieved at one hour reaction time (Figure 4.4).

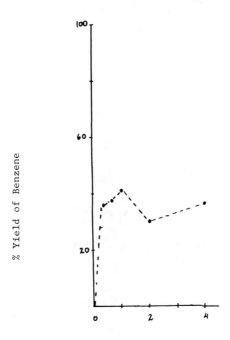

Reaction Time, hours

Fig. 4.4. Time Course for Decarboxylation of Benzoic Acid: Base Case

346

The decreased yield of benzene at the long reaction time (i.e. two hours) indicates increasing complexity of reaction under the strong oxidizing conditions.

The persulfate/silver ion method of decarboxylation effectively eliminates the use of quinoline, a solvent which is not exceedingly desirable from a process standpoint. The method is simple, fast, and energy conservative. It does, however, call for the use of acetonitrile in the reaction process, another solvent which is not so desirable in a process. The acetonitrile serves as the proton donor (see decarboxylation scheme, Figure 4.2) and this plays an important role in the reaction. Because this work addresses an eventual commercial process it was important to determine how much acetonitrile is necessary for the decarboxylation.

Benzoic acid was subjected to persulfate/silver ion decarboxylation in 50:50 acetonitrile/water, 25:75 acetonitrile/water, and 100% water. Relative reactant concentrations are shown in Figure 4.5.

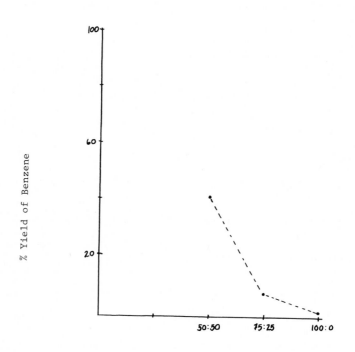

Water: Acetonitrile

Fig. 4.5. Decarboxylation of Benzoic Acid: Acetonitrile/Water Variations

A reduction in the amount of acetonitrile had a profound effect on the efficiency of reaction. Whereas in the 50:50 case there was a 36% yield of benzene, in the 25:75 case the yield dropped to 6%. There was negligible decarboxylation under totally aqueous conditions. These results illustrate the need for a donor solvent if reaction is to be effective. However, in the decarboxylation of lignite hydrolysate it may not be necessary to add 50% acetonitrile to the process stream. Some of the products of the alkaline hydrolysis may be effective proton donors themselves (e.g. the phenolic compounds).

4.4.2 Decarboxylation of Lignite Hydrolysate

Lignite hydrolysate was subjected to decarboxylation using the persulfate/silver ion method. Rather than using the entire hydrolysis slurry for the experiment work, only the solubilized organics were used. The hydrolysis slurry was centrifuged, and the resulting supernatant was decanted and acidified. The organic material which was precipitated was isolated upon centrifugation. This material was used because it was assumed to be richest in the carboxylic acid-like product most susceptible to decarboxylation.

The organic hydrolysate was combined with sodium persulfate and silver nitrate in 50% acetonitrile. The reaction mixture was heated to 100°C in a Parr bomb and cooled. The product was analyzed via GC.

The 6% yield of water-insoluble organic material represents 4% overall conversion of the input lignite volatile solids. This in itself is certainly not so overwhelmingly encouraging for start-up of a full-scale process. It does indicate, though, that there may be some promise for the production of BTX-type liquid fuels from lignite.

The work shown here represents a limited and isolated effort in decarboxylation of a lignite hydrolysate. The products of the hydrolysis are exceedingly complex. Future work should be directed towards two research areas. One is the extensive identification of the major components of the products of lignite breakdown. If these materials are subject to rigorous characterization and identification they can be fitted more adequately into the decarboxylation scheme. The mechanistic details for decarboxylation of the lignite hydrolysate will be better understood in light of the product identifications. The second area for future research concerns the investigation of the role of the catalyst in the decarboxylation. From a process viewpoint, it is important to identify the form the catalyst must take (e.g. fixed-bed), the lifetime of the catalyst (for purposes of regeneration), and any improvements in catalyst performance. This is a cumbersome task, but one which must be accomplished if the direct liqeufaction of lignite is to be considered a viable means of liquid fuels production.

5. THE PILOT-SCALE FACILITY FOR DIRECT LIQUEFACTION OF LIGNITE

The laboratory phase of the liquefaction program illustrated that lignite could be solubilized under conditions of alkaline hydrolysis. The hydrolysate contained aromatic organic acids which are capable of decarboxylation to water-insoluble organics. The experimental program was used to identify some of the necessary processing conditions, such as temperature and alkali concentration, and to characterize the nature of the reacted material. Under batch conditions, maximum lignite solubilization was observed at 250°C, 8% volatile solids loading, and 20% sodium carbonate (weight per weight of lignite volatile solids). The results demonstrated the feasibility of an aqueous treatment of lignite and provided insight into the preliminary design of a pilot-scale facility.

Here, "pilot-scale" is taken to mean an expanded scale laboratory facility. In this light, the pilot phase of the research would concentrate on the more detailed analysis of the alkaline hydrolysis of the lignite in an effort to "optimize" the process. The work would focus on the development of a kinetic model which identifies the reaction time-temperature profiles, as well as the reactant concentrations corresponding to maximum solubilization of the lignite under the "pilot conditions".

5.1. The Reactor System for Lignite Pretreatment

The pretreatment apparatus which was utilized in the laboratory program at Dynatech is a batch reactor system. It is neither sufficiently large for the production of enough extracted material for use in subsequent processing steps, nor is it sufficiently versatile for achieving the short retention times necessary for a detailed kinetic study. Therefore, it is proposed to build a continuous apparatus for the pilot-scale work. This satisfies the scaled-up throughout requirements for investigation of subsequent processing steps, and yields meaningful results for the development of a predictive kinetic model. The apparatus is schematicized in Figure 5.1.

Because limited information is currently available which would suggest an appropriate retention time for a continuous system, some preliminary experiments are necessary. For short retention times, in the order of seconds, a plug flow reaction system would seem to be appropriate. For longer reaction times, in the order of minutes, a CSTR would probably be appropriate.   The batch system currently in use has a retention time in the order of hours and experiments have indicated that shorter retention times are necessary.   Maximum solubilization under batch conditions was achieved when the reaction just reached 250°C – longer reaction times resulted in rapid repolymerization of the reaction intermediates (see Section 3).

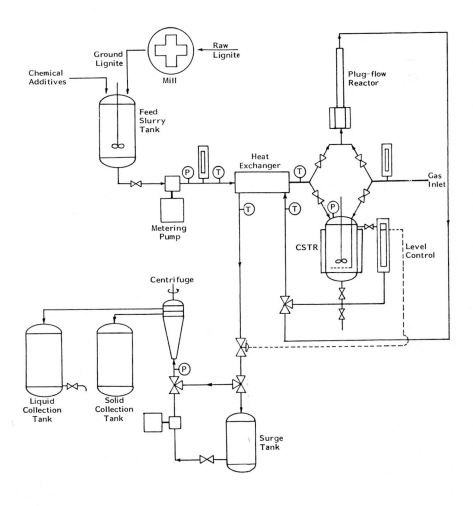

Fig. 5.1. Preliminary Design for Laboratory Hydrolysis Unit

The equipment differences between a plug flow reactor and a CSTR are small, except for the reaction chamber itself. The feed tanks, pumps, valving, reactant stream quench and instrumentation are the same for both systems. The plug flow reaction chamber is a jacketed tube which rapidly heats the reaction mixture, and maintains the elevated temperature for a specified time. The CSTR chamber is a jacketed vessel in which the entering stream is heated by mixing with the reacting mixture. In both systems the residence time is adjusted by varying the flow rate of the reactants.

The parameters which would be investigated in this pilot-scale program are temperature, mean residence time, coal concentration and particle size, and chemical additive concentrations. These parameters would be varied, with the objective of developing a predictive model to describe the dynamics of the process, to be used to approximate the optimal design conditions. To cover the broad range of residence times, both the plug flow (1 - 30 sec residence times) and CSTR (2 - 60 min residence times) systems could be used.

The kinetic model which would result from the pilot program would indicate the optimal time-temperature profile. This would give some indication as to whether a staged reactor configuration would be appropriate for the overall process design. Staging would be an attractive addition to the pilot-scale facility since it would offer an advantage in materials handling and potentially add to the yield of desirable product.

6. ACKNOWLEDGEMENT

This project was carried out under US Department of Energy Contract No. DE-AC01-81ER10914-3 from the Division of Advanced Energy Projects, Office of Basic Energy Sciences, and the sponsorship of Burlington Northern, Inc., Seattle, Washington. The authors wish to express their special thanks to Dr. Ryszard Gajewski, Director, Division of Advanced Energy Projects, and to Mr. David S. Gleason and Mr. Henry J. Sandri, Jr., of Burlington Northern, Inc.

7. REFERENCES

Aueritt, P. U.S. Geol. Survey Bull, 1275, 1967, cited in Ind. Chem.,
     7th ed., Kent, J.A. ed., Van Nostrand, 1974, p. 25
Buckles, R.E. and Wheeler, N.G., 1963. Organic Synthesis Coll. Vol. IV
     857-859
deRiel, S.R., Heneghan, E.P., Houmere, D.M., and Trantolo, D.J. 1984.
     Dynatech Technical Report No. 2282 to the Dept. of Energy,
     Division of Advanced Energy Projects
Entel, J. 1955. J. Am. Chem. Soc., 7, 611
Fisher, F. 1919. Ges. Abhandl. Kenntnis Khole, 4, 13
Fisher, F. and Walter 1921. Ges. Abhandl. Kenntnis Khole, 6, 79
Fisher, F. and Friedrich 1921. Ges. Abhandl. Kenntnis Khole, 6, 108
Fristad, W.E., Fry, M.A. and Klang, J.A. 1983. J. Org. Chem. 48,
     3575-3577
Helsel, R.W. and Dence, C.W. 1975. Chem. Eng. Prog. 55-59
Juettner, B., Smith, R.C. and Howard, H.C. 1935. J. Am. Chem. Soc., 57,
     2322
Kerkovius and Dimrath, 1913. Ann, 399, 120, in Juettner et al.
Levy, P.F. deRiel, S.R., Heneghan, E.P., Cheng, L.K. and Sanderson,
     J.E. 1983. Dynatech Technical Report No. 2260 to the Dept. of
     Energy, Division of Advanced Energy Projects
Montgomery, R. and Holly, E. 1956. Fuel, 36, 63-75
Montgomery, R. and Holly, E. 1956. Fuel, 35, 60-65
National Coal Association, cited in Ind. Chem., 7th ed. Kent, J.A., ed.
     Van Nostrand, 1974
Sondreal, E., Willson, W. and Steinberg, V. 1982. Fuel, 61, 925-932
Wise, D.L., Kitchell, H.P., and Trantolo, D.J. 1983. Dynatech Technical
     Report No. 2244 to Meridian Land and Mineral Co., Billings, MT

APPENDIX A
COAL ANALYSIS REPORTS

The following reports detail the ultimate, proximate and petrographic analysis of the original lignite samples and the pretreated lignite samples from this experimental program. The original lignite samples were obtained from the Beulah mine and are identified as R455 and A790. The pretreated lignite samples were residues from a 300°C/0.5 hour hydrolysis. These are identified as R486 and A817.

R455

R455 is the original lignite sample from which the solid residues were leached. This coal consists of several components or macerals which have been measured quantitatively (by volume %) and are presented in the enclosed petrographic analysis (Table I). It should be noted that this sample has a total vitrinite content of 63.21% and an average vitrinite reflectance in oil of 0.32. The ash was found to be 12.96 (weight%) on a dry basis. This sample exhibits a noticeable banding in its microstructure.

R486

R486 is the solid residue taken from Dynatech's experiment; a high temperature basic leach of the original lignite sample. Remnants of identifiable coal macerals are still present in the residue, however, in much reduced amounts. Total vitrinite is only 6.24% as compared to 63.21% in the original lignite (see Table I). Exinite and semi-fusinite totals also dropped significantly in the residue. These data are based on a quantitative petrographic analysis of the residue. A "residual matrix" was found to be present in the residue which accounted for 64.38% of the total sample (by volume). The matrix is organic and is somewhat similar to coal in its general appearance. However, it is lighter in color than the vitrinite in the original lignite and its reflectance is considerably higher (0.45). The reflectance of the remnant vitrinite particles in the residue was found to be the same as that of the vitrinite in the original lignite. There is a little difference in ash between the two samples. The residue has an ash of 16.36% (dry basis) compared to 12.96% for the lignite. Banding, which was present in the original sample, is noticeably absent in the residue.

TABLE I

Sample	V2	V3	V	Matrix	VT*	E	R	SF	Total	SF	M	F	MM	Total	Avg. Ro
R455 Lignite	15.97	39.11	--		8.13	7.77	0.74	2.40	74.12	4.80	8.50	5.00	7.58	25.88	.32
R486 Residue	--	--	6.24**	64.38		1.62	--	2.17***	--	--	11.65	4.24	9.70	--	.45

Abbreviations for entities:

V2	=	vitrinoid type 2 (average reflectance in oil 0.20 to 0.29, etc.)
E	=	exinoids
R	=	resinoids
SF	=	semi-fusinoids
M	=	micrinoids
F	=	fusinoids
MM	=	mineral matter calculated from following formula:

$$MM = \frac{\left[100 - \dfrac{1.08\ Ash,\% + 0.55\ Sulfur,\%}{1.35\ (avg.\ sp.\ gr.\ of\ coal\ entities)}\right] + 1.08\ Ash,\% + 0.55\ Sulfur,\%}{2.8} \times \frac{1.08\ Ash,\% + 0.55\ Sulfur,\%}{2.8\ (avg.\ sp.\ gr.\ of\ mineral\ matter)} \times 100$$

*VT = telinite or cellular vitrinite
**Includes all vitrinoid types
***Includes all semi-fusinoid types

A790

SAMPLE: LIGNITE BEULAH ZAP

Coal Analysis Report - Ultimate and Proximate Analysis

	AS RECEIVED	DRY BASIS
% MOISTURE	38.25	-
% ASH	4.77	7.73
% VOLATILE MATTER	26.24	42.49
% FIXED CARBON	30.74	49.78
% SULFUR	0.34	0.55
% CHLORINE	0.03	0.05
% CARBON	42.36	68.60
% HYDROGEN*	2.51	4.07
% NITROGEN	0.52	0.84
% OXYGEN* (By difference)	11.21	18.16
BTU/Lb	6986	11313
MAF - BTU/Lb	-	12260

(* excluding moisture)

Coal Analysis Report

ASH ANALYSIS

% SiO_2	20.5
% Al_2O_3	11.3
% Fe_2O_3	3.22
% CaO	28.0
% MgO	8.55
% Na_2O	3.05
% K_2O	0.53
% TiO_2	0.63
% MnO	0.16
% P_2O_5	1.08
% SO_3	20.2

A817
SAMPLE: RESIDUAL LIGNITE SOLIDS

Coal Analysis Report - Ultimate and Proximate Analysis

	AS RECEIVED	DRY BASIS
% MOISTURE	3.64	-
% ASH	13.11	13.61
% VOLATILE MATTER	-	-
% FIXED CARBON	-	-
% SULFUR	0.16	0.16
% CHLORINE	0.03	0.03
% CARBON	63.94	66.36
% HYDROGEN*	3.59	3.73
% NITROGEN	0.88	0.91
% OXYGEN* (By difference)	14.65	15.20
BTU/Lb	-	-
MAF - BTU/Lb	-	-

(* excluding moisture)

Energy Recovery from Lignin, Peat and Lower Rank Coals, edited by D.J. Trantolo and D.L. Wise
Elsevier Science Publishers B.V., Amsterdam 1989 — Printed in The Netherlands

Chapter 11

PROCESS AND COST COMPUTER MODEL
OF THE DIRECT LIQUEFACTION OF LIGNITE TO A BTX-TYPE FUEL*

Ernest E. Kern,
Houston Lighting & Power Company,
Houston, Texas 77001, USA

1.   ABSTRACT

The ASPEN PLUS** process simulator and economic evaluation system was
used to develop a computer model of a process for the direct liquefaction of
lignite. The liquefaction process was developed by the Dynatech R/D Company
of Cambridge, Massachusetts with research funds from the US Department of
Energy and Meridian Land and Mineral company of Billings, Montana.

The purpose of this study was to augment the DOE laboratory work with
an economic evaluation that was not in the original scope of the project.
Houston Lighting & Power Company funded the computer study.

The results of the base case indicate that the fuel can be produced at
about $9 per million Btu's or $1.16 per gallon in 1984 dollars. This price
is not competitive in today's market but is significantly less than other
synfuel price projections.

2.   SUMMARY

This study examines the economics of a process for the direct liquefaction of lignite. This process was developed by the Dynatech R/D Company
as part of a Department of Energy funded research project.   The economic
evaluation and process simulation were done with the ASPEN PLUS computer
code developed at the Massachusetts Institute of Technology with funds from
DOE. ASPEN PLUS was designed to simulate synthetic fuel processes and other
processes that include multiphase streams and complex substances such as
lignite.   ASPEN PLUS also has a costing system to perform equipment sizing

---

*Benzene, Toluene and Xylene (BTX)
**ASPEN PLUS is a trademark of Aspen Technology, Inc.

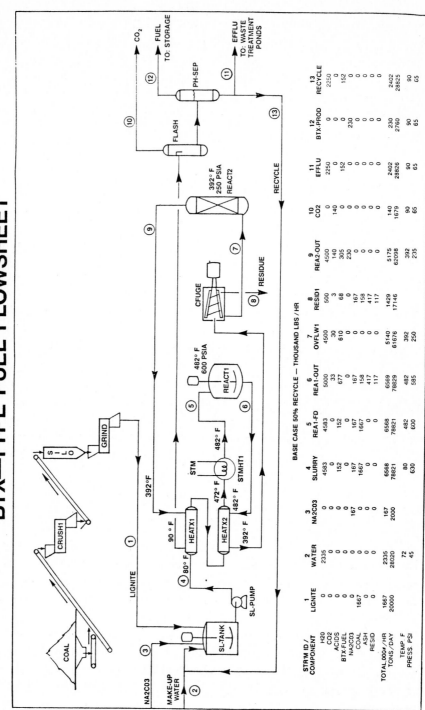

Figure 1.

BTX—TYPE FUEL FLOWSHEET

and costing, capital investment estimation, operating cost estimation and profitability analysis of a process such as the one under study.

ASPEN PLUS flow sheet simulator was used to calculate a heat and material balance for the liquefaction process, power requirements for crushing, grinding and pumping, and heat duties for the heat exchangers. These data were then used in the cost model to size the equipment and determine their cost. In some cases where cost models were not in the program, vendor quotes were used, such as in the case of the coal handling equipment and the centrifuges. The profitability of a project is evaluated by the interest rate of return. In this case the product prices were calculated given a user-specified IROR of 15%. ASPEN PLUS calculates the product price so that the revenues reduce the investment balance to zero at the end of the project's economic life. The economic life was specified as 15 years.

The direct liquefaction of lignite to a BTX-type fuel is a two-stage chemical treatment. The first step is an alkaline hydrolysis of a portion of the organic matter in the lignite making that portion water soluble. These solubles are primarily organic acids in the C6 to C20 range. The second step takes the organic acids and converts them to immiscible aromatic compounds by removing the acid groups (decarboxylation). Because of the complexity of the organic compounds and the uncertainty of the composition of the process streams, model compounds were used to characterize the products of the two reactions. The model compounds were benzoic acid for the first reaction and toluene for the product of the second reaction.

The reaction conditions, yield, retention times were estimated from laboratory tests conducted by Dynatech over the last two years. No refining of the product fuel was included in this study; however, an intermediate flashing step or distillation may be desirable to separate the light hydrocarbons for diesel fuel and the heavy components for boiler fuel.

Figure 1 shows the base case heat and material balances for a 20,000 ton-per-day plant producing 15,000 barrels per day of fuel. The total capital investment required is estimated to be $645 million at 50% debt. In order to achieve a 15% interest rate of return on equity, it was calculated to require an initial selling price of $8.95 per million Btu's or $1.16 per gallon.

This price is not competitive in today's market but is significantly less than, for instance, the Great Plains Syn-Gas estimated price of $16/MM Btu and more for other liquefaction processes. Currently, spot-market prices (July 27, 1984) for No. 2 heating oil and unleaded gasoline are $0.72 and $0.73 per gallon respectively (approximately $4.80 per million Btu's).

TABLE 1

BTX-TYPE FUEL PROJECT
BASE CASE PARAMETERS

LIGNITE FEED RATE	20,000 TPD
LIGNITE HEATING VALUE	6,000 BTU/LB
LIGNITE PRICE	$12/TON
RECYCLE PROCESS WATER	50% OF EFFLUENT STREAM
REACTOR-1 CONVERSION	63% OF VOLATILE SOLIDS
REACTOR-2 CONVERSION	50% OF THE ACIDS
BTX - FUEL HEATING VALUE	18,000 BTU/LB
INTEREST RATE OF RETURN ON EQUITY	15%
INFLATION	7%/YR
PROJECT START	MARCH 1985
COMMERCIAL PRODUCTION	OCTOBER 1988
PERCENT DEBT	50%

Table I: Base case parameters for an economic evaluation of the direct liquefaction of lignite to BTX-Type Fuel. The plant is assumed to be a mine-mouth operation located in the lignite fields of east Texas. Reactor yields are based on Dynatech R/D Company laboratory work.

3.   RESULTS AND CONCLUSIONS

A mine mouth plant was chosen as the basis for the economic evaluation of the direct liquefaction of lignite. The plant would be located in the lignite fields of east Texas. Some of the design data and cost information was obtained from a lignite-fired electric generating station presently being constructed by Houston Lighting & Power in that area. The power plant will burn 30,000 tons per day of lignite as compared with 20,000 tons per day for the process being studied. The costs were factored accordingly.

The process design, conditions and yields were obtained from an interim report by Dynatech R/D Company, dated January 17, 1984, and verbal communic- ations with Dr. Debra Trantolo and Steve Haralampu.

The process consists of the following unit operations:

I. Coal Handling (Conveying and Storage)

II. Crushing and Grinding

III. Slurry Preparation and Pumping

IV. Heat Exchange

V. Alkaline Hydrolysis - Reactor No. 1

VI. Solids Separation (Centrifuge)

VII. Decarboxylation - Reactor No. 2

VIII. Separations (Gas-Liquid and Liquid-Liquid)

IX. Waste Treatment

The flowsheet diagram is shown in Figure 1 along with the base case heat and material balance. The heat and material balance was calculated with ASPEN PLUS; these data were then used in the costing model to size equipment and make the economic calculations. The base case parameters for the economic evaluation are shown in Table I. The whole simulation and economic evaluation boils down to one number which is the present value of the initial selling price of the product. For the base case that price is $8.95 per million Btu (MKB). A sensitivity analysis of the initial selling price of the BTX-fuel was done with some key parameters, the results of which are shown in Table IV and Figures 2 to 5. The price ranges from a low of $8.65 per million Btu to a high of $13.49. The parameter showing the most effect is the slurry solids concentration because of the higher capital costs for the additional equipment required to handle the large volumes of water. When going from a solids concentration of 27% to 8%, the volume of the slurry more than triples, causing the equipment cost to double and an accompanying increase in the initial selling price of 50%.

A 50% increase in lignite price from $12/ton to $18/ton increases the product price by 12%. The Btu yield for the base case, that is, the Btu value of the product fuel divided by the Btu content of the lignite, is 41% (99.4 billion Btu/day divided by 240 billion Btu/day).

The financing of the project by differing debt ratios also has a significant effect on the price of the fuel. As is shown in the plot of accumulative cash flow in Figure 6 for percent debt of 0, 50, 80%, the risks of the project are shared in the construction years, but profits also have to be shared in the following years.

The overall heat transfer coefficient used to size the heat exchangers has a significant effect because the heat exchangers are the single most expensive item in the equipment list. Lowering the heat transfer coefficient from 350 to 150 Btu/hr·ft²·°F increases the total number of heat exchangers from 48 to 128 and the total equipment cost from $123 million to $167 million resulting in an 18% increase in the initial selling price of the fuel.

BASE CASE RESULTS

INVESTMENT $ MILLION

      Physical Plant $ 389
      Interest During Construction (10%)   37
      Start-up   63
      Tax Credit   -51
      Working Capital   89
      Contingency   117

            TOTAL $ 644

SCHEDULE

      Project Start   March 1985
      Commercial Production   October 1988

REVENUE

      Capacity Production Rate (Tons/Day)   2760
      Normal Production Rate (Tons/Day)   2346
      Initial Selling Price ($/Ton 1988)   437
      Initial Selling Price ($MKB 1988)   12.14
      Initial Selling Price ($MKB 1984)   8.95
      Initial Selling Price ($/Gal 1984)   1.16
      Return on Investment   17.4%
      Payout Time (Years)   5.8

CAPITAL INVESTMENT $ MILLION

      Total Equipment Cost   123
      Total Field Construction Cost   310
      Total Depreciable Cost   506
      Total Capital   668

ANNUAL OPERATING COST 1988

      Total Raw Materials   148
      Total Utilities   25
      Waste Treatment   2.9
      Catalyst   2.9
      Labor   32
      Supplies   12
      General Works   28
      Depreciation   34

      Gross Operating Cost $285

TABLE II: Base case results for the economics of the BTX-Type Fuel computer
model. The ASPEN PLUS flowsheet simulator was used to calculate a heat and
material balance. This information was then fed into the costing model to
determine capital costs, operating costs and initial selling price. The
economic feasibility of the process is then based on the calculated selling
price of the fuel compared to current market price. Current spot market
prices for heating oil and gasoline are $0.72 per gallon or approximately
$4.80 per million Btu's.

TABLE III

BTX-TYPE FUEL
MAJOR EQUIPMENT
BASE CASE

ITEM	NO. OF UNITS	DESIGN P, T OR MOTOR SIZE	SIZE EACH	$ x 1000 UNIT PRICE	$ x 1000 TOTAL	$ x 1000 INSTALLED
COAL HDL EQUIP	1	-	20,000 TPD	-	4,669	24,746
COAL SILOS	8	-	800 TON	77	612	3,266
CRUSHERS	2	750HP MOTOR	1,500 TPH	147	294	878
GRINDERS	8	750HP MOTOR	100 TPH	184	1,470	4,390
SLURRY TANKS	16	W/75HP AGIT'RS	20,000 GAL	51	821	3,045
SLURRY PUMPS	16	300HP MOTOR	575 GPM	31	499	1,408
HEAT EXCHANGERS	32	650PSI,500 F	73 MBH	499	15,981	41,441
STEAM HEAT EXCH	16	800PSI,550 F	4.8 MBH	16	257	667
REACTORS No. 1	16	650PSI,500 F	46,000 GAL	495	7,925	26,987
CENTRIFUGES	4	750HP MOTOR	100 TPH	662	2,646	6,648
REACTORS No. 2	8	275PSI,450 F	34,000 GAL	382	3,058	8,477
FLASH TANKS	8	75PSI,150 F	30,000 GAL	81	651	2,361
PHASE SEP.	8	75PSI,150 F	30,000 GAL	81	651	2,36L
STORAGE TANKS	6	ATM	400,000 GAL	99	592	3,159
WASTE WATER TRM	1	-	-	-	-	9,027
				TOTAL:	$40,126	$138,861

TABLE IV
BTX-TYPE FUEL
ASPEN-PLUS COST MODEL

CASE	LIGNITE PRICE $/TON	PERCENT DEBT %	SLURRY SOLIDS %	EQUIP. COST	INVESTMENT PHYSICAL COST	$MM TOTAL COST	INITIAL SELLING PRICE $/TON	$/MKB	PRESENT VALUE $/MKB	COMMER. PRODUCTION DATE
A (BASE CASE)	12	50	27	123	389	644	437	12.14	8.95	OCT 1988
B	12	0	27	123	389	607	462	12.83	9.46	OCT 1988
C	12	80	27	123	389	666	422	11.72	8.65	OCT 1988
D	18	50	27	123	389	664	491	13.64	10.06	OCT 1988
E	12	50	13	157	512	823	498	13.83	10.03	DEC 1988
F	12	50	8	235	813	1273	681	18.92	13.49	APR 1989
G (U=150)*	12	50	27	167	543	873	526	14.61	10.60	JAN 1989

Table IV: Results of the case study on the direct liquefaction of lignite using the ASPEN PLUS costing model. The effect of lignite price, slurry solids concentration, percent debt and heat transfer coefficient on the initial selling price of the product was determined assuming a 15% interest rate of return on equity.

*Heat Transfer Coefficient U=350 $Btu/hr \cdot Ft^2 \cdot {}^\circ F$ in all other cases

364

TABLE V

SENSITIVITY OF INITIAL SELLING PRICE OF BTX-FUEL

% Debt	1st Quarter of Commercial Operation Cash Flow Balance $MM	Initial Selling Price $/MKB (1984 $)
0	-749	9.46
50 *	-483	8.95
80	-323	8.65

% Slurry Concentration	Equip Cost S MM	Initial Selling Price $/MKB (1984 $)
8	235	13.49
13	157	10.03
27 *	123	8.95

Lignite Price $/Ton	Raw Material Cost, $MM	Initial Selling Price $/MKB (1984 $)
12 *	148	8.95
18	190	10.06

Overall Heat Transfer Coefficient	Heat Exchanger Installed Cost $MM	Initial Selling Price $/MKB (1984 $)
150	94	10.60
350 *	42	8.95

*Base Case

Table V: Sensitivity of the initial selling price of BTX-Fuel as a function of percent debt, slurry solids concentration, lignite price and heat transfer coefficient as calculated by the ASPEN PLUS costing model.

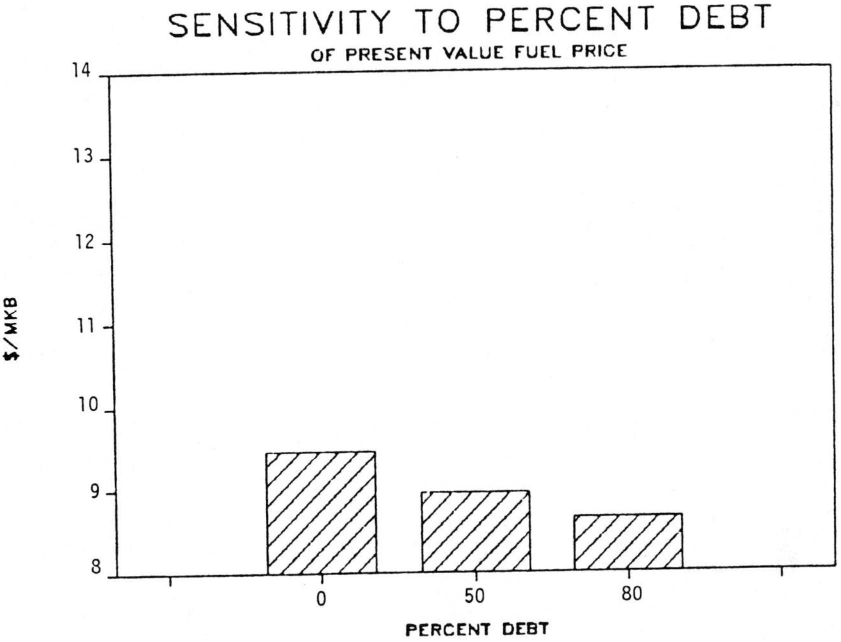

SENSITIVITY TO PERCENT DEBT
OF PRESENT VALUE FUEL PRICE

Figure 2: Sensitivity to percent debt of the calculated present value initial selling price of the product from the "Direct Liquefaction of Lignite to BTX-Type Fuel". Prices were determined by the ASPEN PLUS cost model using the input from the process simulation and vendor quotes. The base case assumed 50% debt and 15% interest rate of return on equity giving a price of $8.95 per million Btu's.

366

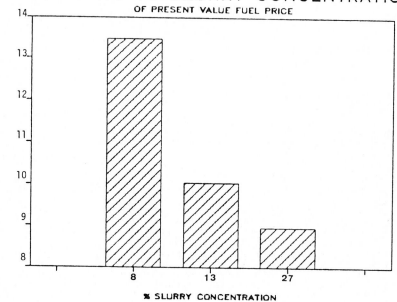

Figure 3: Sensitivity of the calculated present value initial selling price
to slurry concentration for the product from the "Direct Liquefaction of
Lignite to BTX-Type Fuel". The feed to the reactors is in the form of a
finely ground coal and water mixture in the range of 8 to 27% solids. At
the lower concentrations the volume of water that has to be handled becomes
so large that the equipment costs escalate very rapidly. A reduction
in solids concentration from 27 to 8 percent doubled the equipment cost
raising the fuel price from $8.95 to $13.49.

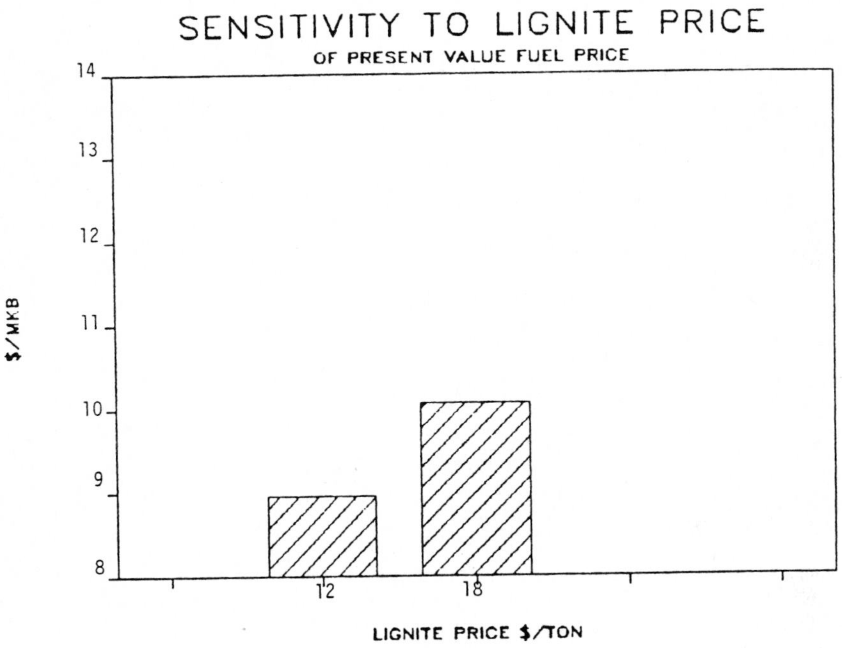

SENSITIVITY TO LIGNITE PRICE
OF PRESENT VALUE FUEL PRICE

LIGNITE PRICE $/TON

Figure 4: Sensitivity of the BTX-Fuel price to lignite price. The lignite constitutes 30% of the annual operating costs and therefore has a significant effect on the cost of the product. Approximately 40% of the Btu value of the lignite put into the process is recovered as Btu heating value in the product fuel. The rest is lost to the residue and waste effluent streams and in the conversion of the organic acid groups to CO_2 in reactor No. 2. This is a significantly lower conversion efficiency than in the coal gasification process. For example, the thermal efficiency of the Shell coal gasification process is 80%.

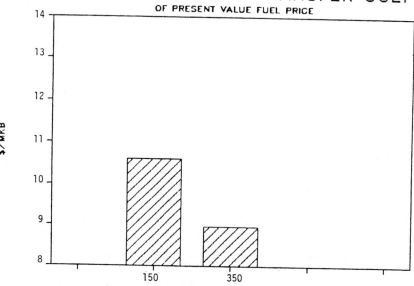

Figure 5: The heat transfer coefficient has a significant impact on the initial selling price because the heat exchangers are the single most expensive items on the equipment list. Lowering heat transfer coefficient from 350 to 150 Btu/hr·Ft² · °F increases the total equipment cost from $123 million to $167 million resulting in a 18% increase in the initial selling price of the fuel.

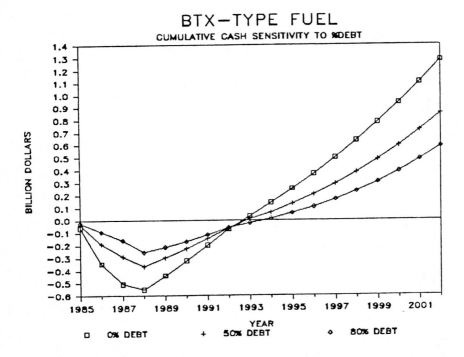

Figure 6: Cumulative cash flow shows a payout time of five to six years, depending on the amount of debt financed.

4.   CONCEPTUAL PROCESS DESIGN

A two-step chemical process has been proposed by Dynatech R&D Company
to convert lignite or peat to a liquid aromatic fuel. The fuel is characterized as being a Benzene, Toleune, Xylene (BTX) type fuel.   In actuality,
it consists of a whole range of hydrocarbons from C6 to C20. The first step
in the process is the alkaline hydrolysis of the so-called volatile solids
of the lignite to produce water soluble organic acids which, in the second
step, are converted to immiscible aromatics by decarboxylation (removing the
acid groups).

A detailed flowsheet of the process is shown in Figure 1. The plant is
assumed to be a mine-mouth operation to eliminate lignite transportation
costs. However, since lignite is fed to the first reactor as a slurry of
about 20% solids, it may be practical to produce a more concentrated slurry
at the mine and transport the slurry by pipeline to a more convenient site.

A conceptual design, broken down into 9 unit operations, follows:

I.   Coal Handling

The coal handling facility includes all the conveyor systems from
the mine to the storage silos plus the equipment for stack-out and
reclamation of lignite in the dead and active storage piles. The full
installed costs were obtained from the Houston Lighting & Power's
Limestone Electric Generation Station (LEGS) project, located in
Limestone County about 125 miles northwest from downtown Houston. The
contractor for the coal handling is McNally Pittsburg Inc. The total
field erected costs including equipment is $25,746,603. This unit is
sized for 30,000 tons per day. The price was scaled down to 20,000
tons per day capacity as follows:

$$\text{Estimated BTX Coal Handling Cost} = (20{,}000/30{,}000)^{0.6} \times 25{,}746{,}603$$
$$= \$20{,}200{,}000$$

The power requirement was estimated at 6,000 kw.

II.   Crushing and Grinding

The crusher cost was estimated from the LEGS project in which two
crushers were installed, each capable of handling full load. The cost
of each, with drive, was $120,000.

The grinders are sized for 110 tons per hour each, requiring eight
units (no spares). Cost is estimated at $150,000 each.   The power
requirements were calculated by ASPEN PLUS for coal particle size
reduction to 100 mesh (150 microns).

The coal silos were sized for an 8-hour surge capacity, field erected, each containing 800 tons and 33,000 cubic feet in size.

In order to calculate the total installed cost of the crushers and grinders, they were put into the ASPEN COST MODEL as compressors. This was nececcary because no cost model exists for crushers and grinders in ASPEN. This substitution was recommended by ASPEN-TECH.

III. Slurry Preparation and Slurry Pumps

The slurry prep-tanks were sized for half-hour retention time, shop-fabricated with 75 hp agitators each. The agitators were sized and costed by the Lightning Mixing Company. The tanks were deliberately kept small (20,000 gal) so that with vigorous agitation, the coal particles could be kept in suspension without having to resort to chemical additives. If the slurry tanks were to be used for surge capacity, they would be prohibitively large and would require tremendous agitators. However, a penalty must be paid for having small prep-tanks; that is if one grinder fails, two reactor trains will be shut down in less than 30 minutes.

In the base case there are 16 slurry pumps raising the pressure of the slurry to go through three heat exchangers and into the first reactor at 600 psia. Subsequent process equipment operates at lower pressures so that no additional pumps are required. The power requirement was calculated by ASPEN PLUS. For the base case 16 pumps (one pump for each reactor) are required at 575 GPM, using a 300 hp motor.

IV. Heat Exchangers

The lignite/water slurry must be heated to 482°F (250°C) in a series of three heat exchangers; the first one exchanges heat between the feed stream at 80°F and the effluent stream from the Reactor No. 2 entering at 392°F (200°C). The second exchanges heat with the effluent from Reactor No. 1 bringing the feed temperature to 472°F. A third heat exchanger, using steam at 750 psig, brings the temperature of the slurry to the required 482°F.

An overall heat transfer coefficient (U) of 350 Btu/hr ft^2 °F was used for all three heat exchangers. Since the heat exchangers are the single most costly items for the entire process (see Table III), the initial selling price of the product fuel is very sensitive to the heat transfer coefficient that is assumed. To test the sensitivity, a case was run using U = 150. This increases the selling price by 18%, and more than doubled the cost of the heat exchangers (see Table III).

V. Reactor No. 1 - Alkaline Hydrolysis

The alkaline hydrolysis reaction is carried out in a continuously stirred tank reactor (CSTR) at 482°F (250°C) and 600 psia with a resident time of one hour. The reactors are 15 ft diameter by 30 ft tangent-to-tangent in height. They are designed for 650 psi and 500°F, 1/8 inch corrosion allowance and 75% full. Each reactor has a 75 hp agitator. The calculated shell thickness is 4.5 inches and the shell weight of each vessel is 367,000 lbs.

VI. Solids Separation

Initially, a hydrocyclone was used for removing the solids from the stream prior to Reactor No. 2 because of the simplicity of the design for handling the pressures and temperatures required. However, the design engineers at Dorr-Oliver said that hydrocyclones would not be practical to remove the extremely small residue and ash particles present in this mixture. It was therefore decided to use a centrifuge similar in design to the centrifuge used in the dewatering of pipeline coal slurry in a test for the ETSI pipeline project.

Four 100 TPH (solids) centrifuges using 750 hp motors are required at a quoted price from the Bird Machinery Co. of $540,000 each (1984 price).

VII. Reactor No. 2 - Decarboxylation

Reactor No. 2 carries out a reaction to remove the carboxylic acid groups from the water soluble organic acids produced in Reactor No. 1. The reactor is designed as a fixed bed catalytic reactor, 12 ft in diameter and 36 ft tangent-to-tangent in height, made of carbon steel with 1/8 inch corrosion allowance. Eight are required for the base case. Retention time is twenty minutes.

The reaction is carried out at 250 psia and 392°F (200°C) using a silver/copper co-catalyst.

VIII. Separations

Effluent stream (9) from Reactor No. 2 is cooled by heat exchange with the incoming slurry feed stream (4) to 90°F. At this point, the stream consists of a three phase mixture (dispersion) of CO_2 gas, immiscible hydrocarbon liquid and a water phase with dissolved un-reacted organic acids. The CO_2 flashed off in the Flash Tanks can either be vented to the atmosphere or recovered as a product. No credit was taken for the sale of CO_2 and no provision was made for clean-up of the stream prior to exhaust to the atmosphere. Most likely, the two will cancel each other.

From the flash tank the liquid goes to the phase separator to separate the aqueous phase from the organic (organic is lighter). 50 percent of the bottoms, containing unreacted acids, is recycled to the slurry prep tank; the other half goes to a waste water treatment plant.

IX. <u>Waste Water Treatment Plant</u>

The aqueous effluent stream from the plant contains about 6% organics which must be treated in some way. For this study it was assumed that the organics would be treated in a conventional waste water treatment plant. However, because of the high content of organics it may be desirable to treat the effluent stream in a process called wet oxidaion. In this process, air is bubbled through the waste water in a pressurized reactor at 600 psia 600°F. Heat and fuel gas are by-products and could be used somewhere else in the plant.

The organics in the waste water may also be recovered as a product by extraction or precipitation.

The optimization of the waste water treatment is not in the scope of this study. For this study the waste treatment is handled as an off-site utility.

APPENDIX I

ASPEN PLUS Process Simulation
for the Direct Liquefaction of Lignite
to a BTX-Type Fuel

```
BASE CASE - INPUT FILE
PROCES SIMULATION

TITLE 'BTX-TYPE FUEL'
DESCRIPTION "SIMULATION OF DIRECT LIQUEFACTION OF PEAT AND LIGNITE
            TO BTX-TYPE FUEL WITHOUT OXIDATION IN FIRST REACTOR"

IN-UNITS ENG
OUT-UNITS ENG
HISTORY MSG-LEVEL PROPERTIES=2
;    COMPONENTS

DATABANKS SOLIDS
PROPERTIES SYSOP0
;
COMPONENTS  H2O WATER / CO2 CARBON-DIOXIDE /
ACIDS BENZOIC-ACID / FUEL TOLUENE / NACO3 CALCITE /CAT2 ACETONITRILE/
COAL / ASH / RESID
ATTR-COMPS  COAL  PROXANAL ULTANAL SULFANAL /
            ASH   PROXANAL ULTANAL SULFANAL /
            RESID PROXANAL ULTANAL SULFANAL

NC-PROPS COAL   ENTHALPY  HCOALGEN / DENSITY DCOALIGT
NC-PROPS ASH    ENTHALPY  HCOALGEN / DENSITY DCOALIST
NC-PROPS RESID  ENTHALPY  HCOALGEN / DENSITY DCOALIST
;
; PROXANAL: MOISTURE / MF FIXED CARBON / MF VOLATILE MATTER / MF ASH
;
; DEFINING THE STREAMS, SUBSTREAMS AND PARTICLE SIZE DISTRIBUTION
;
DEF-STREAM-CLASS MIXCINCP SUBS=MIXED CISOLID NCPSD
DEF-STREAMS CONVEN ALL /MIXNCPSD SECT1 / MIXCINCP SECT2
DEF-SUBS-ATTR  PSD  PSD
IN-UNITS  LENGTH= MU
INTERVALS 15
SIZE-LIMITS  0/44/63/88/125/177/250/354/500/707/1000/
             2000/4000/8000/16000/50000
;
;    FEED STREAMS
;
STREAM LIGNITE
  SUBSTREAM NCPSD TEMP= 85 PRES= 15 MASS-FLOW= 1666666
  MASS-FRAC  COAL 1.0/ ASH 1D-10/ RESID 1D-10
  SUBS-ATTR PSD PSD FRAC= 0 0 0 0 0 0 0 .1 .1 .2 .35 .25
  COMP-ATTR COAL  PROXANAL (31.70  46.91 41.60 11.49)/
                  ULTANAL (11.49 63.3 4.29 0.72 0.0 1.22 18.65)/
                  SULFANAL (0.56 0.03 0.63)
  COMP-ATTR ASH  PROXANAL ( 0.00   0.00  0.00 100.0)/
                 ULTANAL (11.49 63.3 4.29 0.72 0.0 1.22 18.65)/
                 SULFANAL (0.56 0.03 0.63)
  COMP-ATTR RESID  PROXANAL (0 98.0 1.0 1.0)/
                   ULTANAL (11.49 63.3 4.29 0.72 0.0 1.22 18.65)/
                   SULFANAL (0.56 0.03 0.63)
;

STREAM WATER      TEMP=72  PRES=45
  MASS-FLOW  H2O 10.417D+06
STREAM FCAT1
  SUBSTREAM CISOLID PRES= 15 TEMP= 72
  MASS-FLOW NA2CO3 1.6667D+05
STREAM FCAT2 PRES=250 TEMP= 72
  MASS-FLOW CAT2 5880
;
FLOWSHEET SECT1
  BLOCK CRUSH1  IN = LIGNITE           OUT = SCR1
  BLOCK GRIND   IN = SCR1              OUT = GR-LIG
FLOWSHEET SECT2
  BLOCK CHGR1   IN = GR-LIG            OUT = DSTR3
  BLOCK SL-TANK IN = DSTR3 WATER FCAT1 OUT = SLURRY
  BLOCK SL-PUMP IN = SLURRY            OUT = PSLURRY
  BLOCK HEATX1  IN = PSLURRY           OUT = HX1-OUT
  BLOCK STMHT   IN = HX1-OUT           OUT = REA1-FD
  BLOCK SPLIT1  IN = REA1-FD           OUT = WAT-1 DSTR1
  BLOCK REACT1  IN = DSTR1             OUT = DSTR2
  BLOCK COMB1   IN = DSTR2 WAT-1       OUT = REA1-OUT'
  BLOCK HEATX2  IN = REA1-OUT          OUT = HX2-OUT
  BLOCK SEPA    IN = HX2-OUT           OUT = OVFLW1 RESID1
FLOWSHEET SECT3
  BLOCK CHGR2   IN = OVFLW1            OUT = DSTR4
  BLOCK PBOOST  IN = DSTR4             OUT = REA2-FD
  BLOCK REACT2  IN = REA2-FD FCAT2     OUT = REA2-OUT
  BLOCK HEATX3  IN = REA2-OUT          OUT = HX-OUT
  BLOCK FLASH1  IN = HX3-OUT           OUT = RCO2 FL-BOT
  BLOCK PH-SEP  IN = FL-BOT            OUT = BTX-PROD EFFLU
;
BLOCK CRUSH1  CRUSHER
  IN-UNITS  LENGTH= MU
  PARAM  DOUT=2500  TYPE= 1
  BWI   NCPSD  11.37
;
BLOCK GRIND   CRUSHER
  IN-UNITS  LENGTH= MU
  PARAM  DOUT =150      TYPE = 3
  BWI   SSID = NCPSD              BWI = 11.37
;
BLOCK CHGR1   CLCHNG
;
BLOCK SL-TANK  MIXER
;
BLOCK SL-PUMP  PUMP
  PARAM  PRES = 630      TYPE = 2      NPHASE= 1
;
BLOCK HEATX1  HEATER
  PARAM  TEMP= 472 PRES= -20
BLOCK STMHT   HEATER
  PARAM  TEMP= 482  PRES= -10
;
```

```
BLOCK SPLIT1        SEP
  FRAC  SUBS= MIXED STREAM= WAT-1  COMPS= H2O    FRACS= 1.0
  FRAC  SUBS= CISOLID STREAM= WAT-1 COMPS=NA2CO3 FRACS=1.0
;
BLOCK REACT1        RYIELD
  PARAM         TEMP = 482   PRES= -15
  MASS-YIELD    MIXED H2O 25.0 / ACIDS 31.5/ CO2 2.0/
                NCPSD COAL 9.5 / ASH 25.0 / RESID 7.0
;
BLOCK COMB1         MIXER
;
BLOCK HEATX2        HEATER
  PARAM         TEMP=392    PRES= -10
;
BLOCK SEPA          CFUGE
  DIAMETER DIA=3.667 RPS=1200
  CAKE-PROPS WET= 0.5  SPRES=1.0  MRES=1.0
  RATIOS H:R=3
;
BLOCK CHGR2         CLCHNG
;
BLOCK PBOOST        HEATER
  PARAM  PRES= 250
;
BLOCK REACT2        RSTOIC
  PARAM         TEMP = 392 PRES = -15
  STOIC 1  MIXED ACIDS -1 / CO2 1 / FUEL 1
  CONV  1  MIXED ACIDS 0.5
;
BLOCK HEATX3        HEATER
  PARAM         TEMP= 90     PRES=-10
;
BLOCK FLASH1        FLASH2
  PARAM  PRES=15
;
BLOCK PH-SEP        SEP
  FRAC  STREAM=BTX-PROD  COMPS=FUEL   FRACS=1.0
;
STREAM-REPORT
  STANDARD OPTIONS = MASSFLOW  MASSFRAC STDVOLFLOW EXCL-STREAMS= DSTR1 &
           WAT-1 DSTR2 DSTR3 DSTR4
RUN-CONTROL  MAX-TIME = 10
```

TABLE OF CONTENTS

ASPEN PLUS   VER: IBM-VS2   REL: DEC83F   INST: UCC   4/10/84   PAGE   1
BTX-TYPE FUEL
RUN CONTROL SECTION

RUN CONTROL INFORMATION

TYPE OF RUN: NEW

INPUT FILE NAME: JCLINPUT

OUTPUT PROBLEM DATA FILE NAME: ASPEN   VERSION NO.   1

CALLING PROGRAM NAME: ASPEN

SIMULATION REQUESTED FOR ENTIRE FLOWSHEET

DESCRIPTION

    SIMULATION OF DIRECT LIQUEFACTION OF PEAT AND LIGNITE TO BTX-TYPE
    FUEL WITHOUT OXIDATION IN FIRST REACTOR

BLOCK STATUS

****
*   * * *
* ALL UNIT OPERATION BLOCKS WERE COMPLETED NORMALLY
*   * * *
****

ASPEN PLUS   VER: IBM-VS2   REL: DEC83F   INST: UCC   4/10/84   PAGE   2
BTX-TYPE FUEL
FLOWSHEET SECTION

FLOWSHEET CONNECTIVITY BY STREAMS

FLOWSHEET SECTION SECT1

STREAM	SOURCE	DEST	STREAM	SOURCE	DEST
LIGNITE	----	CRUSH1	SCR1	CRUSH1	GRIND
GR-LIG	GRIND	CHGR1			

FLOWSHEET SECTION SECT2

STREAM	SOURCE	DEST	STREAM	SOURCE	DEST
FCAT1	----	SL-TANK	WATER	----	SL-TANK
DSTR3	CHGR1	SL-TANK	SLURRY	SL-TANK	SL-PUMP
FSLURRY	SL-PUMP	HEATX1	HX1-OUT	HEATX1	STMHT
REA1-FD	STMHT	SPLIT1	WAT-1	SPLIT1	COMB1
DSTR1	SPLIT1	REACT1	DSTR2	REACT1	COMB1
REA1-OUT	COMB1	HEATX2	HX2-COUT	HEATX2	SEPA
OVFLW1	SEPA	CHGR2	RESID1	SEPA	----

FLOWSHEET SECTION SECT3

STREAM	SOURCE	DEST	STREAM	SOURCE	DEST
FCAT2	----	REACT2	REA2-FD	CHGR2	REACT2
REA2-OUT	REACT2	HEATX3	HX3-COUT	HEATX3	PH-SEP
RCO2	PH-SEP	----	BTX-PROD	PH-SEP	----
FFFLU	PH-SEP				

FLOWSHEET CONNECTIVITY BY BLOCKS

FLOWSHEET SECTION SECT1

BLOCK	INLETS	OUTLETS
CRUSH1	LIGNITE	SCR1
GRIND	SCR1	GR-LIG

FLOWSHEET SECTION SECT2

BLOCK	INLETS	OUTLETS
CHGR1	GR-LIG	DSTR3
SL-TANK	DSTR3 WATER FCAT1	SLURRY
SL-PUMP	SLURRY	FSLURRY
HEATX1	FSLURRY	HX1-OUT
STMHT	HX1-OUT	REA1-FD
SPLIT1	REA1-FD	WAT-1 DSTR1
REACT1	DSTR1	DSTR2
COMB1	DSTR2 WAT-1	REA1-OUT
HEATX2	REA1-OUT	HX2-COUT
SEPA	HX2-COUT	OVFLW1 RESID1

FLOWSHEET SECTION SECT3

BLOCK	INLETS	OUTLETS
CHGR2	OVFLW1	REA2-FD
REACT2	REA2-FD FCAT2	REA2-OUT
HEATX3	REA2-OUT	HX3-COUT
PH-SEP	HX3-COUT	RCO2 BTX-PROD EFFLU

ASPEN PLUS   VER: IBM-VS2   REL: DEC83F   INST: UCC   4/10/84   PAGE   3

BTX-TYPE FUEL
FLOWSHEET SECTION

COMPUTATIONAL SEQUENCE

SEQUENCE USED WAS:
CRUSH1 GRIND CHGR1 SL-TANK SL-PUMP HEATX1 STMHT SPLIT1 REACT1 COMB1
HEATX2 SEPA CHGR2 REACT2 HEATX3 PH-SEP

OVERALL FLOWSHEET BALANCE

*** MASS AND ENERGY BALANCE ***

	IN	OUT	RELATIVE DIFF.
CONVENTIONAL COMPONENTS (LBMOL/HR)			
H2O	254418.	279859.	-0.909092D-01
CO2	0.0000000D+00	2802.69	-1.00000
ACIDS	0.0000000D+00	2759.26	-1.00000
FUEL	0.0000000D+00	1969.55	-1.00000
NA2CO3	1665.22	0.0000000D+00	1.00000
CAT2	143.229	143.229	0.0000000D+00
SUBTOTAL (LBMOL/HR)	256226.	287554.	-0.108885
(LB/HR)	0.475588D+07	0.568934D+07	-0.164071
NON-CONVENTIONAL COMPONENTS (LB/HR)			
COAL	0.166667D+07	174167.	0.895500
ASH	0.166667D-03	453334.	-1.00000
RESID	0.166667D-03	128334.	-1.00000
SUBTOTAL (LB/HR)	0.166667D+07	760834.	0.543499
TOTAL BALANCE			
MASS (LB/HR)	0.642255D+07	0.645018D+07	-0.428312D-02
ENTHALPY(BTU/HR)	-0.367923D+11	-0.358831D+11	-0.247136D-01

ASPEN PLUS   VER: IBM-VS2   REL: DEC83F   INST: UCC   4/10/84   PAGE   4

BTX-TYPE FUEL
PHYSICAL PROPERTIES SECTION

COMPONENTS

ID	TYPE	FORMULA	NAME OR ALIAS
H2O	C	H2O	WATER
CO2	C	CO2	CARBON-DIOXIDE
ACIDS	C	C7H6O2	BENZOIC-ACID
FUEL	C	C7H8	TOLUENE
NA2CO3	C	CACO3-2	CALCITE
CAT2	C	C2H3N	ACETONITRILE
COAL	NC		MISSING
ASH	NC		MISSING
RESID	NC		MISSING

ID	ATTRIBUTE TYPES		
COAL	PROXANAL	ULTANAL	SULFANAL
ASH	PROXANAL	ULTANAL	SULFANAL
RESID	PROXANAL	ULTANAL	SULFANAL

```
ASPEN PLUS  VER: IBM-VS2   REL: DEC83F  INST: UCC          4/10/84   PAGE   5
                           BTX-TYPE FUEL
                           U-O-S BLOCK SECTION

BLOCK: CRUSH1   MODEL: CRUSHER
-----
        INLET = LIGNITE        OUTLET = SCR1
PROPERTY OPTION SET SYSOP0

        *** MASS AND ENERGY BALANCE ***
                    IN            OUT          RELATIVE DIFF.
CONV. COMP. (LBMOL/HR)
        (LB/HR)     0.0000000D+00  0.0000000D+00  0.0000000D+00
                    0.0000000D+00  0.0000000D+00  0.0000000D+00
NONCONV. COMP (LB/HR)  0.166667D+07  0.166667D+07  0.139698D-15
TOTAL BALANCE
MASS(LB/HR)         0.166667D+07  0.166667D+07  0.139698D-15
ENTHALPY(BTU/HR)    -0.464206D+10  -0.464206D+10  -0.410684D-15

        *** INPUT DATA ***

OPERATING MODE: 0 = PRIMARY, 1 = SECONDARY            0
CRUSHER TYPE: 1 = GYRATORY/JAW, 2 = SINGLE ROLL
              3 = MULTIPLE ROLL, 4 = CAGE MILL         1
DIAMETER OF SOLIDS OUTLET ,FT                     0.0082021
BOND WORK INDEX FOR SUBSTREAM NCPSD  KWHR/TON     11.3700

        *** RESULTS ***

POWER REQUIREMENT ,HP                             1,608.00
PARTICLE DIAMETER WHICH IS
   LARGER THAN 80% OF INLET MASS  FT              0.14175
PARTICLE DIAMETER WHICH IS
   LARGER THAN 80% OF OUTLET MASS  FT             0.010754

BLOCK: GRIND   MODEL: CRUSHER
-----
        INLET = SCR1        OUTLET = GR-LIG
PROPERTY OPTION SET SYSOP0

        *** MASS AND ENERGY BALANCE ***
                    IN            OUT          RELATIVE DIFF.
CONV. COMP. (LBMOL/HR)
        (LB/HR)     0.0000000D+00  0.0000000D+00  0.0000000D+00
                    0.0000000D+00  0.0000000D+00  0.0000000D+00
NONCONV. COMP (LB/HR)  0.166667D+07  0.166667D+07  0.419095D-15
TOTAL BALANCE
MASS(LB/HR)         0.166667D+07  0.166667D+07  0.419095D-15
ENTHALPY(BTU/HR)    -0.464206D+10  -0.464206D+10  -0.205442D-15

        *** INPUT DATA ***

OPERATING MODE: 0 = PRIMARY, 1 = SECONDARY            0
CRUSHER TYPE: 1 = GYRATORY/JAW, 2 = SINGLE ROLL
              3 = MULTIPLE ROLL, 4 = CAGE MILL         3
DIAMETER OF SOLIDS OUTLET ,FT                    0.00049215
BOND WORK INDEX FOR SUBSTREAM NCPSD  KWHR/TON    11.3700
```

```
ASPEN PLUS  VER: IBM-VS2   REL: DEC83F  INST: UCC          4/10/84   PAGE   6
                           BTX-TYPE FUEL
                           U-O-S BLOCK SECTION

BLOCK: GRIND   MODEL: CRUSHER (CONTINUED)
        *** RESULTS ***

POWER REQUIREMENT ,HP                             6,048.05
PARTICLE DIAMETER WHICH IS
   LARGER THAN 80% OF INLET MASS  FT              0.010754
PARTICLE DIAMETER WHICH IS
   LARGER THAN 80% OF OUTLET MASS  FT             0.00077496

BLOCK: SL-TANK   MODEL: MIXER
-----
INLET STREAM(S):    DSTR3    SLURRY    WATER    FCAT1
OUTLET STREAM:      SLURRY
PROPERTY OPTION SET:   SYSOP0
FREE WATER OPTION SET: SYSOP12
SOLUBLE WATER OPTION:  ORGANIC OPTION SET

        *** MASS AND ENERGY BALANCE ***
                    IN            OUT          RELATIVE DIFF.
CONV. COMP. (LBMOL/HR)  256083.       256083.     0.0000000D+00
        (LB/HR)     0.4750000D+07  0.4750000D+07  0.0000000D+00
NONCONV. COMP (LB/HR)  0.166667D+07  0.166667D+07  0.0000000D+00
TOTAL BALANCE
MASS(LB/HR)         0.641667D+07  0.641667D+07  0.0000000D+00
ENTHALPY(BTU/HR)    -0.367956D+11  -0.367956D+11  0.682254D-06

        *** INPUT DATA ***

TWO  PHASE      FLASH
MAXIMUM NO. ITERATIONS                           30
CONVERGENCE TOLERANCE                            0.000100000
OUTLET PRESSURE: MINIMUM OF INLET STREAM PRESSURES

BLOCK: SL-PUMP   MODEL: PUMP
-----
INLET STREAM(S):       SLURRY
OUTLET STREAM:         PSLURRY
PROPERTY OPTION SET:   SYSOP0
FREE WATER OPTION SET: SYSOP12
SOLUBLE WATER OPTION:  ORGANIC OPTION SET

        *** MASS AND ENERGY BALANCE ***
                    IN            OUT          RELATIVE DIFF.
CONV. COMP. (LBMOL/HR)  256083.       256083.     0.0000000D+00
        (LB/HR)     0.4750000D+07  0.4750000D+07  0.0000000D+00
NONCONV. COMP (LB/HR)  0.166667D+07  0.166667D+07  0.0000000D+00
TOTAL BALANCE
MASS(LB/HR)         0.641667D+07  0.641667D+07  0.0000000D+00
ENTHALPY(BTU/HR)    -0.367956D+11  -0.367327D+11  -0.351115D-03
```

ASPEN PLUS   VER: IBM-VS2   REL: DEC83F   INST: UCC   4/10/84   PAGE   7
                              BTX-TYPE FUEL
                              U-O-S BLOCK SECTION

BLOCK: SL-PUMP   MODEL: PUMP (CONTINUED)

                 *** INPUT DATA ***
REQUIRED EXIT PRESSURE (PSI)   630.000
PUMP EFFICIENCY   MISSING
DRIVER EFFICIENCY   1.000000

FLASH SPECIFICATIONS:
LIQUID PHASE CALCULATION
NO FLASH PERFORMED
MAXIMUM NUMBER OF ITERATIONS   30
TOLERANCE   0.0001000000

                 *** RESULTS ***
VOLUMETRIC FLOW RATE (CUFT/HR)   73,789.8
PRESSURE CHANGE (PSI)   615.000
FLUID POWER (HP)   3,300.42
BRAKE POWER (HP)   5,077.57
ELECTRICITY (KW)   3,786.34
PUMP EFFICIENCY USED   0.65000
NET WORK (HP)   -5,077.57

BLOCK: HEATX1   MODEL: HEATER

INLET STREAM(S):   FSLURRY
OUTLET STREAM: HX1-OUT
PROPERTY OPTION SET: SYSOP0
FREE WATER OPTION SET: SYSOP12
SOLUBLE WATER OPTION: ORGANIC OPTION SET

             *** MASS AND ENERGY BALANCE ***
                        IN   OUT   RELATIVE DIFF.
CONV. COMP. (LBMOL/HR) 256083.   256083.   0.0000000D+00
            (LB/HR)   0.4750000D+07 0.4750000D+07 0.0000000D+00
NONCONV. COMP(LB/HR)   0.1666667D+07 0.1666667D+07 0.0000000D+00
TOTAL BALANCE
MASS(LB/HR)   0.641667D+07   0.641667D+07 0.000000D+00
ENTHALPY(BTU/HR)   -0.367827D+11 -0.344360D+11 -0.637984D-01

                 *** INPUT DATA ***
TWO PHASE TP FLASH
SPECIFIED TEMPERATURE F   472.000
PRESSURE DROP   PSI   20.0000
MAXIMUM NO. ITERATIONS   30
CONVERGENCE TOLERANCE   0.0001000000

ASPEN PLUS   VER: IBM-VS2   REL: DEC83F   INST: UCC   4/10/84   PAGE   8
                              BTX-TYPE FUEL
                              U-O-S BLOCK SECTION

BLOCK: HEATX1   MODEL: HEATER (CONTINUED)

                 *** RESULTS ***
OUTLET TEMPERATURE   F   472.00
OUTLET PRESSURE   PSI   610.00
HEAT DUTY   BTU/HR   0.23467D+10
VAPOR FRACTION   0.000000D+00

V-L PHASE EQUILIBRIUM :

COMP   F(I)   X(I)   Y(I)   K(I)
H2O   1.0000   1.0000   1.0000   0.85118

BLOCK: STMHT   MODEL: HEATER

INLET STREAM(S):   HX1-OUT
OUTLET STREAM: REA1-FD
PROPERTY OPTION SET: SYSOP0
FREE WATER OPTION SET: SYSOP12
SOLUBLE WATER OPTION: ORGANIC OPTION SET

             *** MASS AND ENERGY BALANCE ***
                        IN   OUT   RELATIVE DIFF.
CONV. COMP. (LBMOL/HR) 256083.   256083.   0.0000000D+00
            (LB/HR)   0.4750000D+07 0.4750000D+07 0.0000000D+00
NONCONV. COMP(LB/HR)   0.1666667D+07 0.1666667D+07 0.0000000D+00
TOTAL BALANCE
MASS(LB/HR)   0.641667D+07   0.641667D+07 0.000000D+00
ENTHALPY(BTU/HR)   -0.344360D+11 -0.343660D+11 -0.220808D-02

                 *** INPUT DATA ***
TWO PHASE TP FLASH
SPECIFIED TEMPERATURE F   482.000
PRESSURE DROP   PSI   10.00000
MAXIMUM NO. ITERATIONS   30
CONVERGENCE TOLERANCE   0.0001000000

                 *** RESULTS ***
OUTLET TEMPERATURE   F   482.00
OUTLET PRESSURE   PSI   600.00
HEAT DUTY   BTU/HR   0.76037D+08
VAPOR FRACTION   0.000000D+00

V-L PHASE EQUILIBRIUM :

COMP   F(I)   X(I)   Y(I)   K(I)
H2O   1.0000   1.0000   1.0000   0.95118

ASPEN PLUS   VER: IBM-VS2   REL: DEC83F   INST: UCC   4/10/84 PAGE 9
                            BTX-TYPE FUEL
                            U-O-S BLOCK SECTION

BLOCK: SPLIT1   MODEL: SEP

INLET STREAM: REA1-FD
OUTLET STREAMS: WAT-1
                        DSTR1
PROPERTY OPTION SET: SYSOP0
FREE WATER OPTION SET: SYSOP12
SOLUBLE WATER OPTION: ORGANIC OPTION SET

        *** MASS AND ENERGY BALANCE ***

	IN	OUT	RELATIVE DIFF.
CONV. COMP. (LBMOL/HR)	256083.	256083.	0.0000000D+00
(LB/HR)	0.4750000D+07	0.4750000D+07	0.0000000D+00
NONCONV. COMP (LB/HR)	0.1666667D+07	0.1666667D+07	-0.139698D-15
TOTAL BALANCE			
MASS (LB/HR)	0.641667D+07	0.641667D+07	-0.362853D-16
ENTHALPY(BTU/HR)	-0.343600D+11	-0.343600D+11	0.0000000D+00

        *** INPUT DATA ***

FLASH SPECS FOR STREAM WAT-1
TWO   PHASE TP FLASH
SPECIFIED TEMPERATURE F   482.000
PRESSURE DROP   PSI   0.0
MAXIMUM NO. ITERATIONS   30
CONVERGENCE TOLERANCE   0.0001000000

FLASH SPECS FOR STREAM DSTR1
TWO   PHASE TP FLASH
SPECIFIED TEMPERATURE F   482.000
PRESSURE DROP   PSI   0.0
MAXIMUM NO. ITERATIONS   30
CONVERGENCE TOLERANCE   0.0001000000

FRACTION OF FEED
    SUBSTREAM= MIXED
    STREAM= WAT-1   CPT= H2O   FRACTION=   1.000000

        *** RESULTS ***

COMPONENT = H2O
    STREAM   SUBSTREAM   SPLIT FRACTION
    WAT-1   MIXED   1.000000

COMPONENT = NA2CO3
    STREAM   SUBSTREAM   SPLIT FRACTION
    DSTR1   CISOLID   1.000000

ASPEN PLUS   VER: IBM-VS2   REL: DEC83F   INST: UCC   4/10/84 PAGE 10
                            BTX-TYPE FUEL
                            U-O-S BLOCK SECTION

BLOCK: SPLIT1   MODEL: SEP (CONTINUED)

COMPONENT = COAL
    STREAM   SUBSTREAM   SPLIT FRACTION
    DSTR1   NCFSD   1.000000

COMPONENT = ASH
    STREAM   SUBSTREAM   SPLIT FRACTION
    DSTR1   NCFSD   1.000000

COMPONENT = RESID
    STREAM   SUBSTREAM   SPLIT FRACTION
    DSTR1   NCFSD   1.000000

BLOCK: REACT1   MODEL: RYIELD

INLET STREAM: DSTR1
OUTLET STREAM: DSTR2
PROPERTY OPTION SET: SYSOP0
FREE WATER OPTION SET: SYSOP12
SOLUBLE WATER OPTION: ORGANIC OPTION SET

        *** MASS AND ENERGY BALANCE ***

	IN	OUT	RELATIVE DIFF.
CONV. COMP. (LBMOL/HR)	1665.22	31003.8	-0.946290
(LB/HR)	166670.	0.1072500D+07	-0.844597
NONCONV. COMP (LB/HR)	0.1666667D+07	760834.	0.543499
TOTAL BALANCE			
MASS (LB/HR)	0.183334D+07	0.183334D+07	0.126998D-15
ENTHALPY(BTU/HR)	-0.525924D-10	-0.459267D+10	-0.128471

        *** INPUT DATA ***

TWO   PHASE TP FLASH
SPECIFIED TEMPERATURE F   482.000
PRESSURE DROP   PSI   15.0000
MAXIMUM NO. ITERATIONS   30
CONVERGENCE TOLERANCE   0.0001000000

MOLE-YIELD
    SUBSTREAM MIXED   :
    NONE SPECIFIED
    SUBSTREAM CISOLID :
    NONE SPECIFIED
    SUBSTREAM NC   :
    NONE SPECIFIED

MASS-YIELD
    SUBSTREAM MIXED   :
    H2O   25.0   CO2   2.00   ACIDS   31.5
    SUBSTREAM CISOLID :
    NONE SPECIFIED
    SUBSTREAM NC   :
    COAL   9.50   ASH   25.0   RESID   7.00

ASPEN PLUS   VER: IBM-VS2   REL: DEC83F   INST: UCC   4/10/84   PAGE   12
                                          BTX-TYPE FUEL
                                          U-O-S BLOCK SECTION

BLOCK: HEATX2   MODEL: HEATER (CONTINUED)

*** MASS AND ENERGY BALANCE ***

	IN	OUT	RELATIVE DIFF.
CONV. COMP.(LBMOL/HR)	285421.	285421.	0.000000D+00
(LB/HR)	0.565583D+07	0.565583D+07	0.000000D+00
NONCONV. COMP(LB/HR)	760834.	760834.	0.000000D+00
TOTAL BALANCE			
MASS(LB/HR)	0.641667D+07	0.641667D+07	0.000000D+00
ENTHALPY(BTU/HR)	-0.336843D+11	-0.344040D+11	0.209195D-01

*** INPUT DATA ***

TWO PHASE TP FLASH		
SPECIFIED TEMPERATURE	F	392.000
PRESSURE DROP	PSI	10.00000
MAXIMUM NO. ITERATIONS		30
CONVERGENCE TOLERANCE		0.0001000000

*** RESULTS ***

OUTLET TEMPERATURE	F	392.00
OUTLET PRESSURE	PSI	575.00
HEAT DUTY	BTU/HR	-0.71972D+09
VAPOR FRACTION		0.000000D+00

V-L PHASE EQUILIBRIUM :

COMP	F(I)	X(I)	Y(I)	K(I)
H2O	0.98051	0.98051	0.85670	0.38938
CO2	0.29190D-02	0.29190D-02	0.14308	21.845
ACIDS	0.16568D-01	0.16568D-01	0.21719D-03	0.58420D-02

BLOCK: SEPA   MODEL: SSPLIT
INLET STREAM: HX2-COUT
OUTLET STREAMS: OVFLW1
                   RESID1

*** MASS AND ENERGY BALANCE ***

	IN	OUT	RELATIVE DIFF.
CONV. COMP.(LBMOL/HR)	285421.	285421.	0.101968D-15
(LB/HR)	0.565583D+07	0.565583D+07	0.164666D-15
NONCONV. COMP(LB/HR)	760834.	760834.	0.000000D+00
TOTAL BALANCE			
MASS(LB/HR)	0.641667D+07	0.641667D+07	0.145141D-15
ENTHALPY(BTU/HR)	-0.344040D+11	-0.344040D+11	-0.194039D-15

ASPEN PLUS   VER: IBM-VS2   REL: DEC83F   INST: UCC   4/10/84   PAGE   11
                                          BTX-TYPE FUEL
                                          U-O-S BLOCK SECTION

BLOCK: REACT1   MODEL: RYIELD (CONTINUED)

*** RESULTS ***

OUTLET TEMPERATURE	F	482.00
OUTLET PRESSURE	PSI	585.00
HEAT DUTY	BTU/HR	0.67567D+09
VAPOR FRACTION		0.11223

V-L PHASE EQUILIBRIUM :

COMP	F(I)	X(I)	Y(I)	K(I)
H2O	0.82060	0.82286	0.80276	0.97557
CO2	0.26872D-01	0.58810D-02	0.17292	32.805
ACIDS	0.15252	0.17126	0.47189D-02	0.25219D-01

BLOCK: COMB1   MODEL: MIXER
INLET STREAM(S):   DSTR2   WAT-1
OUTLET STREAM: REA1-OUT
PROPERTY OPTION SET: SYSOP0
FREE WATER OPTION SET: SYSOP12
SOLUBLE WATER OPTION: ORGANIC OPTION SET

*** MASS AND ENERGY BALANCE ***

	IN	OUT	RELATIVE DIFF.
CONV. COMP.(LBMOL/HR)	285421.	285421.	0.000000D+00
(LB/HR)	0.565583D+07	0.565583D+07	0.823292D-16
NONCONV. COMP(LB/HR)	760834.	760834.	0.000000D+00
TOTAL BALANCE			
MASS(LB/HR)	0.641667D+07	0.641667D+07	0.725706D-16
ENTHALPY(BTU/HR)	-0.336843D+11	-0.336843D+11	-0.330604D-07

*** INPUT DATA ***

TWO PHASE FLASH	
MAXIMUM NO. ITERATIONS	30
CONVERGENCE TOLERANCE	0.0001000000
OUTLET PRESSURE: MINIMUM OF INLET STREAM PRESSURES

BLOCK: HEATX2   MODEL: HEATER
INLET STREAM(S):   REA1-OUT   HX2-COUT
OUTLET STREAM:
PROPERTY OPTION SET: SYSOP0
FREE WATER OPTION SET: SYSOP12
SOLUBLE WATER OPTION: ORGANIC OPTION SET

```
ASPEN PLUS   VER: IBM-VS2   REL: DEC83F   INST: UCC        4/10/84   PAGE 13
                            BTX-TYPE FUEL
                            U-O-S BLOCK SECTION

BLOCK:  SEPA   MODEL: SSPLIT (CONTINUED)

                   *** INPUT DATA ***
 FRACTION OF FLOW
  SUBSTRM=         STRM=        FRAC=
   MIXED           OVFLW1       0.83300
   NCPSD           RESID1       1.000000
   CISOLID         RESID1       1.000000

                   *** RESULTS ***

 STRM= OVFLW1  SUBSTRM= MIXED     SPLIT FRACT=    0.83300
                        CISOLID                   0.0
                        NCPSD                     0.0

 STRM= RESID1  SUBSTRM= MIXED     SPLIT FRACT=    0.16700
                        CISOLID                   1.000000
                        NCPSD                     1.000000

BLOCK:  REACT2   MODEL: RSTOIC
 ------------------------------
 INLET STREAM:          REA2-FD
 OUTLET STREAM:         REA2-OUT
 PROPERTY OPTION SET:   SYSOP0
 FREE WATER OPTION SET: SYSOP12
 SOLUBLE WATER OPTION:  ORGANIC OPTION SET

               *** MASS AND ENERGY BALANCE ***
                       IN            OUT         RELATIVE DIFF.
 TOTAL BALANCE
  MOLE (LBMOL/HR)      237899.       239869.      -0.821094D-02
  MASS (LB/HR )        0.471719D+07  0.474482D+07 -0.582253D-02
  ENTHALPY(BTU/HR )   -0.278519D+11 -0.273078D+11 -0.195247D-01

               *** INPUT DATA ***
 SIMULTANEOUS REACTIONS
 STOICHIOMETRY MATRIX:

 REACTION # 1:
  SUBSTREAM MIXED :
  CO2   1.00     ACIDS  -1.00     FUEL   1.00

 REACTION CONVERSION SPECS: NUMBER=   1
  REACTION # 1:
   SUBSTREAM:MIXED   KEY COMP:ACIDS   CONV FRAC: 0.5000
```

```
ASPEN PLUS   VER: IBM-VS2   REL: DEC83F   INST: UCC        4/10/84   PAGE 14
                            BTX-TYPE FUEL
                            U-O-S BLOCK SECTION

BLOCK:  REACT2   MODEL: RSTOIC (CONTINUED)
 TWO   PHASE  TP FLASH
 SPECIFIED TEMPERATURE F                     392.000
 PRESSURE DROP        PSI                     15.0000
 MAXIMUM NO. ITERATIONS                       30
 CONVERGENCE TOLERANCE                        0.0001000000

                               *** RESULTS ***
 OUTLET TEMPERATURE      F                    392.00
 OUTLET PRESSURE         PSI                   235.00
 HEAT DUTY               BTU/HR                0.70720D+09
 VAPOR FRACTION                                0.16350

 V-L PHASE EQUILIBRIUM :

 COMP       F(I)          X(I)          Y(I)          K(I)
 H2O        0.97188       0.97945       0.93315       0.95273
 CO2        0.11104D-01   0.11597D-02   0.61993D-01   53.450
 ACIDS      0.82109D-02   0.97889D-02   0.13992D-03   0.14294D-01
 FUEL       0.82109D-02   0.90024D-02   0.41618D-02   0.46230D-01
 CAT2       0.59712D-03   0.60269D-03   0.56859D-03   0.94341

BLOCK:  HEATX3   MODEL: HEATER
 -----------------------------
 INLET STREAM(S):            REA2-OUT
 OUTLET STREAM:              HX3-COUT
 PROPERTY OPTION SET:        SYSOP0
 FREE WATER OPTION SET:      SYSOP12
 SOLUBLE WATER OPTION:       ORGANIC OPTION SET

               *** MASS AND ENERGY BALANCE ***
                       IN            OUT         RELATIVE DIFF.
 TOTAL BALANCE
  MOLE (LBMOL/HR)      239869.       239869.       0.0000000D+00
  MASS (LB/HR )        0.474482D+07  0.474482D+07  0.0000000D+00
  ENTHALPY(BTU/HR )   -0.273078D+11 -0.293342D+11  0.690790D-01

               *** INPUT DATA ***
 TWO   PHASE  TP FLASH
 SPECIFIED TEMPERATURE F                  90.0000
 PRESSURE DROP        PSI                 10.00000
 MAXIMUM NO. ITERATIONS                   30
 CONVERGENCE TOLERANCE                    0.0001000000
```

ASPEN PLUS   VER: IBM-VS2   REL: DEC83F   INST: UCC   4/10/84   PAGE 15
                                    BTX-TYPE FUEL
                                    U-O-S BLOCK SECTION

BLOCK: HEATX3   MODEL: HEATER (CONTINUED)

*** RESULTS ***

OUTLET TEMPERATURE   F   90.000
OUTLET PRESSURE   PSI   225.00
HEAT DUTY   BTU/HR   -0.20264D+10
VAPOR FRACTION   0.00000D+00

V-L PHASE EQUILIBRIUM :

COMP	F(I)	X(I)	Y(I)	K(I)
H2O	0.97188	0.97188	0.53376D-01	0.31476D-02
CO2	0.11104D-01	0.11104D-01	0.94601	4.8826
ACIDS	0.82109D-02	0.82109D-02	0.23924D-06	0.16698D-05
FUEL	0.82109D-02	0.82109D-02	0.50912D-03	0.55536D-02
CAT2	0.59712D-03	0.59712D-03	0.10775D-03	0.10340D-01

BLOCK: PH-SEP   MODEL: SEP

INLET STREAM: HX3-COUT
OUTLET STREAMS: RCO2
                  BTX-PROD
                  EFFLU

PROPERTY OPTION SET: SYSOP0
FREE WATER OPTION SET: SYSOP12
SOLUBLE WATER OPTION: ORGANIC OPTION SET

*** MASS AND ENERGY BALANCE ***

	IN	OUT	RELATIVE DIFF.
TOTAL BALANCE			
MOLE (LBMOL/HR)	239869.	239869.	0.000000D+00
MASS (LB/HR)	0.474492D+07	0.474492D+07	-0.981410D-16
ENTHALPY(BTU/HR)	-0.293342D+11	-0.293342D+11	0.000000D+00

*** INPUT DATA ***

FLASH SPECS FOR STREAM RCO2
TWO   PHASE TP FLASH
SPECIFIED TEMPERATURE   F   90.0000
PRESSURE DROP   PSI   0.0
MAXIMUM NO. ITERATIONS   30
CONVERGENCE TOLERANCE   0.0001000000

ASPEN PLUS   VER: IBM-VS2   REL: DEC83F   INST: UCC   4/10/84   PAGE 16
                                    BTX-TYPE FUEL
                                    U-O-S BLOCK SECTION

BLOCK: PH-SEP   MODEL: SEP (CONTINUED)

FLASH SPECS FOR STREAM BTX-PROD
TWO   PHASE TP FLASH
SPECIFIED TEMPERATURE   F   90.0000
PRESSURE DROP   PSI   0.0
MAXIMUM NO. ITERATIONS   30
CONVERGENCE TOLERANCE   0.0001000000

FLASH SPECS FOR STREAM EFFLU
TWO   PHASE TP FLASH
SPECIFIED TEMPERATURE   F   90.0000
PRESSURE DROP   PSI   0.0
MAXIMUM NO. ITERATIONS   30
CONVERGENCE TOLERANCE   0.0001000000

FRACTION OF FEED
    SUBSTREAM= MIXED   CPT= CO2   FRACTION= 1.00000
       STREAM= RCO2
       STREAM= BTX-PROD   CPT= FUEL   FRACTION= 1.00000

*** RESULTS ***

COMPONENT = H2O
STREAM	SUBSTREAM	SPLIT FRACTION
EFFLU	MIXED	1.000000

COMPONENT = CO2
STREAM	SUBSTREAM	SPLIT FRACTION
RCO2	MIXED	1.000000

COMPONENT = ACIDS
STREAM	SUBSTREAM	SPLIT FRACTION
EFFLU	MIXED	1.000000

COMPONENT = FUEL
STREAM	SUBSTREAM	SPLIT FRACTION
BTX-PROD	MIXED	1.000000

COMPONENT = CAT2
STREAM	SUBSTREAM	SPLIT FRACTION
EFFLU	MIXED	1.000000

SUBSTREAM ATTR PSD TYPE: PSD

INTERVAL	LOWER LIMIT		UPPER LIMIT	
1	0.0	FT	1.4436-04	FT
2	1.4436-04	FT	2.0669-04	FT
3	2.0669-04	FT	2.8871-04	FT
4	2.8871-04	FT	4.1010-04	FT
5	4.1010-04	FT	5.8071-04	FT
6	5.8071-04	FT	8.2021-04	FT
7	8.2021-04	FT	0.0012	FT
8	0.0012	FT	0.0016	FT
9	0.0016	FT	0.0023	FT
10	0.0023	FT	0.0033	FT
11	0.0033	FT	0.0066	FT
12	0.0066	FT	0.0131	FT
13	0.0131	FT	0.0262	FT
14	0.0262	FT	0.0525	FT
15	0.0525	FT	0.1640	FT

LIGNITE SCR1 GR-LIG

	LIGNITE	SCR1	GR-LIG
STREAM ID	LIGNITE	SCR1	GR-LIG
FROM :	-----	-----	GRIND
TO :	CRUSH1	GRIND	CHGR1
CLASS:	MIXNCPSD	MIXNCPSD	MIXNCPSD
TOTAL STREAM:			
LB/HR	1.6667+06	1.6667+06	1.6667+06
SUBSTREAM: MIXED			
PHASE:	MIXED	MIXED	MIXED
COMPONENTS: LB/HR			
H2O	0.0	0.0	0.0
CO2	0.0	0.0	0.0
ACIDS	0.0	0.0	0.0
FUEL	0.0	0.0	0.0
NA2CO3	0.0	0.0	0.0
CAT2	0.0	0.0	0.0
COMPONENTS: MASS FRAC			
H2O	MISSING	MISSING	MISSING
CO2	MISSING	MISSING	MISSING
ACIDS	MISSING	MISSING	MISSING
FUEL	MISSING	MISSING	MISSING
NA2CO3	MISSING	MISSING	MISSING
CAT2	MISSING	MISSING	MISSING
COMPONENTS: STD CUFT/HR			
H2O	0.0	0.0	0.0
CO2	MISSING	MISSING	MISSING
ACIDS	MISSING	MISSING	MISSING
FUEL	0.0	0.0	0.0
NA2CO3	MISSING	MISSING	MISSING
CAT2	MISSING	MISSING	MISSING
TOTAL FLOW:			
LBMOL/HR	0.0	0.0	0.0
LB/HR	0.0	0.0	0.0
CUFT/HR	0.0	0.0	0.0

SUBSTREAM: NCPSD	STRUCTURE: NON CONVENTIONAL		
COMPONENTS: LB/HR			
COAL	1.6667+06	1.6667+06	1.6667+06
ASH	1.6667-04	1.6667-04	1.6667-04
RESID	1.6667-04	1.6667-04	1.6667-04
COMPONENTS: MASS FRAC			
COAL	1.0000	1.0000	1.0000
ASH	1.0000-10	1.0000-10	1.0000-10
RESID	1.0000-10	1.0000-10	1.0000-10
TOTAL FLOW:			
LB/HR	1.6667+06	1.6667+06	1.6667+06
STATE VARIABLES:			
TEMP F	85.0000	85.0000	85.0000
PRES PSI	15.0000	15.0000	15.0000
VFRAC	0.0	0.0	0.0
LFRAC	0.0	0.0	0.0
SFRAC	1.0000	1.0000	1.0000

LIGNITE SCR1 GR-LIG (CONTINUED)

STREAM ID		LIGNITE	SCR1	GR-LIG
RESID	PROXANAL			
	MOISTURE	0.0	0.0	0.0
	FC	98.0000	98.0000	98.0000
	VM	1.0000	1.0000	1.0000
	ASH	1.0000	1.0000	1.0000
	ULTANAL			
	ASH	11.4900	11.4900	11.4900
	CARBON	63.3000	63.3000	63.3000
	HYDROGEN	4.2900	4.2900	4.2900
	NITROGEN	0.7200	0.7200	0.7200
	CHLORINE	0.0	0.0	0.0
	SULFUR	1.2200	1.2200	1.2200
	OXYGEN	18.6500	18.6500	18.6500
	SULFANAL			
	PYRITIC	0.5600	0.5600	0.5600
	SULFATE	0.0300	0.0300	0.0300
	ORGANIC	0.6300	0.6300	0.6300
SUBSTREAM ATTRIBUTES:				
PSD				
	FRAC1	0.0	0.0154	0.1675
	FRAC2	0.0	0.0063	0.0756
	FRAC3	0.0	0.0125	0.1091
	FRAC4	0.0	0.0138	0.1793
	FRAC5	0.0	0.0226	0.2264
	FRAC6	0.0	0.0212	0.2442
	FRAC7	0.0	0.0290	0.0019
	FRAC8	0.0	0.0424	0.0
	FRAC9	0.0	0.0522	0.0
	FRAC10	0.0	0.1163	0.0
	FRAC11	0.1000	0.3423	0.0
	FRAC12	0.1000	0.3075	0.0
	FRAC13	0.0500	0.0036	0.0
	FRAC14	0.0500	0.0	0.0
	FRAC15	0.2500	0.0	0.0

LIGNITE SCR1 GR-LIG (CONTINUED)

STREAM ID		LIGNITE	SCR1	GR-LIG
ENTHALPY:				
	BTU/LB	-2785.2393	-2785.2393	-2785.2393
	BTU/HR	-4.6421+09	-4.6421+09	-4.6421+09
DENSITY:				
	LB/CUFT	89.4015	89.4015	89.4015
AVG MW		1.0000	1.0000	1.0000
COMPONENT ATTRIBUTES:				
COAL	PROXANAL			
	MOISTURE	31.7000	31.7000	31.7000
	FC	46.9100	46.9100	46.9100
	VM	41.6000	41.6000	41.6000
	ASH	11.4900	11.4900	11.4900
	ULTANAL			
	ASH	11.4900	11.4900	11.4900
	CARBON	63.3000	63.3000	63.3000
	HYDROGEN	4.2900	4.2900	4.2900
	NITROGEN	0.7200	0.7200	0.7200
	CHLORINE	0.0	0.0	0.0
	SULFUR	1.2200	1.2200	1.2200
	OXYGEN	18.6500	18.6500	18.6500
	SULFANAL			
	PYRITIC	0.5600	0.5600	0.5600
	SULFATE	0.0300	0.0300	0.0300
	ORGANIC	0.6300	0.6300	0.6300
ASH	PROXANAL			
	MOISTURE	0.0	0.0	0.0
	FC	0.0	0.0	0.0
	VM	0.0	0.0	0.0
	ASH	100.0000	100.0000	100.0000
	ULTANAL			
	ASH	11.4900	11.4900	11.4900
	CARBON	63.3000	63.3000	63.3000
	HYDROGEN	4.2900	4.2900	4.2900
	NITROGEN	0.7200	0.7200	0.7200
	CHLORINE	0.0	0.0	0.0
	SULFUR	1.2200	1.2200	1.2200
	OXYGEN	18.6500	18.6500	18.6500
	SULFANAL			
	PYRITIC	0.5600	0.5600	0.5600
	SULFATE	0.0300	0.0300	0.0300
	ORGANIC	0.6300	0.6300	0.6300

388

FCAT1 WATER SLURRY PSLURRY HX1-OUT

STREAM ID	FCAT1	WATER	SLURRY	PSLURRY	HX1-OUT
FROM :			SL-TANK	SL-PUMP	HEATX1
TO :	SL-TANK	SL-TANK	SL-PUMP	HEATX1	STMHT
CLASS:	MIXCINCP	MIXCINCP	MIXCINCP	MIXCINCP	MIXINCP
TOTAL STREAM:					
LB/HR	1.6667+05	4.5833+06	6.4167+06	6.4167+06	6.4167+06
BTU/HR	-8.6454+08	-3.1289+10	-3.6796+10	-3.6783+10	-3.4436+10
SUBSTREAM: MIXED					
PHASE:	MIXED	LIQUID	LIQUID	LIQUID	LIQUID
COMPONENTS: LB/HR					
H2O	0.0	4.5833+06	4.5833+06	4.5833+06	4.5833+06
CO2	0.0	0.0	0.0	0.0	0.0
ACIDS	0.0	0.0	0.0	0.0	0.0
FUEL	0.0	0.0	0.0	0.0	0.0
NA2CO3	0.0	0.0	0.0	0.0	0.0
CAT2	1.6667+05	0.0	0.0	0.0	0.0
COMPONENTS: MASS FRAC					
H2O	MISSING	1.0000	1.0000	1.0000	1.0000
CO2	MISSING	0.0	0.0	0.0	0.0
ACIDS	MISSING	0.0	0.0	0.0	0.0
FUEL	MISSING	0.0	0.0	0.0	0.0
NA2CO3	MISSING	0.0	0.0	0.0	0.0
CAT2	MISSING	0.0	0.0	0.0	0.0
COMPONENTS: STD CUFT/HR					
H2O	0.0	7.3561+04	7.3561+04	7.3561+04	7.3561+04
CO2	MISSING	MISSING	MISSING	MISSING	MISSING
ACIDS	MISSING	MISSING	MISSING	MISSING	MISSING
FUEL	MISSING	MISSING	MISSING	MISSING	MISSING
NA2CO3	MISSING	MISSING	MISSING	MISSING	MISSING
CAT2	0.0	MISSING	MISSING	MISSING	MISSING
TOTAL CUFT/HR	0.0	7.3561+04	7.3561+04	7.3561+04	7.3561+04
TOTAL FLOW:					
LBMOL/HR	0.0	2.5442+05	2.5442+05	2.5442+05	2.5442+05
LB/HR	0.0	4.5833+06	4.5833+06	4.5833+06	4.5833+06
CUFT/HR	0.0	7.3736+04	7.3790+04	7.3893+04	9.9016+04
STATE VARIABLES:					
TEMP F	MISSING	72.0000	72.0000	75.9618	472.0000
PRES PSI	MISSING	45.0000	15.0000	630.0000	610.0000
VFRAC	MISSING	0.0	0.0	0.0	0.0
LFRAC	MISSING	1.0000	1.0000	1.0000	1.0000
SFRAC	MISSING	0.0	0.0	0.0	0.0
ENTHALPY:					
BTU/LBMOL	MISSING	-1.2298+05	-1.2298+05	-1.2291+05	-1.1465+05
BTU/LB	MISSING	-6826.6936	-6825.3383	-6822.8829	-6363.3276
BTU/HR	MISSING	-3.1289+10	-3.1289+10	-3.1272+10	-2.9170+10
ENTROPY:					
BTU/LBMOL-R	MISSING	-39.0659	-39.0199	-38.9319	-29.0299
BTU/LB-R	MISSING	-2.1685	-2.1660	-2.1611	-1.5559
DENSITY:					
LBMOL/CUFT	MISSING	3.4504	3.4479	3.4430	2.5695
LB/CUFT	MISSING	62.1586	62.1133	62.0225	46.2890
AVG MW	MISSING	18.0150	18.0150	18.0150	18.0150

FCAT1 WATER SLURRY PSLURRY HX1-OUT (CONTINUED)

STREAM ID	FCAT1	WATER	SLURRY	PSLURRY	HX1-OUT
SUBSTREAM: CISOLID STRUCTURE: CONVENTIONAL					
COMPONENTS: LB/HR					
H2O	0.0	0.0	0.0	0.0	0.0
CO2	0.0	0.0	0.0	0.0	0.0
ACIDS	0.0	0.0	0.0	0.0	0.0
FUEL	0.0	0.0	0.0	0.0	0.0
NA2CO3	0.0	0.0	0.0	0.0	0.0
CAT2	1.6667+05	0.0	1.6667+05	1.6667+05	1.6667+05
COMPONENTS: MASS FRAC					
H2O	0.0	MISSING	0.0	0.0	0.0
CO2	0.0	MISSING	0.0	0.0	0.0
ACIDS	0.0	MISSING	0.0	0.0	0.0
FUEL	0.0	MISSING	0.0	0.0	0.0
NA2CO3	0.0	MISSING	0.0	0.0	0.0
CAT2	1.0000	MISSING	1.0000	1.0000	1.0000
COMPONENTS: STD CUFT/HR					
H2O	0.0	0.0	0.0	0.0	0.0
CO2	MISSING	MISSING	MISSING	MISSING	MISSING
ACIDS	MISSING	MISSING	MISSING	MISSING	MISSING
FUEL	0.0	0.0	0.0	0.0	0.0
NA2CO3	MISSING	MISSING	MISSING	MISSING	MISSING
CAT2	MISSING	MISSING	MISSING	MISSING	MISSING
TOTAL CUFT/HR	0.0	0.0	0.0	0.0	0.0
TOTAL FLOW:					
LBMOL/HR	1665.2180	0.0	1665.2180	1665.2180	1665.2180
LB/HR	1.6667+05	0.0	1.6667+05	1.6667+05	1.6667+05
CUFT/HR	985.1861	0.0	985.1861	985.1861	985.1861
STATE VARIABLES:					
TEMP F	72.0000	MISSING	73.3581	75.9618	472.0000
PRES PSI	600.0000	MISSING	15.0000	630.0000	610.0000
VFRAC	0.0	MISSING	0.0	0.0	0.0
LFRAC	0.0	MISSING	0.0	0.0	0.0
SFRAC	1.0000	MISSING	1.0000	1.0000	1.0000
ENTHALPY:					
BTU/LBMOL	-5.1918+05	MISSING	-5.1915+05	-5.1910+05	-5.0995+05
BTU/LB	-5187.1338	MISSING	-5186.8647	-5186.3477	-5094.9785
BTU/HR	-8.6454+08	MISSING	-8.6449+09	-8.6441+08	-8.4918+08
ENTROPY:					
BTU/LBMOL-R	-63.0943	MISSING	-63.0437	-62.9468	-50.2970
BTU/LB-R	-0.6304	MISSING	-0.6299	-0.6299	-0.5025
DENSITY:					
LBMOL/CUFT	1.6903	MISSING	1.6903	1.6903	1.6903
LB/CUFT	169.1762	MISSING	169.1762	169.1762	169.1762
AVG MW	100.0890	MISSING	100.0890	100.0890	100.0890
SUBSTREAM: NCPSD STRUCTURE: NON CONVENTIONAL					
COMPONENTS: LB/HR					
COAL	0.0	0.0	0.0	1.6667+06	1.6667+06
ASH	0.0	0.0	0.0	1.6667-04	1.6667-04
RESID	0.0	0.0	0.0	1.6667-04	1.6667-04

389

ASPEN PLUS   VER: IBM-VS2   REL: DEC83F   INST: UCC   4/10/84   PAGE 23
BTX-TYPE FUEL
STREAM SECTION

FCAT1 WATER SLURRY PSLURRY HX1-OUT (CONTINUED)

STREAM ID	FCAT1	WATER	SLURRY	PSLURRY	HX1-OUT
COMPONENTS: MASS FRAC					
COAL	0.0	0.0	1.0000	1.0000	1.0000
ASH	0.0	0.0	1.0000-10	1.0000-10	1.0000-10
RESID	0.0	0.0	1.0000-10	1.0000-10	1.0000-10
TOTAL FLOW:					
LB/HR	0.0	0.0	1.6667+06	1.6667+06	1.6667+06
STATE VARIABLES:					
TEMP F	MISSING	MISSING	73.3581	75.9618	472.0000
PRES PSI	MISSING	MISSING	15.0000	630.0000	610.0000
VFRAC	MISSING	MISSING	0.0	0.0	0.0
LFRAC	MISSING	MISSING	1.0000	1.0000	1.0000
SFRAC					
ENTHALPY:					
BTU/LB	MISSING	MISSING	-2788.8708	-2788.0607	-2650.2182
BTU/HR	MISSING	MISSING	-4.6481+09	-4.6468+09	-4.4170+09
DENSITY:					
LB/CUFT	MISSING	MISSING	89.4015	89.4015	89.4015
AVG MW	1.0000	1.0000	1.0000	1.0000	1.0000
COMPONENT ATTRIBUTES:					
COAL PROXANAL					
MOISTURE	MISSING	MISSING	31.7000	31.7000	31.7000
FC	MISSING	MISSING	46.9100	46.9100	46.9100
VM	MISSING	MISSING	41.6000	41.6000	41.6000
ASH	MISSING	MISSING	11.4900	11.4900	11.4900
ULTANAL					
ASH	MISSING	MISSING	11.4900	11.4900	11.4900
CARBON	MISSING	MISSING	63.3000	63.3000	63.3000
HYDROGEN	MISSING	MISSING	4.2900	4.2900	4.2900
NITROGEN	MISSING	MISSING	0.7200	0.7200	0.7200
CHLORINE	MISSING	MISSING	0.0	0.0	0.0
SULFUR	MISSING	MISSING	1.2200	1.2200	1.2200
OXYGEN	MISSING	MISSING	18.6500	18.6500	18.6500
SULFANAL					
PYRITIC	MISSING	MISSING	0.5600	0.5600	0.5600
SULFATE	MISSING	MISSING	0.0300	0.0300	0.0300
ORGANIC	MISSING	MISSING	0.6300	0.6300	0.6300
ASH PROXANAL					
MOISTURE	MISSING	MISSING	0.0	0.0	0.0
FC	MISSING	MISSING	0.0	0.0	0.0
VM	MISSING	MISSING	0.0	0.0	0.0
ASH	MISSING	MISSING	100.0000	100.0000	100.0000

ASPEN PLUS   VER: IBM-VS2   REL: DEC83F   INST: UCC   4/10/84   PAGE 24
BTX-TYPE FUEL
STREAM SECTION

FCAT1 WATER SLURRY PSLURRY HX1-OUT (CONTINUED)

STREAM ID	FCAT1	WATER	SLURRY	FSLURRY	HX1-OUT
ULTANAL					
ASH	MISSING	MISSING	11.4900	11.4900	11.4900
CARBON	MISSING	MISSING	63.3000	63.3000	63.3000
HYDROGEN	MISSING	MISSING	4.2900	4.2900	4.2900
NITROGEN	MISSING	MISSING	0.7200	0.7200	0.7200
CHLORINE	MISSING	MISSING	0.0	0.0	0.0
SULFUR	MISSING	MISSING	1.2200	1.2200	1.2200
OXYGEN	MISSING	MISSING	18.6500	18.6500	18.6500
SULFANAL					
PYRITIC	MISSING	MISSING	0.5600	0.5600	0.5600
SULFATE	MISSING	MISSING	0.0300	0.0300	0.0300
ORGANIC	MISSING	MISSING	0.6300	0.6300	0.6300
RESID PROXANAL					
MOISTURE	MISSING	MISSING	0.0	0.0	0.0
FC	MISSING	MISSING	98.0000	98.0000	98.0000
VM	MISSING	MISSING	1.0000	1.0000	1.0000
ASH	MISSING	MISSING	1.0000	1.0000	1.0000
ULTANAL					
ASH	MISSING	MISSING	11.4900	11.4900	11.4900
CARBON	MISSING	MISSING	63.3000	63.3000	63.3000
HYDROGEN	MISSING	MISSING	4.2900	4.2900	4.2900
NITROGEN	MISSING	MISSING	0.7200	0.7200	0.7200
CHLORINE	MISSING	MISSING	0.0	0.0	0.0
SULFUR	MISSING	MISSING	1.2200	1.2200	1.2200
OXYGEN	MISSING	MISSING	18.6500	18.6500	18.6500
SULFANAL					
PYRITIC	MISSING	MISSING	0.5600	0.5600	0.5600
SULFATE	MISSING	MISSING	0.0300	0.0300	0.0300
ORGANIC	MISSING	MISSING	0.6300	0.6300	0.6300
SUBSTREAM ATTRIBUTES:					
PSD					
FRAC1	MISSING	MISSING	0.1675	0.1675	0.1675
FRAC2	MISSING	MISSING	0.0756	0.0756	0.0756
FRAC3	MISSING	MISSING	0.1091	0.1091	0.1091
FRAC4	MISSING	MISSING	0.1753	0.1753	0.1753
FRAC5	MISSING	MISSING	0.2264	0.2264	0.2264
FRAC6	MISSING	MISSING	0.2442	0.2442	0.2442
FRAC7	MISSING	MISSING	0.0019	0.0019	0.0019
FRAC8	MISSING	MISSING	0.0	0.0	0.0
FRAC9	MISSING	MISSING	0.0	0.0	0.0
FRAC10	MISSING	MISSING	0.0	0.0	0.0
FRAC11	MISSING	MISSING	0.0	0.0	0.0
FRAC12	MISSING	MISSING	0.0	0.0	0.0
FRAC13	MISSING	MISSING	0.0	0.0	0.0
FRAC14	MISSING	MISSING	0.0	0.0	0.0
FRAC15	MISSING	MISSING	0.0	0.0	0.0

ASPEN PLUS   VER: IBM-VS2   REL: DEC83F   INST: UCC   4/10/84   PAGE 25

BTX-TYPE FUEL
STREAM SECTION

REA1-FD REA1-OUT HX2-COUT OVFLW1 RESID1

	REA1-FD	REA1-OUT	HX2-COUT	OVFLW1	RESID1
STREAM ID	REA1-FD	REA1-OUT	HX2-COUT	OVFLW1	RESID1
FROM	STMHT	COMB1	HEATX2	SEFA	SEFA
TO	SPLIT1	HEATX2	SEFA	CHGR2	CHGR2
CLASS	MIXCINCP	MIXCINCP	MIXCINCP	MIXCINCP	MIXCINCP
TOTAL STREAM:					
LB/HR	6.4167+06	6.4167+06	6.4167+06	4.7113+06	1.7054+06
BTU/HR	-3.4360+10	-3.3684+10	-3.3484+10	-2.7855+10	-6.5489+09
SUBSTREAM: MIXED					
PHASE	LIQUID	MIXED	LIQUID	LIQUID	LIQUID
COMPONENTS: LB/HR					
H2O	4.5833+06	5.0417+06	5.0417+06	4.1997+06	8.4196+05
CO2	0.0	3.6667+04	3.6667+04	3.0543+04	6123.3422
ACIDS	0.0	5.7750+05	5.7750+05	4.8106+05	9.6413+04
FUEL	0.0	0.0	0.0	0.0	0.0
NA2CO3	0.0	0.0	0.0	0.0	0.0
CAT2	0.0	0.0	0.0	0.0	0.0
COMPONENTS: MASS FRAC					
H2O	1.0000	0.8914	0.8914	0.8914	0.8914
CO2	0.0	0.0065	0.0065	0.0065	0.0065
ACIDS	0.0	0.1021	0.1021	0.1021	0.1021
FUEL	0.0	0.0	0.0	0.0	0.0
NA2CO3	0.0	0.0	0.0	0.0	0.0
CAT2	0.0	0.0	0.0	0.0	0.0
COMPONENTS: STD CUFT/HR					
H2O	7.3561+04	8.0917+04	8.0917+04	6.7404+04	1.3513+04
CO2	MISSING	MISSING	MISSING	MISSING	MISSING
ACIDS	MISSING	MISSING	MISSING	MISSING	MISSING
FUEL	0.0	0.0	0.0	0.0	0.0
NA2CO3	MISSING	MISSING	MISSING	MISSING	MISSING
CAT2	MISSING	MISSING	MISSING	MISSING	MISSING
TOTAL CUFT/HR	7.3561+04	8.0917+04	8.0917+04	6.7404+04	1.3513+04
TOTAL FLOW:					
LBMOL/HR	2.5442+05	2.8542+05	2.8542+05	2.3776+05	4.7665+04
LB/HR	4.5833+06	5.6558+06	5.6558+06	4.7113+06	9.4452+05
CUFT/HR	1.0012+05	1.9805+05	1.1132+05	9.2146+04	1.8674+04
STATE VARIABLES:					
TEMP F	482.0000	479.6346	392.0000	392.0000	392.0000
PRES PSI	600.0000	585.0000	575.0000	575.0000	575.0000
VFRAC	0.0	0.0158	0.0	0.0	0.0
LFRAC	1.0000	0.9842	1.0000	1.0000	1.0000
SFRAC	0.0	0.0	0.0	0.0	0.0
ENTHALPY:					
BTU/LBMOL	-1.1455+05	-1.1470+05	-1.1716+05	-1.1716+05	-1.1716+05
BTU/LB	-6349.2361	-5788.3595	-5912.4006	-5912.4006	-5912.4008
BTU/HR	-2.9101+10	-3.2738+10	-2.7855+10	-2.7855+10	-5.5844+09
ENTROPY:					
BTU/LBMOL-R	-27.7705	-29.0274	-30.5323	-30.5323	-30.5323
BTU/LB-R	-1.5415	-1.4144	-1.5408	-1.5408	-1.5408
DENSITY:					
LBMOL/CUFT	2.5411	1.4413	2.5525	2.5525	2.5525
LB/CUFT	45.7779	28.5606	50.5797	50.5797	50.5797
AVG MW	18.0150	19.8157	19.8157	19.8157	19.8157

ASPEN PLUS   VER: IBM-VS2   REL: DEC83F   INST: UCC   4/10/84   PAGE 26

BTX-TYPE FUEL
STREAM SECTION

REA1-FD REA1-OUT HX2-COUT OVFLW1 RESID1 (CONTINUED)

	REA1-FD	REA1-OUT	HX2-COUT	OVFLW1	RESID1
STREAM ID	REA1-FD	REA1-OUT	HX2-COUT	OVFLW1	RESID1
SUBSTREAM: CISOLID STRUCTURE: CONVENTIONAL					
COMPONENTS: LB/HR					
H2O	0.0	0.0	0.0	0.0	0.0
CO2	0.0	0.0	0.0	0.0	0.0
ACIDS	0.0	0.0	0.0	0.0	0.0
FUEL	0.0	0.0	0.0	0.0	0.0
NA2CO3	1.6667+05	1.6667+05	0.0	0.0	0.0
CAT2	0.0	0.0	0.0	0.0	0.0
COMPONENTS: MASS FRAC					
H2O	0.0	0.0	0.0	MISSING	MISSING
CO2	0.0	0.0	0.0	MISSING	MISSING
ACIDS	0.0	0.0	0.0	MISSING	MISSING
FUEL	0.0	0.0	0.0	MISSING	MISSING
NA2CO3	1.0000	1.0000	MISSING	MISSING	MISSING
CAT2	0.0	0.0	0.0	MISSING	MISSING
COMPONENTS: STD CUFT/HR					
H2O	0.0	0.0	0.0	0.0	0.0
CO2	MISSING	MISSING	MISSING	MISSING	MISSING
ACIDS	MISSING	MISSING	MISSING	MISSING	MISSING
FUEL	0.0	0.0	0.0	0.0	0.0
NA2CO3	MISSING	MISSING	MISSING	MISSING	MISSING
CAT2	MISSING	MISSING	MISSING	MISSING	MISSING
TOTAL CUFT/HR	0.0	0.0	0.0	0.0	0.0
TOTAL FLOW:					
LBMOL/HR	1665.2180	MISSING	MISSING	MISSING	MISSING
LB/HR	1.6667+05	MISSING	MISSING	MISSING	MISSING
CUFT/HR	985.1861	MISSING	MISSING	MISSING	MISSING
STATE VARIABLES:					
TEMP F	482.0000	MISSING	MISSING	MISSING	MISSING
PRES PSI	600.0000	585.0000	575.0000	575.0000	575.0000
VFRAC	MISSING	MISSING	MISSING	MISSING	MISSING
LFRAC	MISSING	MISSING	MISSING	MISSING	MISSING
SFRAC	1.0000	MISSING	MISSING	MISSING	MISSING
ENTHALPY:					
BTU/LBMOL	-5.0970+05	MISSING	MISSING	MISSING	MISSING
BTU/LB	-5092.4548	MISSING	MISSING	MISSING	MISSING
BTU/HR	-8.4876+08	MISSING	MISSING	MISSING	MISSING
ENTROPY:					
BTU/LBMOL-R	-50.0273	MISSING	MISSING	MISSING	MISSING
BTU/LB-R	-0.4998	MISSING	MISSING	MISSING	MISSING
DENSITY:					
LBMOL/CUFT	1.6903	MISSING	MISSING	MISSING	MISSING
LB/CUFT	169.1762	MISSING	MISSING	MISSING	MISSING
AVG MW	100.0890	MISSING	MISSING	MISSING	MISSING
SUBSTREAM: NCPSD STRUCTURE: NON CONVENTIONAL					
COMPONENTS: LB/HR					
COAL	1.6667+06	1.6667+06	MISSING	0.0	1.7417+05
ASH	1.6667+04	1.6667+04	MISSING	0.0	4.5833+05
RESID	1.6667+04	1.6667+04	MISSING	0.0	1.2833+05

ASPEN PLUS   VER: IBM-VS2   REL: DEC83F   INST: UCC   4/10/84   PAGE 27
BTX-TYPE FUEL
STREAM SECTION

REA1-FD REA1-OUT HX2-COUT OVFLW1 RESID1 (CONTINUED)

STREAM ID	REA1-FD	REA1-OUT	HX2-COUT	OVFLW1	RESID1
COMPONENTS: MASS FRAC					
COAL	1.0000				
ASH	1.0000-10	0.2289	0.2289	0.0	0.2289
RESID	1.0000-10	0.6024	0.6024	0.0	0.6024
		0.1687	0.1687	0.0	0.1687
TOTAL FLOW:					
LB/HR	1.6667+06	7.6083+05	7.6083+05	0.0	7.6083+05
STATE VARIABLES:					
TEMP F	482.0000	479.6346	392.0000	392.0000	392.0000
PRES PSI	600.0000	585.0000	575.0000	575.0000	575.0000
VFRAC	0.0	0.0	0.0	0.0	0.0
LFRAC	0.0	0.0	0.0	0.0	0.0
SFRAC	1.0000	1.0000	1.0000	1.0000	1.0000
ENTHALPY:					
BTU/LB	-2646.3498	-1243.9472	-1267.6669	-1267.6669	-1267.6669
BTU/HR	-4.4106+09	-9.4644+08	-9.6448+08	0.0	-9.6448+08
DENSITY:					
LB/CUFT	89.4015	89.4015	89.4015	89.4015	89.4015
AVG MW	1.0000	1.0000	1.0000	1.0000	1.0000
COMPONENT ATTRIBUTES:					
COAL PROXANAL					
MOISTURE	31.7000	31.7000	31.7000	31.7000	31.7000
FC	46.9100	46.9100	46.9100	46.9100	46.9100
VM	41.6000	41.6000	41.6000	41.6000	41.6000
ASH	11.4900	11.4900	11.4900	11.4900	11.4900
ULTANAL					
ASH	11.4900	11.4900	11.4900	11.4900	11.4900
CARBON	63.3000	63.3000	63.3000	63.3000	63.3000
HYDROGEN	4.2900	4.2900	4.2900	4.2900	4.2900
NITROGEN	0.7200	0.7200	0.7200	0.7200	0.7200
CHLORINE	0.0	0.0	0.0	0.0	0.0
SULFUR	1.2200	1.2200	1.2200	1.2200	1.2200
OXYGEN	18.6500	18.6500	18.6500	18.6500	18.6500
SULFANAL					
PYRITIC	0.5600	0.5600	0.5600	0.5600	0.5600
SULFATE	0.0300	0.0300	0.0300	0.0300	0.0300
ORGANIC	0.6300	0.6300	0.6300	0.6300	0.6300
ASH PROXANAL					
MOISTURE	0.0	0.0	0.0	0.0	0.0
FC	0.0	0.0	0.0	0.0	0.0
VM	0.0	0.0	0.0	0.0	0.0
ASH	100.0000	100.0000	100.0000	100.0000	100.0000

ASPEN PLUS   VER: IBM-VS2   REL: DEC83F   INST: UCC   4/10/84   PAGE 28
BTX-TYPE FUEL
STREAM SECTION

REA1-FD REA1-OUT HX2-COUT OVFLW1 RESID1 (CONTINUED)

STREAM ID	REA1-FD	REA1-OUT	HX2-COUT	OVFLW1	RESID1
ULTANAL					
ASH	11.4900	11.4900	11.4900	11.4900	11.4900
CARBON	63.3000	63.3000	63.3000	63.3000	63.3000
HYDROGEN	4.2900	4.2900	4.2900	4.2900	4.2900
NITROGEN	0.7200	0.7200	0.7200	0.7200	0.7200
CHLORINE	0.0	0.0	0.0	0.0	0.0
SULFUR	1.2200	1.2200	1.2200	1.2200	1.2200
OXYGEN	18.6500	18.6500	18.6500	18.6500	18.6500
SULFANAL					
PYRITIC	0.5600	0.5600	0.5600	0.5600	0.5600
SULFATE	0.0300	0.0300	0.0300	0.0300	0.0300
ORGANIC	0.6300	0.6300	0.6300	0.6300	0.6300
PROXANAL					
MOISTURE	0.0	0.0	0.0	0.0	0.0
FC	98.0000	98.0000	98.0000	98.0000	98.0000
VM	1.0000	1.0000	1.0000	1.0000	1.0000
ASH	1.0000	1.0000	1.0000	1.0000	1.0000
RESID ULTANAL					
ASH	11.4900	11.4900	11.4900	11.4900	11.4900
CARBON	63.3000	63.3000	63.3000	63.3000	63.3000
HYDROGEN	4.2900	4.2900	4.2900	4.2900	4.2900
NITROGEN	0.7200	0.7200	0.7200	0.7200	0.7200
CHLORINE	0.0	0.0	0.0	0.0	0.0
SULFUR	1.2200	1.2200	1.2200	1.2200	1.2200
OXYGEN	18.6500	18.6500	18.6500	18.6500	18.6500
SULFANAL					
PYRITIC	0.5600	0.5600	0.5600	0.5600	0.5600
SULFATE	0.0300	0.0300	0.0300	0.0300	0.0300
ORGANIC	0.6300	0.6300	0.6300	0.6300	0.6300
SUBSTREAM ATTRIBUTES:					
PSD					
FRAC1	0.1675	0.1675	0.1675	0.1675	0.1675
FRAC2	0.0756	0.0756	0.0756	0.0756	0.0756
FRAC3	0.1091	0.1091	0.1091	0.1091	0.1091
FRAC4	0.1753	0.1753	0.1753	0.1753	0.1753
FRAC5	0.2264	0.2264	0.2264	0.2264	0.2264
FRAC6	0.2442	0.2442	0.2442	0.2442	0.2442
FRAC7	0.0019	0.0019	0.0019	0.0019	0.0019
FRAC8	0.0	0.0	0.0	0.0	0.0
FRAC9	0.0	0.0	0.0	0.0	0.0
FRAC10	0.0	0.0	0.0	0.0	0.0
FRAC11	0.0	0.0	0.0	0.0	0.0
FRAC12	0.0	0.0	0.0	0.0	0.0
FRAC13	0.0	0.0	0.0	0.0	0.0
FRAC14	0.0	0.0	0.0	0.0	0.0
FRAC15	0.0	0.0	0.0	0.0	0.0

BTX-TYPE FUEL
STREAM SECTION

FCAT2 REA2-FD REA2-OUT HX3-COUT RCO2

STREAM ID	FCAT2	REA2-FD	REA2-OUT	HX3-COUT	RCO2
FROM :		CHGR2	REACT2	HEATX3	PH-SEP
TO :	REACT2	REACT2	HEATX3	PH-SEP	----
SUBSTREAM: MIXED					
PHASE:	LIQUID	LIQUID	MIXED	LIQUID	VAPOR
COMPONENTS: LB/HR					
H2O	0.0	0.0	4.1997+06	4.1997+06	0.0
CO2	0.0	3.0543+04	1.1722+05	1.1722+05	1.1722+05
ACIDS	0.0	4.8106+05	2.4053+05	2.4053+05	0.0
FUEL	0.0	0.0	1.8148+05	1.8148+05	0.0
NA2CO3	0.0	0.0	0.0	0.0	0.0
CAT2	5880.0000	0.0	0.0	0.0	0.0
COMPONENTS: MASS FRAC					
H2O	0.0	0.0	0.8914	0.8851	0.0
CO2	0.0	0.0065	0.0247	0.0247	1.0000
ACIDS	0.0	0.1021	0.0507	0.0507	0.0
FUEL	0.0	0.0	0.0382	0.0382	0.0
NA2CO3	0.0	0.0	0.0	0.0	0.0
CAT2	1.0000	0.0	0.0	0.0012	0.0
COMPONENTS: STD CUFT/HR					
H2O	0.0	0.0	6.7404+04	6.7404+04	0.0
CO2	MISSING	MISSING	MISSING	MISSING	MISSING
ACIDS	MISSING	MISSING	MISSING	MISSING	MISSING
FUEL	0.0	0.0	3341.6207	3341.6207	0.0
NA2CO3	MISSING	MISSING	MISSING	MISSING	MISSING
CAT2	MISSING	MISSING	MISSING	MISSING	MISSING
TOTAL CUFT/HR	0.0	6.7404+04	7.0745+04	7.0745+04	0.0
TOTAL FLOW:					
LBMOL/HR	143.2295	2.3776+05	2.3987+05	2.3987+05	2663.5591
LB/HR	5880.0000	4.7113+06	4.7448+06	4.7448+06	1.1722+05
CUFT/HR	120.8646	9.3146+04	7.6522+04	7.6522+04	6.9829+04
STATE VARIABLES:					
TEMP F	72.0000	250.0000	392.0000	392.0000	90.0000
PRES PSI	250.0000	575.0000	225.0000	225.0000	225.0000
VFRAC	0.0	0.0	0.0	0.1635	1.0000
LFRAC	1.0000	1.0000	1.0000	0.8365	0.0
SFRAC	0.0	0.0	0.0	0.0	0.0
ENTHALPY:					
BTU/LBMOL	-2.2857+04	-1.1716+05	-1.1384+05	-1.1229+05	-1.6917+05
BTU/LB	556.7673	-5912.4006	-5755.2907	-6182.3621	-3843.9924
BTU/HR	3.2758+06	-2.7855+10	-2.7308+10	-2.9334+10	-4.5061+08
ENTROPY:					
BTU/LBMOL-R	-37.7482	-30.5323	-27.3578	-38.6056	-4.5007
BTU/LB-R	-0.9195	-1.5408	-1.3330	-1.9517	-0.1032
DENSITY:					
LBMOL/CUFT	1.1850	2.5525	3.1346		0.0381
LB/CUFT	48.6495	50.5797	62.0058		1.6787
AVG MW	41.0530	19.8157	19.7809		44.0100

BTX-TYPE FUEL
STREAM SECTION

BTX-PROD EFFLU

STREAM ID	BTX-PROD	EFFLU
FROM :	PH-SEP	PH-SEP
TO :	----	----
SUBSTREAM: MIXED		
PHASE:	LIQUID	LIQUID
COMPONENTS: LB/HR		
H2O	0.0	4.1997+06
CO2	0.0	0.0
ACIDS	0.0	2.4053+05
FUEL	1.8148+05	0.0
NA2CO3	0.0	0.0
CAT2	0.0	5880.0000
COMPONENTS: MASS FRAC		
H2O	0.0	0.9446
CO2	0.0	0.0
ACIDS	0.0	0.0541
FUEL	1.0000	0.0
NA2CO3	0.0	0.0
CAT2	0.0	0.0013
COMPONENTS: STD CUFT/HR		
H2O	0.0	6.7404+04
CO2	MISSING	MISSING
ACIDS	MISSING	MISSING
FUEL	3341.6207	0.0
NA2CO3	MISSING	MISSING
CAT2	MISSING	MISSING
TOTAL CUFT/HR	3341.6207	6.7404+04
TOTAL FLOW:		
LBMOL/HR	1969.5482	2.3524+05
LB/HR	1.8148+05	4.4461+06
CUFT/HR	3397.3356	7.1768+04
STATE VARIABLES:		
TEMP F	90.0000	90.0000
PRES PSI	225.0000	225.0000
VFRAC	0.0	0.0
LFRAC	1.0000	1.0000
SFRAC	0.0	0.0
ENTHALPY:		
BTU/LBMOL	5802.7647	-1.2283+05
BTU/LB	62.9770	-6498.9285
BTU/HR	1.1429+07	-2.8895+10
ENTROPY:		
BTU/LBMOL-R	-80.5099	-38.8250
BTU/LB-R	-0.8738	-2.0542
DENSITY:		
LBMOL/CUFT	0.5797	3.2777
LB/CUFT	53.4172	61.9512
AVG MW	92.1410	18.9007

APPENDIX II

Process Model of
1000 Ton/Day Demonstration Plant for
Production of BTX-Type Fuel

In the following is presented the results of the computer-based process model for production of BTX-type fuel from Texas lignite, especially presented for a 1000 ton/day demonstration plant. The objective of carrying out this particular process model for 1000 tons/day was to evaluate the costs that might be required to initiate a demonstration scale process. Overall, the total capital requirement for this projected demonstration plant of 1000 tons/day was seen as less costly than other lignite liquefaction processes.

ASPEN PLUS   VER: IBM-VS2   REL: DEC83F   INST: UCC   1/22/85   PAGE   1
BTX-TYPE FUEL
RUN CONTROL SECTION

RUN CONTROL INFORMATION

TYPE OF RUN: NEW

INPUT FILE NAME: JCLINPUT

OUTPUT PROBLEM DATA FILE NAME: ASPEN   VERSION NO.   1

CALLING PROGRAM NAME: ASPEN

COST ONLY RUN

DESCRIPTION

ECONOMICS OF THE CONVERSION OF LIGNITE TO BTX-TYPE FUEL *** 27%
SOLIDS *** LIGNITE FEED RATE = 1,000 TONS/DAY *** LIGNITE PRICE =
$12/TON *** 2ND REACT.CONV. = 50% *** RECYCLE = 50% *** DEET =
50% *** IRR = 15%

TABLE OF CONTENTS

Page 2

ASPEN PLUS   VER: IEM-VS2   REL: DEC83F   INST: UCC   1/22/85   PAGE 2
BTX-TYPE FUEL
ECONOMIC EVALUATION SECTION

EXECUTIVE SUMMARY

INVESTMENT
PHYSICAL PLANT	$	35,714,679
INTEREST DURING CONSTRUCTION (10.00%)		1,889,168
STARTUP		5,118,498
TAX CREDIT		-4,642,908
WORKING CAPITAL		6,352,029
CONTINGENCY ALLOWANCE		10,714,404
TOTAL	$	55,145,869

SCHEDULE
PROJECT START	MAR, 1985
MECHANICAL COMPLETION	OCT, 1986
COMMERCIAL PRODUCTION	FEB, 1987

REVENUE
CAPACITY PRODUCTION RATE	(TONS/DAY)	158
NORMAL PRODUCTION RATE	(TONS/DAY)	117
INITIAL SELLING PRICE	($/TON)	640 * CALCULATED

COSTS
INITIAL UNIT COST	($/TON)	422
OPERATING RATE 100%	($/TON)	503
OPERATING RATE 75%	($/TON)	665
OPERATING RATE 50%		

DEBT
AMOUNT TO BE FINANCED (50.00%)	$	23,211,541
INTEREST RATE, LONG TERM		12.00%

PROFITABILITY
INTEREST RATE OF RETURN ON EQUITY		15.00%
PAYOUT TIME -- YEARS		6.00
RETURN ON INVESTMENT		19.67%
DISCOUNT RATE		17.00%
VENTURE WORTH	$	28,187,164
VENTURE WORTH/INITIAL INVESTMENT		1.02
BREAK-EVEN FRACTION		0.39
BREAK-EVEN VOLUME	(TONS/DAY)	51

Page 3

ASPEN PLUS   VER: IEM-VS2   REL: DEC83F   INST: UCC   1/22/85   PAGE 3
BTX-TYPE FUEL
ECONOMIC EVALUATION SECTION

CAPITAL INVESTMENT REPORT
ASPEN METHOD

** LOCATION: HOUSTON   PROJECT COMPLETION: 1986, QUARTER IV **

	MATERIAL COST	LABOR COST	LABOR HOUR
PROCESS UNITS	$ 11,665,794	$ 5,046,830	190,446
UTILITY UNITS	1,454,444	0	0
RECVNG, SHIPPING & STOR	0	0	0
SERVICE BUILDING	913,872	947,529	35,756
SERVICE SYST & DISTRBUTN	718,960	325,950	12,300
ADDITIONAL DIRECT	0	0	0
TOTAL DIRECT COST	$ 14,753,070	$ 6,320,309	238,502
SITE DEVELOPMENT	147,531	78,191	2,951
FREIGHT	295,061		
SALES TAX	442,592		
SUBTOTAL	$ 15,638,254	$ 6,392,500	241,453
CONTRACTOR FLD INDIRECTS	1,727,595	4,670,905	146,884
FIELD CONSTRUCTION	$ 17,365,849	$ 11,069,405	388,337

TOTAL FIELD CONSTRUCTION COST	$	28,435,254
CONTRACTOR ENGINEERING & HOME OFFICE	3,412,230	
OWNER'S COST	1,592,374	
FEES, PERMITS & INSURANCE	2,274,820	
ADDITIONAL DEPRECIABLE COST	0	
SUBTOTAL		7,279,425
TOTAL DIRECT & INDIRECTS	$	35,714,679
PROCESS CONTINGENCY	5,357,202	
PROJECT DEFINITION CONTINGENCY	5,357,202	
TOTAL CONTINGENCY		10,714,404
TOTAL DEPRECIABLE CAPITAL	$	46,429,083
LAND	928,582	
ROYALTY & EXPENSES	0	
ADDITIONAL NON-DEPRECIABLE	0	
SUBTOTAL (NON-DEPRECIABLE COSTS)		928,582
TOTAL FIXED INVESTMENT	$	47,357,664
WORKING CAPITAL		6,352,029
START UP COST		5,118,498
TOTAL CAPITAL	$	58,828,191

ASPEN PLUS   VER: IBM-VS2   REL: DEC83F   INST: UCC   1/22/85   PAGE 4
                                    BTX-TYPE FUEL
                        ECONOMIC EVALUATION SECTION

ANNUAL OPERATING COST

BASIS:
PRINCIPAL PRODUCT: 1986, QUARTER IV
                        BTX-FUEL
PLANT AVAILABILITY: 350/365.25 DAYS
PRODUCTION CAPACITY: 100.0 %
PRODUCTION PER YEAR: 96600. (1000 LB)

TOTAL RAW-MATERIAL	$	7,114,357
TOTAL UTILITIES	$	1,230,872
WASTE TREATMENT	$	130,582
CATALYST	$	130,692

LABOR
OPERATORS: 7 PER SHIFT, 3 SHIFTS

OPERATING (58799 HR)	$	882,000	
MAINTENANCE		1,671,447	
SUPERVISION		510,689	
FRINGE BENEFITS		1,225,655	
SUBTOTAL, LABOR		$	4,289,791

SUPPLIES			
OPERATING	$	88,200	
MAINTENANCE		1,114,298	
SUBTOTAL, SUPPLIES		$	1,202,498

GENERAL WORKS			
GENERAL & ADMIN	$	1,838,482	
PROPERTY TAX		928,582	
PROPERTY INSURANCE		371,433	
SUBTOTAL, GENERAL WORKS		$	3,138,496

DEPRECIATION (15 YEARS, STRAIGHT LINE)	$	3,095,272
GROSS OPERATING COST	$	20,332,560
LESS: BY-PRODUCT CREDIT	$	0
NET OPERATING COST	$	20,332,560
		==============

SUMMARY:

FIXED COST	$	8,630,785
VARIABLE COST	$	8,606,502
GROSS OPERATING COST	$	20,332,560
GROSS COST EXCL. DEPRECIATION	$	17,237,287

ASPEN PLUS   VER: IBM-VS2   REL: DEC83F   INST: UCC   1/22/85   PAGE 5
                                    BTX-TYPE FUEL
                        ECONOMIC EVALUATION SECTION

CASH FLOW DURING CONSTRUCTION

QUARTER	1985, I	1985, II	1985, III	1985, IV
CASH FLOW:				
CAPITAL EXP	30,210	1,864,102	7,395,423	12,922,364
DEBT REC'VD	15,105	932,051	3,697,711	6,461,182
INTEREST	120	11,739	75,782	210,960
NON-DEPRE	929,582	0	0	0
NET CASH	-943,807	-943,790	-3,773,493	-6,672,142
CUM CASH	-943,807	-1,887,597	-5,661,090	-12,333,232
RETURN	5,561	45,460	127,468	322,373
CUM RETURN	5,561	51,021	178,489	500,862
BALANCE	-949,368	-1,938,618	-5,839,579	-12,834,094

QUARTER	1986, I	1986, II	1986, III
CASH FLOW:			
CAPITAL EXP	13,105,706	7,882,980	2,693,787
DEBT REC'VD	6,555,853	3,941,490	1,346,894
INTEREST	373,089	490,866	541,496
STARTUP COST	0	0	767,775
NET CASH	-6,925,942	-4,432,356	-2,656,164
CUM CASH	-19,259,175	-23,691,531	-26,347,695
RETURN	586,234	814,748	958,441
CUM RETURN	1,087,096	1,901,843	2,860,284
BALANCE	-20,346,270	-25,593,374	-29,207,979

ASPEN PLUS   VER: IBM-VS2   REL: DEC83F   INST: UCC   1/22/85   PAGE   6
BTX-TYPE FUEL
ECONOMIC EVALUATION SECTION

CASH FLOW DURING OPERATION

QUARTER	1986, IV
CASH FLOW:	
SALES REVN	2,598,483
OPER COST	2,163,006
SALES & ADMN	311,818
DEBT REC'VD	23,214,541
DEBT RET'RD	23,601,450
INTEREST	433,213
DEPRECIATION	1,081,346
INCOME TAX	-1,511,813
WORKING CAP	6,352,029
STARTUP COST	2,388,632
NET CASH	3,069,626
CUM CASH	-24,498,215
RETURN	668,180
CUM RETURN	3,879,839
BALANCE	-28,378,054

QUARTER	1987, I	1987, II	1987, III	1987, IV
CASH FLOW:				
SALES REVN	3,953,081	6,073,447	6,990,698	7,109,949
OPER COST	3,290,588	4,141,459	4,212,107	4,283,959
SALES & ADMN	474,370	824,814	838,884	853,194
DEBT RET'RD	386,909	386,909	386,909	386,909
INTEREST	642,477	644,887	635,768	622,649
DEPRECIATION	1,622,019	1,622,019	1,622,019	1,622,019
INCOME TAX	-2,820,089	-143,893	-126,432	-108,749
WORKING CAP	6,352,029	0	0	0
STARTUP COST	1,194,316	0	0	0
NET CASH	-7,079,332	1,406,180	1,045,463	1,071,987
CUM CASH	-31,577,547	-30,171,367	-29,125,905	-28,053,918
RETURN	1,225,498	1,302,530	1,305,307	1,314,234
CUM RETURN	5,105,336	6,407,866	7,713,173	9,027,407
BALANCE	-36,682,884	-36,579,233	-36,839,078	-37,081,324

ASPEN PLUS   VER: IBM-VS2   REL: DEC83F   INST: UCC   1/22/85   PAGE   7
BTX-TYPE FUEL
ECONOMIC EVALUATION SECTION

CASH FLOW DURING OPERATION (CONTINUED)

QUARTER	1988, I	1988, II	1988, III	1988, IV
CASH FLOW:				
SALES REVN	7,231,235	7,354,589	7,480,047	7,607,646
OPER COST	4,357,037	4,431,762	4,506,954	4,583,836
SALES & ADMN	867,748	882,551	897,606	912,917
DEBT RET'RD	386,909	386,909	386,909	386,909
INTEREST	611,531	600,412	589,293	578,174
DEPRECIATION	2,297,978	2,297,978	2,297,978	2,297,978
INCOME TAX	-361,224	-343,085	-324,713	-306,104
NET CASH	1,369,233	1,396,441	1,423,999	1,451,913
CUM CASH	-26,684,684	-25,288,243	-23,864,245	-22,412,332
RETURN	1,317,623	1,315,262	1,311,824	1,307,258
CUM RETURN	10,345,030	11,660,292	12,972,116	14,279,375
BALANCE	-37,029,714	-36,948,535	-36,836,361	-36,691,707

QUARTER	1989, I	1989, II	1989, III	1989, IV
CASH FLOW:				
SALES REVN	7,737,421	7,869,410	8,003,650	8,140,181
OPER COST	4,662,030	4,741,557	4,822,441	4,904,705
SALES & ADMN	928,491	944,229	960,438	976,822
DEBT RET'RD	386,909	386,909	386,909	386,909
INTEREST	567,056	555,937	544,818	533,699
DEPRECIATION	2,187,712	2,187,712	2,187,712	2,187,712
INCOME TAX	-243,147	-224,050	-204,703	-185,103
NET CASH	1,436,083	1,464,728	1,493,746	1,523,149
CUM CASH	-20,976,249	-19,511,521	-18,017,773	-16,494,625
RETURN	1,302,308	1,296,947	1,290,047	1,282,452
CUM RETURN	15,581,683	16,878,629	18,168,977	19,451,429
BALANCE	-36,557,932	-36,390,151	-36,186,750	-35,946,053

ASPEN PLUS   VER: IBM-VS2   REL: DEC83F   INST: UCC   1/22/85   PAGE   E
BTX-TYPE FUEL
ECONOMIC EVALUATION SECTION

CASH FLOW DURING OPERATION (CONTINUED)

QUARTER	1990, I	1990, II	1990, III	1990, IV
CASH FLOW:				
SALES REVN	8,279,040	8,420,269	8,563,906	8,709,994
OPER COST	4,988,372	5,073,466	5,160,012	5,248,034
SALES & ADMN	993,485	1,010,432	1,027,669	1,045,199
DEBT RET'RD	386,909	386,909	386,909	386,909
INTEREST	522,581	511,462	500,343	489,224
DEPRECIATION	2,177,056	2,177,056	2,177,056	2,177,056
INCOME TAX	-160,981	-140,859	-120,469	-99,808
NET CASH	1,548,675	1,578,859	1,609,443	1,640,435
CUM CASH	-14,945,949	-13,367,091	-11,757,648	-10,117,213
RETURN	1,773,276	1,262,760	1,250,763	1,237,217
CUM RETURN	20,724,705	21,987,465	23,238,228	24,475,445
BALANCE	-35,670,654	-35,354,555	-34,995,876	-34,592,658

QUARTER	1991, I	1991, II	1991, III	1991, IV
CASH FLOW:				
SALES REVN	8,858,573	9,009,687	9,163,379	9,319,693
OPER COST	5,337,558	5,428,608	5,521,213	5,615,396
SALES & ADMN	1,063,029	1,081,162	1,099,606	1,118,361
DEBT RET'RD	386,909	386,909	386,909	386,909
INTEREST	478,106	466,987	455,868	444,750
DEPRECIATION	2,166,400	2,166,400	2,166,400	2,166,400
INCOME TAX	-74,608	-53,788	-31,883	-10,086
NET CASH	1,667,580	1,689,409	1,731,667	1,724,362
CUM CASH	-8,449,633	-6,750,224	-5,019,557	-3,254,195
RETURN	1,222,129	1,205,426	1,186,960	1,166,648
CUM RETURN	25,697,574	26,903,000	28,089,960	29,256,608
BALANCE	-34,147,207	-33,653,224	-33,108,517	-32,510,804

ASPEN PLUS   VER: IBM-VS2   REL: DEC83F   INST: UCC   1/22/85   PAGE   9
BTX-TYPE FUEL
ECONOMIC EVALUATION SECTION

CASH FLOW DURING OPERATION (CONTINUED)

QUARTER	1992, I	1992, II	1992, III	1992, IV
CASH FLOW:				
SALES REVN	9,478,673	9,640,366	9,804,816	9,972,072
OPER COST	5,711,187	5,808,611	5,907,697	6,008,474
SALES & ADMN	1,137,441	1,155,844	1,176,578	1,196,649
DEBT RET'RD	386,909	386,909	386,909	386,909
INTEREST	433,631	422,512	411,393	400,275
DEPRECIATION	63,935	63,935	63,935	63,935
INCOME TAX	852,992	875,766	893,085	921,096
NET CASH	956,514	999,104	898,085	1,058,670
CUM CASH	-2,297,681	-1,307,577	-283,424	775,245
RETURN	1,159,633	1,166,378	1,172,142	1,176,878
CUM RETURN	30,416,241	31,582,619	32,754,762	33,931,639
BALANCE	-32,713,922	-32,890,197	-33,058,187	-33,156,393

QUARTER	1993, I	1993, II	1993, III	1993, IV
CASH FLOW:				
SALES REVN	10,142,181	10,315,191	10,491,157	10,670,117
OPER COST	6,110,970	6,215,214	6,321,573	6,429,067
SALES & ADMN	1,217,062	1,237,623	1,250,855	1,280,414
DEBT RET'RD	386,909	386,909	386,909	386,909
INTEREST	399,156	378,037	366,918	355,800
DEPRECIATION	63,935	63,935	63,935	63,935
INCOME TAX	944,423	968,073	992,059	1,016,361
NET CASH	1,093,661	1,128,135	1,165,101	1,201,366
CUM CASH	1,868,906	2,998,041	4,163,142	5,364,707
RETURN	1,183,024	1,184,124	1,194,524	1,194,300
CUM RETURN	35,112,160	36,295,185	37,479,509	38,693,869
BALANCE	-33,243,254	-33,297,143	-33,316,566	-33,299,160

ASPEN PLUS   VER: IBM-VS2   REL: DEC82F   INST: UCC   1/22/85   PAGE   10
BTX-TYPE FUEL
ECONOMIC EVALUATION SECTION

CASH FLOW DURING OPERATION (CONTINUED)

QUARTER	1994, I	1994, II	1994, III	1994, IV
CASH FLOW:				
SALES REVN	10,852,133	11,037,255	11,225,534	11,417,025
OPER COST	6,528,708	6,650,279	6,763,723	6,879,102
SALES & ADMN	1,302,256	1,324,471	1,347,064	1,370,043
DEBT RET'RD	386,909	386,909	386,909	386,909
INTEREST	344,681	333,562	322,443	311,325
DEPRECIATION	63,935	63,935	63,935	63,935
INCOME TAX	1,041,010	1,066,003	1,091,348	1,117,049
NET CASH	1,239,540	1,276,031	1,314,047	1,352,598
CUM CASH	6,603,249	7,879,280	9,193,327	10,545,925
RETURN	1,103,088	1,180,281	1,176,250	1,170,640
CUM RETURN	39,846,936	41,027,058	42,203,543	43,374,091
BALANCE	-33,243,698	-33,148,058	-33,010,221	-32,828,166

QUARTER	1995, I	1995, II	1995, III	1995, IV
CASH FLOW:				
SALES REVN	11,611,782	11,809,862	12,011,301	12,216,217
OPER COST	6,996,449	7,115,798	7,237,192	7,360,659
SALES & ADMN	1,393,414	1,417,183	1,441,359	1,465,946
DEBT RET'RD	386,909	386,909	386,909	386,909
INTEREST	300,206	289,087	277,968	266,850
DEPRECIATION	63,935	63,935	63,935	63,935
INCOME TAX	1,143,112	1,169,345	1,196,051	1,223,259
NET CASH	1,391,693	1,431,241	1,471,551	1,512,634
CUM CASH	11,937,618	13,368,958	14,840,510	16,353,144
RETURN	1,163,245	1,154,257	1,142,498	1,130,694
CUM RETURN	44,537,336	45,691,593	46,835,091	47,965,785
BALANCE	-32,599,719	-32,322,634	-31,994,581	-31,613,131

ASPEN PLUS   VER: IBM-VS2   REL: DEC82F   INST: UCC   1/22/85   PAGE   11
BTX-TYPE FUEL
ECONOMIC EVALUATION SECTION

CASH FLOW DURING OPERATION (CONTINUED)

QUARTER	1996, I	1996, II	1996, III	1996, IV
CASH FLOW:				
SALES REVN	12,424,607	12,636,553	12,852,114	13,071,352
OPER COST	7,486,201	7,613,904	7,743,786	7,875,894
SALES & ADMN	1,490,953	1,516,386	1,542,254	1,568,562
DEBT RET'RD	386,909	386,909	386,909	386,909
INTEREST	255,731	244,612	233,493	222,375
DEPRECIATION	53,279	53,279	53,279	53,279
INCOME TAX	1,255,378	1,283,348	1,311,721	1,340,501
NET CASH	1,549,436	1,591,393	1,633,951	1,677,121
CUM CASH	17,902,290	19,493,673	21,127,623	22,804,744
RETURN	1,116,403	1,099,967	1,081,406	1,060,622
CUM RETURN	49,082,378	50,182,345	51,263,752	52,324,374
BALANCE	-31,180,098	-30,688,673	-30,136,129	-29,517,629

QUARTER	1997, I	1997, II	1997, III	1997, IV
CASH FLOW:				
SALES REVN	13,294,330	13,521,111	13,751,762	13,986,346
OPER COST	8,010,449	8,146,877	8,285,951	8,427,196
SALES & ADMN	1,595,320	1,622,533	1,650,211	1,678,362
DEBT RET'RD	386,909	386,909	386,909	386,909
INTEREST	211,256	200,137	189,019	177,900
DEPRECIATION	53,279	53,279	53,279	53,279
INCOME TAX	1,369,696	1,399,314	1,429,361	1,459,844
NET CASH	1,720,914	1,765,341	1,810,411	1,856,136
CUM CASH	24,525,659	26,290,999	28,101,410	29,957,546
RETURN	1,037,911	1,011,966	983,877	953,122
CUM RETURN	53,361,885	54,373,851	55,357,728	56,310,856
BALANCE	-28,836,226	-28,082,852	-27,256,318	-26,353,310

ASPEN PLUS   VER: IBM-VS2   REL: DEC83F   INST: UCC   1/22/85   PAGE 12

BTX-TYPE FUEL
ECONOMIC EVALUATION SECTION

CASH FLOW DURING OPERATION (CONTINUED)

QUARTER	1998, I	1998, II	1998, III	1998, IV
CASH FLOW:				
SALES REVN	14,224,933	14,467,589	14,714,385	14,965,391
OPER COST	8,570,951	8,717,159	8,865,861	9,017,099
SALES & ADMN	1,706,992	1,736,111	1,765,726	1,795,847
DEBT RET'RD	386,909	386,909	386,909	386,909
INTEREST	166,781	155,662	144,544	133,425
DEPRECIATION	53,279	53,279	53,279	53,279
INCOME TAX	1,490,772	1,522,151	1,553,990	1,586,296
NET CASH	1,902,528	1,949,597	1,997,355	2,045,814
CUM CASH	31,860,074	33,809,671	35,807,026	37,852,840
RETURN	919,597	883,162	847,690	801,049
CUM RETURN	57,230,454	58,113,615	58,957,306	59,758,354
BALANCE	-25,370,380	-24,303,944	-23,150,280	-21,905,514

QUARTER	1999, I	1999, II	1999, III	1999, IV
CASH FLOW:				
SALES REVN	15,220,678	15,480,320	15,744,792	16,012,968
OPER COST	9,170,918	9,327,360	9,486,471	9,648,296
SALES & ADMN	1,826,481	1,857,638	1,889,227	1,921,556
DEBT RET'RD	386,909	386,909	386,909	386,909
INTEREST	122,306	111,187	100,069	88,950
DEPRECIATION	53,279	53,279	53,279	53,279
INCOME TAX	1,619,078	1,652,342	1,686,099	1,720,355
NET CASH	2,094,986	2,144,883	2,195,518	2,246,902
CUM CASH	39,947,827	42,092,710	44,288,228	46,535,130
RETURN	755,096	705,686	652,669	595,985
CUM RETURN	60,513,450	61,219,137	61,871,805	62,467,690
BALANCE	-20,565,624	-19,126,427	-17,583,578	-15,932,561

ASPEN PLUS   VER: IBM-VS2   REL: DEC83F   INST: UCC   1/22/85   PAGE 13

BTX-TYPE FUEL
ECONOMIC EVALUATION SECTION

CASH FLOW DURING OPERATION (CONTINUED)

QUARTER	2000, I	2000, II	2000, III	2000, IV
CASH FLOW:				
SALES REVN	16,286,126	16,563,943	16,846,499	17,133,876
OPER COST	9,812,882	9,980,275	10,150,524	10,323,677
SALES & ADMN	1,954,335	1,987,673	2,021,580	2,056,065
DEBT RET'RD	386,909	386,909	386,909	386,909
INTEREST	77,831	66,712	55,594	44,475
DEPRECIATION	53,279	53,279	53,279	53,279
INCOME TAX	1,755,119	1,790,401	1,826,209	1,862,552
NET CASH	2,299,049	2,351,972	2,405,684	2,460,198
CUM CASH	48,834,179	51,186,150	53,591,834	56,052,032
RETURN	535,171	470,358	401,268	327,718
CUM RETURN	63,002,861	63,473,219	63,874,487	64,202,205
BALANCE	-14,168,683	-12,287,069	-10,282,653	-8,150,173

QUARTER	2001, I	2001, II	2001, III	2001, IV
CASH FLOW:				
SALES REVN	17,426,154	17,723,419	18,025,754	6,111,082
OPER COST	10,499,784	10,672,894	10,961,061	3,692,111
SALES & ADMN	2,091,139	2,126,810	2,163,091	244,443
DEBT RET'RD	386,909	386,909	386,909	386,909
INTEREST	33,356	22,237	11,119	-0
DEPRECIATION	53,279	53,279	53,279	53,279
INCOME TAX	1,899,459	1,936,879	1,974,892	-0
SALVAGE VAL				850,499
NET CASH	2,515,528	2,571,689	2,628,683	945,119
CUM CASH	58,567,560	61,139,248	63,767,942	64,711,061
RETURN	249,517	166,467	78,362	16,510
CUM RETURN	64,451,722	64,618,189	64,696,551	64,713,061
BALANCE	-5,884,162	-3,478,940	-928,609	0

ASPEN PLUS   VER: IBM-VS2   REL: DEC83F   INST: UCC   1/22/85   PAGE   14

BTX-TYPE FUEL
ECONOMIC EVALUATION SECTION

INSTALLED COST SUMMARY

PROJECT COST, COMPLETION 1986, QUARTER ,IV
LOCATION: HOUSTON

UNIT:	SL-PREP	REACTORS	TOTAL
EQUIPMENT COST $	4,957,278 $	2,550,031 $	7,507,308
EQUIPMENT SETTING COST	72,400	395,053	467,453
COMMODITY MATERIAL COST $			
PIPING	213,988 $	1,424,260 $	1,638,247
CONCRETE	95,903	395,368	491,272
STEEL	19,871	116,071	135,942
INSTRUMENTS	10,750	134,504	145,254
ELECTRICAL	49,211	96,005	145,216
INSULATION	10,825	125,712	136,537
PAINTING	4,716	22,269	26,985
MISCELLANEOUS	0	0	0
TOTAL COMMODITY MAT'L $	405,265 $	2,314,198 $	2,719,453
COMMODITY LABOR COST $			
PIPING	211,006 $	1,404,413 $	1,615,419
CONCRETE	183,047	672,171	855,217
STEEL	28,873	168,655	197,528
INSTRUMENTS	4,752	59,455	64,207
ELECTRICAL	72,063	140,586	212,649
INSULATION	17,938	208,310	226,248
PAINTING	16,036	75,719	91,755
MISCELLANEOUS	0	0	0
TOTAL COMMODITY LABOR $	513,715 $	2,729,307 $	3,243,022
TOTAL COMMODITY COST $	918,980 $	5,043,495 $	5,962,475
BUILDING MATERIAL $	668,688 $	620,198 $	1,289,886
TESTING MATERIAL	0	0	0
ADDL-INST. MATERIAL	0	0	0
CAPITALIZED SPARES	99,146	51,001	150,146
BUILDING LABOR $	693,315 $	643,039 $	1,336,354
TESTING LABOR	0	0	0
ADDL-INST. LABOR	0	0	0
TOTAL EQUIPMENT COST $	4,957,278 $	2,550,031 $	7,507,308
TOTAL MATERIAL COST	1,173,099	2,985,386	4,158,486
TOTAL LABOR COST	1,279,431	3,767,399	5,046,830
TOTAL INSTALLED COST $	7,409,808 $	9,302,816 $	16,712,624

ASPEN PLUS   VER: IBM-VS2   REL: DEC83F   INST: UCC   1/22/85   PAGE   15

BTX-TYPE FUEL
ECONOMIC EVALUATION SECTION

INSTALLATION LABOR HOURS SUMMARY

PROJECT LABOR HOURS, COMPLETION 1986, QUARTER ,IV
LOCATION: HOUSTON

UNIT:	SL-PREP REACTORS		TOTAL
SETTING LABOR HOURS	2732	14908	17640
COMMODITY LABOR HOURS			
PIPING	7962	52997	60959
CONCRETE	6153	25365	31513
STEEL	1090	6364	7454
INSTRUMENTS	179	2244	2423
ELECTRICAL	2719	5305	8024
INSULATION	677	7861	8538
PAINTING	605	2857	3462
MISCELLANEOUS	0	0	0
TOTAL COMMODITY LABOR	19385	102993	122378
BUILDING LABOR HOURS	26163	24266	50429
TESTING LABOR HOURS	0	0	0
ADDL-INST. LABOR HOURS	0	0	0
TOTAL INSTAL'N LABOR	48280	142166	190446

ASPEN PLUS   VER: IBM-VS2   REL: DEC83F   INST: UCC   1/22/85   PAGE 16
BTX-TYPE FUEL
ECONOMIC EVALUATION SECTION

UNIT: SL-PREP   INSTALLED COST SUMMARY

COAL HANDLING AND SLURRY PREPARATION

PROJECT COST, COMPLETION 1986, QUARTER ,IV
LOCATION: HOUSTON

	MATERIAL COST	LABOR COST	MAT'L & LABOR	LABOR HOURS
EQUIPMENT	$ 4,957,278		4,957,278	
EQUIP SET G		$ 72,400	72,400	2,732
COMMODITY:				
PIPING	$ 213,988	$ 211,006	$ 424,994	7,962
CONCRETE	95,903	163,047	258,950	6,153
STEEL	19,871	28,873	48,744	1,090
INSTRM'NTS	10,750	4,752	15,502	179
ELECTRICAL	49,211	72,063	121,274	2,719
INSULATION	10,825	17,938	28,764	677
PAINTING	4,716	16,036	20,752	605
MISCELL.	0	0	0	0
TOTAL COMMOD'Y	$ 405,265	$ 513,715	$ 918,980	19,385
OTHERS:				
BUILDING	$ 668,688	$ 693,315	$ 1,362,004	26,163
TESTING	0	0	0	0
ADDITIONAL	0	0	0	0
SPARE	99,146		99,146	
TOTAL, OTHER	$ 767,834	$ 693,316	$ 1,461,150	26,163
TOTAL, UNIT	$ 6,130,377	$ 1,279,431	$ 7,409,808	48,280

ASPEN PLUS   VER: IBM-VS2   REL: DEC83F   INST: UCC   1/22/85   PAGE 17
BTX-TYPE FUEL
ECONOMIC EVALUATION SECTION

EQUIPMENT INSTALLATION REPORT
PROJECT COST, COMPLETION 1986, QUARTER ,IV
LOCATION: HOUSTON
UNIT: SL-PREP

COST BLOCK ID	CONVEY	SILO	CRUSH1	GRIND
QUANTITY	1	5	1	1
STANDBY		5		0
BASE EQUIP. COST	$ 3,983,861	$ 124,238	$ 66,695	$ 66,838
ADJUSTED EQUIP. COST	3,983,861	124,238	66,695	66,838
COMMODITY MATL. COST:				
PIPING	0 $	117,597 $	13,889 $	13,918
CONCRETE	0	62,209	4,103	4,112
STEEL	0	5,997	5,050	5,061
INSTRUMENTS	0	0	4,419	4,429
ELECTRICAL	0	4,704	18,623	18,663
INSULATION	0	2,234	1,894	1,898
PAINTING	0	0	631	633
MISCELLANEOUS	0	0	0	0
TOTAL COMM. MAT'L	0 $	192,742 $	48,610 $	48,715
COMMODITY LABOR COST:				
PIPING	0 $	115,959 $	13,695 $	13,725
CONCRETE	0	105,762	6,976	6,991
STEEL	0	8,714	7,338	7,354
INSTRUMENTS	0	0	1,953	1,958
ELECTRICAL	0	0	27,272	27,330
INSULATION	0	7,795	3,138	3,145
PAINTING	0	7,597	2,147	2,151
MISCELLANEOUS	0	0	0	0
TOTAL COMM. LABOR	0 $	245,827 $	62,520 $	62,654
SETTING LABOR COST	0 $	38,453 $	6,626 $	6,640
COMMODITY LABOR HOUR:				
PIPING	0	4,376	517	518
CONCRETE	0	3,991	263	264
STEEL	0	329	277	278
INSTRUMENTS	0	0	74	74
ELECTRICAL	0	0	1,029	1,031
INSULATION	0	294	118	119
PAINTING	0	287	81	81
MISCELLANEOUS	0	0	0	0
TOTAL COMM. LAB HR	0	9,276	2,359	2,364
SETTING LABOR HOUR	0	1,459	250	251
EQUIP. & MATL. COST	$ 3,983,861	$ 316,980	$ 115,306	$ 115,553
LABOR COST	0	284,480	69,146	69,294

ASPEN PLUS   VER: IBM-VS2   REL: DEC83F   INST: UCC   1/22/85   PAGE 1?

BTX-TYPE FUEL
ECONOMIC EVALUATION SECTION

EQUIPMENT INSTALLATION REPORT (CONTINU)
PROJECT COST, COMPLETION 1986, QUARTER .IV
LOCATION: HOUSTON
UNIT: SL-PREP

	SL-TANK	SL-PUMP	TOTAL
COST BLOCK ID			
QUANTITY	1	1	
STANDBY	0	0	
BASE EQUIP. COST $	24,236	17,625	4,283,494
ADJUSTED EQUIP. COST $	48,472	20,571	4,310,676
COMMODITY MATL. COST:			
PIPING $	34,869	5,803	186,076
CONCRETE	12,155	834	83,394
STEEL	1,170	0	17,279
INSTRUMENTS	0	500	9,348
ELECTRICAL	0	5,505	42,792
INSULATION	918	167	9,413
PAINTING	436	0	4,101
MISCELLANEOUS	0	0	0
TOTAL COMM. MAT'L $	49,528	12,809	352,405
COMMODITY LABOR COST:			
PIPING $	34,384	5,722	183,484
CONCRETE	20,632	1,418	141,780
STEEL	1,700	221	25,107
INSTRUMENTS	0		4,132
ELECTRICAL	0	8,062	62,663
INSULATION	1,521	567	15,598
PAINTING	1,482	0	13,944
MISCELLANEOUS	0	0	0
TOTAL COMM. LABOR $	59,718	15,990	446,709
SETTING LABOR COST $	7,540	3,498	62,957
COMMODITY LABOR HOUR:			
PIPING	1,297	216	6,924
CONCRETE	779	54	5,350
STEEL	64	0	947
INSTRUMENTS	0	8	156
ELECTRICAL	0	304	2,765
INSULATION	57	0	589
PAINTING	56	21	526
MISCELLANEOUS	0	0	0
TOTAL COMM. LAB HR	2,254	603	16,857
SETTING LABOR HOUR	285	132	2,376
EQUIP. & MATL. COST $	98,000	33,381	4,663,081
LABOR COST $	67,258	19,488	509,666

ASPEN PLUS   VER: IBM-VS2   REL: DEC83F   INST: UCC   1/22/85   PAGE 10

BTX-TYPE FUEL
ECONOMIC EVALUATION SECTION

UNIT: REACTORS INSTALLED COST SUMMARY
REACTORS, PRODUCT RECOVERY AND WASTE TREATMENT
PROJECT COST, COMPLETION 1986, QUARTER .IV
LOCATION: HOUSTON

	MATERIAL COST	LABOR COST	MAT'L & LABOR	LABOR HOURS
EQUIPMENT $	2,550,031		2,550,031	
EQUIP SET'G $		395,053	395,053	14,908
COMMODITY:				
PIPING $	1,424,260	1,404,413	2,828,673	52,997
CONCRETE	395,368	672,171	1,067,539	25,365
STEEL	116,071	168,655	284,726	6,364
INSTRM'NTS	134,504	59,455	193,958	2,244
ELECTRICAL	96,005	140,586	236,590	5,305
INSULATION	125,712	208,310	334,022	7,861
PAINTING	22,269	75,719	97,988	2,857
MISCELL.	0	0	0	0
TOTAL COMMD'Y $	2,314,188	2,729,307	5,043,495	102,993
OTHERS:				
BUILDING $	620,198	643,039	1,263,237	24,266
TESTING	0	0	0	0
ADDITIONAL	0	0	0	0
SPARE	51,001		51,001	
TOTAL OTHER $	671,199	643,040	1,314,238	24,266
TOTAL UNIT $	5,535,417	3,767,399	9,302,816	142,166

ASPEN PLUS   VER: IBM-VS2   REL: DEC83F   INST: UCC   1/22/85   PAGE 20
BTX-TYPE FUEL
ECONOMIC EVALUATION SECTION

EQUIPMENT INSTALLATION REPORT
PROJECT COST, COMPLETION 1986, QUARTER ,IV
LOCATION: HOUSTON
UNIT: REACTORS

COST BLOCK ID		HEATX1	STMHT1	REACT1	SEPA
QUANTITY		3	1	1	1
STANDBY		0	0	0	0
BASE EQUIP. COST	$	718,920	13,505	426,320	150,065
ADJUSTED EQUIP. COST	$	718,920	13,505	481,518	214,378
COMMODITY MATL. COST:					
PIPING	$	302,819	5,689	271,342	55,584
CONCRETE		37,427	703	44,389	7,102
STEEL		22,456	422	36,318	0
INSTRUMENTS		55,800	1,048	36,318	4,261
ELECTRICAL		11,568	217	16,141	46,874
INSULATION		35,386	665	36,318	0
PAINTING		3,402	64	4,035	1,420
MISCELLANEOUS		0	0	0	0
TOTAL COMM. MAT'L	$	468,859	8,808	444,860	115,242
COMMODITY LABOR COST:					
PIPING	$	298,600	5,609	267,561	54,809
CONCRETE		61,630	1,195	75,466	12,074
STEEL		32,630	611	52,771	0
INSTRUMENTS		24,665	463	16,054	1,894
ELECTRICAL		16,940	318	23,637	68,641
INSULATION		58,675	1,101	60,180	0
PAINTING		11,569	217	13,721	4,830
MISCELLANEOUS		0	0	0	0
TOTAL COMM. LABOR	$	506,670	9,518	509,389	142,238
SETTING LABOR COST	$	50,133	942	77,752	29,784
COMMODITY LABOR HOUR:					
PIPING		11,268	212	10,097	2,068
CONCRETE		2,401	45	2,848	456
STEEL		1,231	23	1,991	0
INSTRUMENTS		931	17	606	71
ELECTRICAL		639	12	892	2,590
INSULATION		2,213	42	2,271	182
PAINTING		437	8	518	0
MISCELLANEOUS		0	0	0	0
TOTAL COMM. LAB HR		19,120	359	19,222	5,367
SETTING LABOR HOUR		1,892	36	2,934	1,124
EQUIP. & MATL. COST	$	1,187,779	22,313	926,379	329,620
LABOR COST	$	556,803	10,460	587,141	172,022

ASPEN PLUS   VER: IBM-VS2   REL: DEC83F   INST: UCC   1/22/85   PAGE 21
BTX-TYPE FUEL
ECONOMIC EVALUATION SECTION

EQUIPMENT INSTALLATION REPORT (CONTINU
PROJECT COST, COMPLETION 1986, QUARTER ,IV
LOCATION: HOUSTON
UNIT: REACTORS

COST BLOCK ID		REACT2	FLASH	PH-SEP	STO-TANK
QUANTITY		1	1	1	1
STANDBY		0	0	0	0
BASE EQUIP. COST	$	132,944	48,166	48,166	459,944
ADJUSTED EQUIP. COST	$	232,820	48,166	48,166	459,944
COMMODITY MATL. COST:					
PIPING	$	110,248	28,723	28,723	435,359
CONCRETE		13,842	5,015	5,015	230,305
STEEL		11,325	4,103	4,103	22,203
INSTRUMENTS		11,325	4,103	4,103	0
ELECTRICAL		5,034	1,824	1,824	0
INSULATION		11,325	4,103	4,103	17,414
PAINTING		1,258	456	456	8,272
MISCELLANEOUS		0	0	0	0
TOTAL COMM. MAT'L	$	164,359	48,327	48,327	713,554
COMMODITY LABOR COST:					
PIPING	$	109,712	28,323	28,323	429,293
CONCRETE		22,557	8,526	8,526	391,545
STEEL		16,456	5,962	5,962	33,262
INSTRUMENTS		5,006	1,814	1,814	0
ELECTRICAL		7,371	2,671	2,671	0
INSULATION		18,767	6,799	6,799	28,856
PAINTING		4,279	1,550	1,550	28,126
MISCELLANEOUS		0	0	0	0
TOTAL COMM. LABOR	$	184,124	55,645	55,645	910,082
SETTING LABOR COST	$	24,246	8,785	8,785	143,098
COMMODITY LABOR HOUR:					
PIPING		4,102	1,069	1,069	16,200
CONCRETE		898	322	322	14,775
STEEL		621	225	225	1,217
INSTRUMENTS		189	68	68	0
ELECTRICAL		278	101	101	1,089
INSULATION		709	257	257	1,061
PAINTING		161	58	58	0
MISCELLANEOUS		0	0	0	0
TOTAL COMM. LAB HR		6,948	2,100	2,100	34,343
SETTING LABOR HOUR		915	331	331	5,400
EQUIP. & MATL. COST	$	397,179	96,493	96,493	1,173,498
LABOR COST	$	208,371	64,429	64,429	1,053,180

ASPEN PLUS   VER: IBM-VS2   REL: DEC83F   INST: UCC   1/22/85   PAGE 22
BTX-TYPE FUEL
ECONOMIC EVALUATION SECTION

EQUIPMENT INSTALLATION REPORT (CONTINU
PROJECT COST, COMPLETION 1986, QUARTER ,IV
LOCATION: HOUSTON
UNIT: REACTORS

COST BLOCK ID	WASTE-W	TOTAL
QUANTITY	1	
STANDBY	0	
BASE EQUIP. COST	$ 0	$ 1,998,031
ADJUSTED EQUIP. COST	$ 0	$ 2,217,418

COMMODITY MATL. COST:	$ 0	$
PIPING		1,238,487
CONCRETE		343,798
STEEL		100,931
INSTRUMENTS		116,960
ELECTRICAL		83,482
INSULATION		109,315
PAINTING		19,364
MISCELLANEOUS		0
TOTAL COMM. MAT'L	$ 0	$ 2,012,337

COMMODITY LABOR COST:	$ 0	$
PIPING		1,221,229
CONCRETE		584,496
STEEL		146,656
INSTRUMENTS		51,700
ELECTRICAL		122,248
INSULATION		181,139
PAINTING		65,843
MISCELLANEOUS		0
TOTAL COMM. LABOR	$ 0	$ 2,373,311
SETTING LABOR COST	$ 0	$ 343,524

COMMODITY LABOR HOUR:	0	
PIPING		46,084
CONCRETE		22,056
STEEL		5,534
INSTRUMENTS		1,951
ELECTRICAL		4,613
INSULATION		6,835
PAINTING		2,485
MISCELLANEOUS		0
TOTAL COMM. LAB HR	0	89,559
SETTING LABOR HOUR	0	12,963

EQUIP. & MATL. COST	$ 0	$ 4,229,755
LABOR COST	$ 0	$ 2,716,835

ASPEN PLUS   VER: IBM-VS2   REL: DEC83F   INST: UCC   1/22/85   PAGE 23
BTX-TYPE FUEL
COST BLOCK SECTION

BLOCK: CONVEY   MODEL: USER
COAL HANDLING EQUIPMENT   *** INPUT PARAMETERS ***

NUMBER OF PIECES OF EQUIPMENT	1
NUMBER OF STAND-BY PIECES	0
CAPACITY SCALE FACTOR	1.00
COST ADJUSTMENT FACTOR	1.00

*** ASSOCIATED UTILITES ***

UTILITY ID FOR ELECTRICITY	POWER	
SPECIFIED RATE OF CONSUMPTION	300.0000	KW

*** COST RESULTS ***

CARBON STEEL COST	$	3,983,900
PURCHASED COST	$	3,983,900

406

ASPEN PLUS   VER: IBM-VS2   REL: DEC83F   INST: UCC   1/22/85   PAGE   24
                                    BTX-TYPE FUEL
                                    COST BLOCK SECTION

BLOCK: SILO   MODEL: TANK
CRUSHED COAL STORAGE SILOS

                    *** INPUT DATA ***

TOTAL VOLUME   1.3334+04   CUFT
TANK TYPE   SHOP-FABRICATED
MATERIAL OF CONSTRUCTION   CARBON STEEL
CAPACITY SCALE FACTOR   1.00
COST ADJUSTMENT FACTOR   1.00

        *** SIZING AND COSTING RESULTS ***

NUMBER OF TANKS   5
VOLUME PER TANK   2666.8000   CUFT
MATERIAL OF CONSTRUCTION FACTOR   1.00

            *** COST RESULTS ***

                        PER PIECE   TOTAL
CARBON STEEL COST $ 24.800 $ 124.200
PURCHASED COST $ 24.800 $ 124.200

ASPEN PLUS   VER: IBM-VS2   REL: DEC83F   INST: UCC   1/22/85   PAGE   25
                                    BTX-TYPE FUEL
                                    COST BLOCK SECTION

BLOCK: CRUSH1   MODEL: COMPR
CRUSHER WITH DRIVE 1 X 75 TONS/HR

                    *** INPUT DATA ***

NUMBER OF COMPRESSORS   1
DRIVER TYPE   MOTOR
INDICATED HORSEPOWER   80.0000   HP
MECHANICAL EFFICIENCY   0.75

            *** USER SUPPLIED COST ***

ADJUSTED COST $ 56.000
INDEX TYPE   EQUIPMENT

            *** ASSOCIATED UTILITES ***

UTILITY ID FOR ELECTRICITY   POWER
SPECIFIED RATE OF CONSUMPTION   60.0000   KW

            *** COST RESULTS ***

CARBON STEEL COST $ 66.700
PURCHASED COST $ 66.700

ASPEN PLUS   VER: IBM-VS2   REL: DEC83F   INST: UCC   1/22/85   PAGE   26
                            BTX-TYPE FUEL
                            COST BLOCK SECTION

BLOCK: GRIND   MODEL: COMPR

GRINDER AND DRIVE

                        *** INPUT DATA ***

NUMBER OF COMPRESSORS   1
INDICATED HORSEPOWER   302.0000   HP
MECHANICAL EFFICIENCY   0.75

                *** USER SUPPLIED COST ***

ADJUSTED COST $ 56.120
INDEX TYPE   EQUIPMENT

                *** ASSOCIATED UTILITES ***

                                        POWER
UTILITY ID FOR ELECTRICITY   225.5000   KW
SPECIFIED RATE OF CONSUMPTION

                *** COST RESULTS ***

CARBON STEEL COST $ 66.800
PURCHASED COST $ 66.800

ASPEN PLUS   VER: IBM-VS2   REL: DEC83F   INST: UCC   1/22/85   PAGE   27
                            BTX-TYPE FUEL
                            COST BLOCK SECTION

BLOCK: SL-TANK   MODEL: TANK

SLURRY PREP-TANK W/AGITATORS 75HP EA.

                        *** INPUT DATA ***

VOLUME FLOW   3670.0000   CUFT/HR
RETENTION TIME REQUIRED   0.5000   HR
TANK TYPE   SHOP-FABRICATED
MATERIAL OF CONSTRUCTION   CARBON STEEL
CAPACITY SCALE FACTOR   1.00
COST ADJUSTMENT FACTOR   2.00

                *** SIZING AND COSTING RESULTS ***

NUMBER OF TANKS   1
TOTAL VOLUME REQUIRED   2202.0000   CUFT
VOLUME PER TANK   2202.0000   CUFT
MATERIAL OF CONSTRUCTION FACTOR   1.00

                *** COST RESULTS ***

CARBON STEEL COST $ 24.200
PURCHASED COST $ 48.500

ASPEN PLUS   VER: IBM-VS2   REL: DEC83F   INST: UCC   1/22/85   PAGE   28
                                 BTX-TYPE FUEL
                              COST BLOCK SECTION
BLOCK: SL-PUMP   MODEL: PUMP

                     *** INPUT DATA ***

MATERIAL OF CONSTRUCTION   CARBON STEEL
MOTOR TYPE   TEFC
SEVERITY OF SERVICE   HEAVY SERVICE
LIQUID VOLUMETRIC FLOW   3670.0000   CUFT/HR
LIQUID DENSITY   84.7000   LB/CUFT
PRESSURE RISE   615.0000   PSI

              *** ASSOCIATED UTILITES ***

UTILITY ID FOR ELECTRICITY   POWER
SPECIFIED RATE OF CONSUMPTION   190.0000   KW

         *** SIZING AND COSTING RESULTS ***

NUMBER OF PUMPS IN UNIT   1
REQUIRED PUMP HEAD   1045.5726   FT-LBF/LB
VOLUMETRIC FLOW PER PUMP   3670.0000   CUFT/HR
MOTOR HORSEPOWER   250.0000
MATERIAL OF CONSTRUCTION FACTOR   1.00
PUMP TYPE FACTOR   2.16
PUMP EFFICIENCY   0.71

              *** COST RESULTS ***

CARBON STEEL COST $ 17,600
PURCHASED COST $ 20,600

ASPEN PLUS   VER: IBM-VS2   REL: DEC83F   INST: UCC   1/22/85   PAGE   29
                                 BTX-TYPE FUEL
                              COST BLOCK SECTION
BLOCK: HEATX1   MODEL: HEATX

                     *** INPUT DATA ***

HEAT EXCHANGER TYPE   FIXED-TUBE
SHELL MATERIAL   CARBON STEEL
TUBE MATERIAL   CARBON STEEL
NUMBER OF SHELL PASSES   1
SHELL PRESSURE   650.0000   PSI
SHELL INLET TEMPERATURE   482.0000   F
SHELL OUTLET TEMPERATURE   90.0000   F
NUMBER OF TUBE PASSES   2
TUBE PRESSURE   650.0000   PSI
TUBE INLET TEMPERATURE   76.0000   F
TUBE OUTLET TEMPERATURE   472.0000   F
HEAT TRANSFER COEFFICIENT   350.0000   BTU/HR-SQFT-R
HEAT DUTY   1.11740+08   BTU/HR

         *** SIZING AND COSTING RESULTS ***

NUMBER OF HEAT EXCHANGERS   3
HEAT TRANSFER AREA PER UNIT   9405.2001   SQFT
MATERIAL OF CONSTRUCTION FACTOR   1.00
PRESSURE FACTOR   2.25
HEAT EXCHANGER TYPE FACTOR   0.75

              *** COST RESULTS ***

                                  PER PIECE   TOTAL
CARBON STEEL COST $ 239,600 $ 718,900
PURCHASED COST $ 239,600 $ 718,900

409

BTX-TYPE FUEL
COST BLOCK SECTION

BLOCK: STMHT1 MODEL: HEATX

*** INPUT DATA ***

```
HEAT EXCHANGER TYPE                 FIXED-TUBE
SHELL MATERIAL                      CARBON STEEL
TUBE MATERIAL                       CARBON STEEL
NUMBER OF SHELL PASSES                      1
SHELL PRESSURE                        800.0000   PSI
SHELL INLET TEMPERATURE               510.0000   F
SHELL OUTLET TEMPERATURE              510.0000   F
NUMBER OF TUBE PASSES                        2
TUBE PRESSURE                         650.0000   PSI
TUBE INLET TEMPERATURE                472.0000   F
TUBE OUTLET TEMPERATURE               482.0000   F
HEAT TRANSFER COEFFICIENT             350.0000   BTU/HR-SQFT-R
HEAT DUTY                             3.8000+06  BTU/HR
```

*** ASSOCIATED UTILITES ***

```
UTILITY ID FOR STEAM                  HPSTEAM
CALCULATED RATE OF CONSUMPTION       5270.4577   LB/HR
```

*** FLOWSHEET REFERENCE DATA ***

```
UTILITY ID - SHELL SIDE               HPSTEAM
```

*** SIZING AND COSTING RESULTS ***

```
NUMBER OF HEAT EXCHANGERS                    1
HEAT TRANSFER AREA PER UNIT           331.5572   SQFT
MATERIAL OF CONSTRUCTION FACTOR          1.00
PRESSURE FACTOR                          1.84
HEAT EXCHANGER TYPE FACTOR               0.55
```

*** COST RESULTS ***

```
CARBON STEEL COST                $      13,500
PURCHASED COST                   $      13,500
```

BTX-TYPE FUEL
COST BLOCK SECTION

BLOCK: REACT1 MODEL: V-VESSEL
REACTORS W/ AGITATORS 75HP EA.

*** INPUT DATA ***

```
NUMBER OF VESSELS                            1
NUMBER OF STANDBY VESSELS                    0
MATERIAL OF CONSTRUCTION            CARBON STEEL
CAPACITY SCALE FACTOR                    1.00
COST ADJUSTMENT FACTOR                   1.14
INNER DIAMETER                         15.0000   FT
VELOCITY RATIO                           0.5000
TANGENT-TO-TANGENT LENGTH              30.0000   FT
PRESSURE                              650.0000   PSI
TEMPERATURE                           500.0000   F
CORROSION ALLOWANCE                     0.0104   FT
PLATFORM ALLOWANCE                      1.4167   FT
```

*** SIZING AND COSTING RESULTS ***

```
TOP SHELL THICKNESS                     0.3750   FT
BOTTOM SHELL THICKNESS                  0.3750   FT
SHELL WEIGHT PER VESSEL               3.6657+05  LB
VOLUME PER VESSEL                    6185.0249   CUFT
MATERIAL OF CONSTRUCTION FACTOR          1.00
```

*** COST RESULTS ***

```
CARBON STEEL COST                $     426.300
PURCHASED COST                   $     481.500
```

ASPEN PLUS   VER: IEM-VS2   REL: DEC83F   INST: UCC   1/22/85   PAGE   32
                                  BTX-TYPE FUEL
                               COST BLOCK SECTION

BLOCK: SEPA   MODEL: PUMP
CENTRIFUGE - 20 TPH EA.

                        *** INPUT DATA ***

NUMBER OF PUMPS   1
MATERIAL OF CONSTRUCTION   CARBON STEEL
MOTOR TYPE   TEFC
SEVERITY OF SERVICE   MODERATE SERVICE
LIQUID VOLUMETRIC FLOW   4012.0000   CUFT/HR
REQUIRED PUMP HEAD   30.0000   FT-LBF/LB

                 *** USER SUPPLIED COST ***

ADJUSTED COST $ 180.0000
INDEX TYPE   EQUIPMENT

               *** ASSOCIATED UTILITES ***

UTILITY ID FOR ELECTRICITY   POWER
SPECIFIED RATE OF CONSUMPTION   112.0000   KW

          *** SIZING AND COSTING RESULTS ***

MATERIAL OF CONSTRUCTION FACTOR   1.00

                  *** COST RESULTS ***

CARBON STEEL COST $ 150.100
PURCHASED COST $ 214.400

ASPEN PLUS   VER: IEM-VS2   REL: DEC83F   INST: UCC   1/22/85   PAGE   33
                                  BTX-TYPE FUEL
                               COST BLOCK SECTION

BLOCK: REACT2   MODEL: V-VESSEL

                        *** INPUT DATA ***

NUMBER OF VESSELS   1
NUMBER OF STANDBY VESSELS   0
MATERIAL OF CONSTRUCTION   CARBON STEEL
CAPACITY SCALE FACTOR   1.00
COST ADJUSTMENT FACTOR   2.00
INNER DIAMETER   8.0000   FT
VELOCITY RATIO   0.5000
TANGENT-TO-TANGENT LENGTH   36.0000   FT
PRESSURE   275.0000   PSI
TEMPERATURE   450.0000   F
CORROSION ALLOWANCE   0.0104   FT
PLATFORM ALLOWANCE   1.4167   FT

          *** SIZING AND COSTING RESULTS ***

TOP SHELL THICKNESS   0.1042   FT
BOTTOM SHELL THICKNESS   0.1042   FT
SHELL WEIGHT PER VESSEL   5.4713+04   LB
VOLUME PER VESSEL   1943.6032   CUFT
MATERIAL OF CONSTRUCTION FACTOR   1.00

                  *** COST RESULTS ***

CARBON STEEL COST $ 132.900
PURCHASED COST $ 232.800

411

BTX-TYPE FUEL
COST BLOCK SECTION

BLOCK: FLASH   MODEL: V-VESSEL

*** INPUT DATA ***

MATERIAL OF CONSTRUCTION	CARBON STEEL	
CAPACITY SCALE FACTOR	1.00	
COST ADJUSTMENT FACTOR	1.00	
VAPOR FLOW	8629.0000	CUFT/HR
LIQUID FLOW	4445.0000	CUFT/HR
VAPOR DENSITY	0.0300	LB/CUFT
LIQUID DENSITY	60.0000	LB/CUFT
VELOCITY RATIO	0.5000	
RETENTION TIME	0.1667	HR
PRESSURE	75.0000	PSI
TEMPERATURE	150.0000	F
CORROSION ALLOWANCE	0.0	FT
PLATFORM ALLOWANCE	1.4167	FT

*** SIZING AND COSTING RESULTS ***

CALCULATED NUMBER OF VESSELS	1	
CALCULATED DIAMETER	9.0000	FT
CALCULATED TANGENT-TO-TANGENT LENGTH	24.8951	FT
TOP SHELL THICKNESS	0.0234	FT
BOTTOM SHELL THICKNESS	0.0234	FT
SHELL WEIGHT PER VESSEL	1.0495+04	LB
VOLUME PER VESSEL	1774.6161	CUFT
MATERIAL OF CONSTRUCTION FACTOR	1.00	

*** COST RESULTS ***

CARBON STEEL COST	$	48.200
PURCHASED COST	$	48.200

BTX-TYPE FUEL
COST BLOCK SECTION

BLOCK: PH-SEP   MODEL: V-VESSEL

*** INPUT DATA ***

MATERIAL OF CONSTRUCTION	CARBON STEEL	
CAPACITY SCALE FACTOR	1.00	
COST ADJUSTMENT FACTOR	1.00	
VAPOR FLOW	0.0	CUFT/HR
LIQUID FLOW	4445.0000	CUFT/HR
VAPOR DENSITY	0.0300	LB/CUFT
LIQUID DENSITY	60.0000	LB/CUFT
VELOCITY RATIO	0.5000	
RETENTION TIME	0.1667	HR
PRESSURE	75.0000	PSI
TEMPERATURE	150.0000	F
CORROSION ALLOWANCE	0.0	FT
PLATFORM ALLOWANCE	1.4167	FT

*** SIZING AND COSTING RESULTS ***

CALCULATED NUMBER OF VESSELS	1	
CALCULATED DIAMETER	9.0000	FT
CALCULATED TANGENT-TO-TANGENT LENGTH	24.8951	FT
TOP SHELL THICKNESS	0.0234	FT
BOTTOM SHELL THICKNESS	0.0234	FT
SHELL WEIGHT PER VESSEL	1.0495+04	LB
VOLUME PER VESSEL	1774.6161	CUFT
MATERIAL OF CONSTRUCTION FACTOR	1.00	

*** COST RESULTS ***

CARBON STEEL COST	$	48.200
PURCHASED COST	$	48.200

412

                                      BTX-TYPE FUEL
                                   COST BLOCK SECTION

BLOCK: STO-TANK MODEL: TANK

                    *** INPUT DATA ***

NUMBER OF TANKS   1
NUMBER OF STANDBYS   0
TOTAL VOLUME   5.6146+05 CUFT
TANK TYPE   FIELD-ERECTED
MATERIAL OF CONSTRUCTION   CARBON STEEL
CAPACITY SCALE FACTOR   1.00
COST ADJUSTMENT FACTOR   1.00

           *** SIZING AND COSTING RESULTS ***

VOLUME PER TANK   5.6146+05 CUFT
MATERIAL OF CONSTRUCTION FACTOR   1.00

                    *** COST RESULTS ***

CARBON STEEL COST $ 459,900
PURCHASED COST $ 459,900

                                      BTX-TYPE FUEL
                                   COST BLOCK SECTION

BLOCK: WASTE-W MODEL: USER
WASTE WATER TREATMENT PLANT

                  *** INPUT PARAMETERS ***

NUMBER OF PIECES OF EQUIPMENT   1
NUMBER OF STAND-BY PIECES   0
CAPACITY SCALE FACTOR   1.00
COST ADJUSTMENT FACTOR   1.00

              *** ASSOCIATED UTILITES ***

UTILITY ID FOR WATER   W-WATER
SPECIFIED RATE OF CONSUMPTION   1.2000+05 LB/HR
UTILITY ID FOR ELECTRICITY   POWER
SPECIFIED RATE OF CONSUMPTION   300.0000 KW

                    *** COST RESULTS ***

CARBON STEEL COST $ 0
PURCHASED COST $ 0

ASPEN PLUS   VER: IBM-VS2   REL: DEC83F   INST: UCC   1/22/85   PAGE 38
BTX-TYPE FUEL
COST BLOCK SECTION

UTILITY USAGE: POWER   (ELECTRICITY)

INPUT DATA:

PRICE   0.0500 $/KWHR
INDEX TYPE   FUEL

RESULT:

INDEXED PRICE (1986 IV Q)   0.0641 $/KWHR

THIS UTILITY IS PURCHASED

USAGE:

COST BLOCK ID	MODEL	USAGE RATE (KW)	ANNUAL COST
CONVEY	USER	300.0000	168.513
CRUSH1	COMPR	60.0000	33.703
GRIND	COMPR	225.5000	126.666
SL-TANK	TANK	56.0000	31.456
SL-PUMP	PUMP	190.0000	106.725
REACT1	V-VESSEL	56.0000	31.456
SEPA	PUMP	112.0000	62.911
WASTE-W	USER	300.0000	168.513
	TOTAL:	1299.5000	$ 729.942

ASPEN PLUS   VER: IBM-VS2   REL: DEC83F   INST: UCC   1/22/85   PAGE 39
BTX-TYPE FUEL
COST BLOCK SECTION

UTILITY USAGE: CW   (WATER)

INPUT DATA:

INLET TEMPERATURE   80.0000   F
RETURN TEMPERATURE   120.0000   F
PRESSURE   14.6959   PSI
COOLING VALUE   40.0000   BTU/LB
PRICE   1.2000-04 $/LB
INDEX TYPE   FUEL

RESULT:

COOLING VALUE   40.0000   BTU/LB
INDEXED PRICE (1986 IV Q)   1.5379-04 $/LB

THIS UTILITY IS PURCHASED

USAGE:

THIS UTILITY IS NOT USED BY ANY COST BLOCKS

ASPEN PLUS   VER: IBM-VS2   REL: DEC83F   INST: UCC   1/22/85   PAGE   40
                            BTX-TYPE FUEL
                            COST BLOCK SECTION

UTILITY USAGE: HPSTEAM   (STEAM)

INPUT DATA:

    HEATING VALUE   721.0000   BTU/LB
    PRICE   0.0080   FUEL
    INDEX TYPE $/LB

RESULT:

    HEATING VALUE (1986 IV Q)   721.0000   BTU/LB
    INDEXED PRICE   0.0103 $/LB

THIS UTILITY IS PURCHASED

USAGE:

COST BLOCK ID	MODEL	USAGE RATE (LB/HR)	ANNUAL COST
STMHT1	HEATX	5270.4577	473,675
	TOTAL:	5270.4577	$ 473,675

ASPEN PLUS   VER: IBM-VS2 : DEC83F   INST: UCC   1/22/85   PAGE   41
                            BTX-TYPE FUEL
                            COST BLOCK SECTION

UTILITY USAGE: W-WATER   (WATER)

INPUT DATA:

    INLET TEMPERATURE   67.9100   F
    RETURN TEMPERATURE   139.9100   F
    PRESSURE   14.6959   PSI
    COOLING VALUE   40.0000   BTU/LB
    PRICE   6.0000-05 $/LB
    INDEX TYPE   FUEL

RESULT:

    COOLING VALUE   40.0000   BTU/LB
    INDEXED PRICE (1986 IV Q) 7.6894-05 $/LB

THIS UTILITY IS GENERATED ON-SITE

    NUMBER OF UNITS   1
    MATERIAL COST $ 1,454,444
    LABOR COST $ 0
    LABOR HOURS   0

    STANDARD CAPACITY COST $ 5,000,000
    STANDARD CAPACITY   2.4300+06   LB/HR
    CAPACITY EXPONENT   0.6000
    MINIMUM CAPACITY   1.0000+04   LB/HR
    MAXIMUM CAPACITY   6.9200+06   LB/HR
    LABOR FACTOR   MISSING
    SCALE FACTOR   1.2500
    EQUIPMENT COST INDEX VALUE   MISSING
    INDEX TYPE   EQUIPMENT
    USAGE:

COST BLOCK ID	MODEL	USAGE RATE (LB/HR)	ANNUAL COST
WASTE-W	USER	1.2000+05	80,886
	TOTAL:	1.2000+05	$ 80,886

SUBJECT INDEX